THE LITTORAL FAUNA OF THE BRITISH ISLES

SHORE COLLECTING

Shore with loose rocks — A rock pool

Netting amongst weeds — Sieving muddy sand

THE LITTORAL FAUNA OF THE BRITISH ISLES

A HANDBOOK FOR COLLECTORS

BY

N. B. EALES, D.Sc.

Formerly Lecturer in Zoology, University of Reading

With a Foreword by

THE LATE

STANLEY KEMP, SC.D., F.R.S.

formerly Director of the Marine Biological Association's Laboratory at Plymouth

FOURTH EDITION

CAMBRIDGE

AT THE UNIVERSITY PRESS

1967

CAMBRIDGE UNIVERSITY PRESS
Cambridge, New York, Melbourne, Madrid, Cape Town,
Singapore, São Paulo, Delhi, Tokyo, Mexico City

Cambridge University Press
The Edinburgh Building, Cambridge CB2 8RU, UK

Published in the United States of America by Cambridge University Press, New York

www.cambridge.org
Information on this title: www.cambridge.org/9780521294706

First edition 1939
Second edition 1950
Reprinted 1952
Third edition 1961
Fourth edition 1967
First paperback edition 2011

A catalogue record for this publication is available from the British Library

ISBN 978-0-521-04862-0 Hardback
ISBN 978-0-521-29470-6 Paperback

TO
THE
STUDENTS
WHO HAVE
WORKED WITH ME
ON
THE SHORE

FOREWORD

One of the chief difficulties which lies in the way of a student who is beginning work on the marine fauna of the British coasts is the identification of the specimens which he collects. The information he wants is as a rule to be found in large monographs, or in papers scattered through long series of zoological journals; and these will almost always include descriptions of very many species that he is never likely to obtain.

This book has been designed by Dr Eales to meet the needs of the student, and since the wealth of our marine fauna is very great it is mainly concerned with animals to be found between tide marks. By the use of the keys which it contains, and with a certain amount of preliminary training in zoology, the student should be able to discover the relationships, and the generic, if not the specific, name of any littoral animal that he has collected. If he has opportunities for dredging or trawling he will find species that are not included, and he will occasionally obtain some that are missing from its pages even when shore collecting. But in all his work he cannot fail to derive much benefit from this book, and the carefully compiled lists of literature will guide him to other sources from which he will be able to verify or complete his identifications.

Only those who have worked on systematic zoology can realise the vast amount of labour which this book has entailed. I feel assured that it will be welcomed most cordially and that it will do much to stimulate interest in our marine fauna.

Continental countries are much ahead of us in the production of such handbooks as this, and I hope the publication of the present volume will lead to a demand for more comprehensive works of the same kind.

STANLEY KEMP

The Plymouth Laboratory

PREFACE TO FIRST EDITION

Most students taking zoology for their degree examination attend a course of marine biology at one of the British Biological Stations. The abundance and varied nature of the marine fauna is a revelation to those who have collected only on land or in fresh water, and the most diligent student is bewildered by its range and wealth of species. If he tries to identify a specimen from books, he frequently acquires the unscientific habit of looking at the pictures and finding "something like it". Even if he specialises in one phylum, he finds far more than he can digest in the short time at his disposal, and may go away with a memory of shapes, colours and names which add little to his scientific knowledge.

Experience of marine collecting in many parts of the British Isles, and of student courses at certain of our Biological Stations, has led to the compilation of this book. Its object is threefold:

(1) To encourage observation of the habitat, habits and structure of the living animal in its natural surroundings.

(2) To supplement this with a closer observation in the laboratory, utilising those characters which are of systematic importance, so that unconsciously the classification of the phylum to which the specimen belongs is acquired.

(3) To provide a preliminary training in systematic work, which prepares the way for the use of advanced monographs on *special groups*. The keys in this book should assist the collector in tracing an animal to its phylum, class, order, family, genus, and even species, after which he should consult the specialist monograph. This method prevents waste of time on numbers of species which do not enter the horizon of the shore collector.

The imperfections and limitations of the keys are obvious. The artificial division "Littoral fauna" has been chosen because the general marine fauna of the British Isles is a subject too wide for the ordinary student to handle, and too big to make a convenient book. On the other hand, the changing nature of the littoral fauna makes it impossible to include everything the student is likely to find, and he should be warned that he must not attempt to make his specimen fit the key, but rather to let the key guide him to the literature. Most of the specimens included here have been collected by the author at Plymouth, Port Erin, Lough Ine, Bangor, Robin Hood's Bay, or on the coasts of west Scotland, south Wales, north Cornwall and the Channel Islands. The fauna lists of Plymouth and Port Erin laboratories have been utilised. The extreme northern and arctic forms have not been described.

Wherever possible the keys are based on external macroscopic characters. Obviously the difficulties are great in such groups as Porifera, Nemertini, Mollusca and Tunicata, to mention only a few, and in many the observation of macroscopic characters must be supplemented by simple microscopic investigation, as in the Tunicates. Microscopic forms like the Peridinians and most of the Copepods, the planktonic forms, and parasitic groups like the Trematodes, Cestodes and Nematodes, have been omitted.

Most of the sketches are original, and illustrate structural points required in classification rather than details of the species. To ensure accuracy of form, many have been drawn under the Edinger projection apparatus.

The author is indebted to Mr W. A. Rogers for the preparation of Fig. 1. She has been assisted throughout by Miss K. M. White, M.Sc., an ardent collector and malacologist, whose co-operation both on the shore and in the laboratory has been invaluable. Prof. F. J. Cole, F.R.S. has readily given advice and help in many ways, especially as regards the literature and facilities for the work in general.

N. B. E.

Zoology Department
University of Reading

PREFACE TO SECOND EDITION

In preparing the second edition certain sections have been brought up to date, in conformity with papers published since 1939. Short descriptions of the eggs or egg capsules have been added for species in which they are readily identifiable. Exercises in the quantitative study of littoral populations have been inserted in the appendix.

1949 N. B. E.

PREFACE TO THIRD EDITION

The publication of monographs on Mysidacea and Tunicata has made alterations necessary in those sections. The Polychaeta have been revised and augmented to include many southern species not formerly included.

1959 N. B. E.

PREFACE TO FOURTH EDITION

The author is indebted to Dr W. J. Rees, of the British Museum (Natural History) for the beautiful photographs of living animals. The zoning of the shore paragraphs have been rewritten, using animals instead of plants as the basis. The explanation of the plates and their reference letters have been rearranged to face the figures to which they refer.

1966 N. B. E.

GENERAL LITERATURE

Fauna Lists:

Plymouth Marine Fauna. 3rd edition, 1957. Marine Biological Association of the United Kingdom.

The Dale Fort Marine Fauna. 1957. Proc. Bristol Naturalists' Society.

Marine Fauna of the Isle of Man. 1937. University Press of Liverpool.

The Fauna, Flora and Geology of the Clyde area. 1901. Publication of the British Association Local Committee. Supplementary List, 1911. *Mar. Biol. Association of the West of Scotland.*

J. H. Barrett & C. M. Yonge. 1958. *Guide to the Seashore.* London: Collins.

Alister Hardy. 1956. *The Open Sea.* New Naturalist, No. 34. London: Collins.

C. M. Yonge. 1949. *The Sea Shore.* New Naturalist, No. 12. London: Collins.

Flatteley & Walton. 1922. *The Biology of the Seashore.* London: Sidgwick and Jackson.

Fowler & Allen. 1928. *Science of the Sea.* Oxford University Press.

Russell & Yonge. 1928. *The Seas.* London: Warne and Co.

Newbigin, revised Elmhirst. 1931. *Life by the Seashore.* London: Allen and Unwin.

Wilson. 1937. *Life of the Shore and Shallow Sea.* London: Ivor Nicholson and Watson.

CONTENTS

PLATES

INTRODUCTION

THE LITTORAL ZONE

The sea shore forms a rich hunting ground for the study of marine life. All the phyla whose members inhabit salt water have representatives in this region, though the littoral animals are often smaller than the roving types (nekton) or the deep-sea forms (benthos). Just as some terrestrial animals confine themselves to special habitats, like the jungle, the prairies, the desert, underground burrows, etc., so on the shore some species are restricted to definite areas. And just as some terrestrial species migrate seasonally, so in the sea similar movements occur. Nudibranchs, Tectibranchs and certain fishes come *inshore* for spawning in spring or autumn. Crabs, on the other hand, go *offshore* for spawning and *inshore* for hatching the egg masses attached to the abdomen.

The diurnal movement of the tides brings about peculiar conditions, however, on the sea shore. The fauna may be deprived of its watery medium twice every day, for a period of some hours, and during this period exposed to the air, the heat of the sun, and at times to fresh water owing to rain. An animal literally "retires into its shell" under these circumstances, burrows into the sand and mud, or takes advantage of the protection afforded by seaweeds and overhanging rocks. Its avian and human foes hunt for it at this time. The tides vary both in time and range from day to day, and also from place to place. Each day the *time* of high tide is approximately 50 minutes later than on the previous day, and since there are two high tides and two low tides every day, the interval between successive high or successive low tides is a little less than 12 hours. The *range* of the tide varies on different days of the lunar month, being greatest immediately after the moon is new or full, and least when the moon is in the first or last quarter. Thus there will be a higher high tide and a lower low tide when the moon is new or full; such a tide is called a *spring tide*, the slack tides of the first and last quarters being *neap tides*. Tides of exceptional range occur near the March and September equinoxes, and these are distinguished as *equinoctial spring tides*. The *height* of the tide also varies from place to place. In Liverpool Bay the difference in vertical height at the equinoctial spring tide between high-water level and low-water level is 31 feet, at Port Erin it is a little less, at Chepstow 42 feet, at Plymouth 19 feet. At Port Erin low water at the equinox occurs in the afternoon, at Plymouth at midday and at Bangor in the early evening. The good tides last about 5 days, increasing up to a maximum at the middle of the period. As some animals retreat into deeper water with the falling tides, the best days for collecting will be those before the maximum. The strip of coast between high-water mark and low-water mark which is uncovered when the tide recedes is the littoral collecting ground and will differ greatly in extent on various parts of our coasts according to the slope of the land, the state of the moon, and to certain peeuliar tidal conditions in restricted areas.

This fluctuating exposure of areas of the shore enables us to classify coastal regions as follows:

A. The **spray-splashed zone,** or **supra-littoral fringe,** above high water spring tide level, covered only at the equinoctial spring tides. The green seaweed *Enteromorpha* lives in the pools, the mollusc *Littorina neritoides* in crevices in the rocks.

B. The **upper shore** or **upper littoral zone,** between high water spring tide and high water neap tide levels. *Pelvetia canaliculata* and *Fucus spiralis* are typical weeds. *Patella vulgata* and *Littorina saxatilis* are on rock and weed respectively. The barnacle *Chthalamus* occurs.

C. The **middle shore,** between high water neap and low water neap tide levels. This can be divided into two:—

(*a*) upper mid-littoral zone

(*b*) lower mid-littoral zone.

Midway between these two is the **mid-tide level.** Characteristic weeds are *Ascophyllum nodosum*, *Fucus vesiculosus* and *F. serratus*. *Balanus balanoides*, *Littorina littoralis* and *Gibbula umbilicalis* extend through (*a*), *G. cineraria* through (*b*).

D. The **lower shore** or **infra-littoral fringe,** from low water neap to low water spring and low water equinoctial spring tide levels. Large brown weeds, such as species of *Laminaria*, and some red weeds are present. *Littorina littorea* and *Gibbula magus* are typical molluscs.

	Tidal level	Zone (Yonge)	Plants	Animals	Zone (Stephenson)		
LITTORAL AREA	HWES HWS	Splash zone	*Enteromorpha*	*Littorina neritoides*	Supra-littoral fringe	A	INTERTIDAL AREA
	HWN	Upper shore	*Pelvetia* *Fucus spiralis*	*Chthamalus*, *L. saxatilis*, *Patella vulgata*	Upper littoral zone	B	
	MTL	Middle Shore: Upper middle (*a*)	*Ascophyllum*, *Fucus vesiculosus* *F. serratus*	*Balanus balanoides* *L. littoralis* *Gibbula umbilicalis*	Upper mid-littoral zone	C(*a*)	
	LWN	Middle Shore: Lower middle (*b*)	ditto	*G. cineraria*	Lower mid-littoral zone	C (*b*)	
	LWS LWES	Lower shore	*Laminaria* spp.	*Littorina littorea* *Gibbula magus*	Infra-littoral fringe	D	
		Sub-littoral zone	Red weeds	Mysids, prawns	Sub-littoral zone	E	

I (Fig. 1)

(*facing p. 3*

E. The **sub-littoral zone,** below the lowest tide mark, but available to the shore collector in deep rock pools, by wading or netting. The seaweeds are chiefly red species. Animals are demersal, e.g. *Hippolyte varians*, prawns and mysids.

Zones A–E constitute the *littoral* area, and are accessible to the shore collector.

F. The shallow water zone or continental shelf, the sloping plateau upon which the British Isles are built. It extends to about the hundred fathom line in depth, and is rich in species.

G. The abysses of the deep sea, possessing a fauna but no flora, since plants cannot subsist in their unlighted depths.

It is with areas A, B, C, D and E that we are concerned here, and the fauna of each zone is as distinctive as the flora. We may, however, classify the shore from another standpoint—that of its geological composition:

1. Pebbly beach.
2. Tidal estuary of stream or river.
3. Dune formation with brackish pools.
4. Sandy shore, ranging from coarse gravelly sand to fine sand.
5. Muddy shore.
6. Rocky shore with movable rocks.
7. Rocky shore with large immovable rocks, reefs with crevices and gullies.
8. Rock pools.

Many animal species are confined to one or more of these areas, though shores may exhibit a mixture of several of them. Thus Port Erin possesses, on a small scale, 2, 4, 6, 7 and 8 in different parts of the bay, Amroth Bay in South Wales 1, 2, 4, 5, 7 and 8. Speaking generally, zones 1, 2, 3 are restricted in species, whereas 4, 5, 6, 7 and 8 are rich, although even here there is considerable variation. Thus the rock may be volcanic tuff, slate, shale, limestone, chalk or sandstone, and its animal population will differ according to the nature of the substratum and the growth of weed. It should be noted that very few animals live where the rock is bare, but a rock covered with weed affords protection to many species.

EXPLANATION OF PLATE I

Scheme to show tidal levels and marine zones, with typical plants and animals. After Yonge and Stephenson. Cf. also Fig. 111. p. 287.

TYPICAL FAUNAS

On a typical rocky coast, with loose and firm rocks, gullies, rock pools and patches of sand or mud, the collector may expect to find many of the following animals. The zones, which are related to the tide levels, are those given on p. 2, but they should not be considered as rigid limits in distribution.

A. Splash zone: *Ligia oceanica*, *Petrobius maritimus*, *Talitrus saltator*, *Orchestia gammarella*, in crevices *Littorina neritoides*, and in pools *Littorina saxatilis*, *Carcinus maenas*, *Jaera marina*, *Gammarus locusta*.

B. Upper shore: *Littorina neritoides*, *L. saxatilis*, *Asterina gibbosa* (in pools in spring), *Patella vulgata*, *Nucella lapillus* *Gammarus marinus*, *G. locusta*, *Balanus balanoides*.

C. Middle shore: *Littorina littoralis* (*obtusata*), *L. littorea*, *Gibbula umbilicalis*, *Ocenebra erinacea*, *Calliostoma zizyphinum*, *Lasaea rubra*, *Modiolus phasiolinus*, *Mytilus edulis*, Chitons, *Leucosolenia botryoides*, *Grantia compressa*, *Sycon coronatum*, *Halichondria panicea*, *Hymeniacidon sanguinea*, *Obelia geniculata*, *Dynamena pumila*, *Plumularia* spp., *Coryne muscoides*, *Clava squamata*, *Clytia johnstoni*, *Actinia equina*, *Anemonia sulcata*, *Lepidonotus* spp., *Nereis pelagica*, *N. diversicolor*, etc., *Cirratulus cirratus*, *Arenicola ecaudata*, *A. branchialis*, *Serpula vermicularis*, *Pomatoceros triqueter*, *Spirorbis borealis*, *Membranipora membranacea*, *Flustrella hispida*, *Idotea* spp., *Carcinus maenas*, *Eupagurus bernhardus* (small), *Portunus puber*, *Botryllus schlosseri*, *Botrylloides leachi*, *Blennius pholis*, *Gobius* spp.

D. Lower shore: *Gromia oviformis*, *Adocia cinerea* (*Reniera*), *Halisarca dujardini*, *Oscarella lobularis*, *Haliclona oculata*, *Pachymatisma johnstoni*, *Ophlitaspongia seriata*, *Spongelia fragilis*.

Leptoplana tremellaris.

Tubularia larynx, *Garveia nutans*, *Depastrum cyathiforme*, *Lucernariopsis campanulata*, *Alcyonium digitatum*, *Tealia felina*, *Bunodactis verrucosa*, *Sagartia* spp., *Eunemertes neesi*, *Amphiporus lactifloreus*, *Tetrastemma* spp., *Lineus* spp., Polynoid worms, *Phyllodoce* spp., Syllid worms, *Amphitrite johnstoni*, *Filograna implexa*, *Sabellaria* spp., *Bugula turbinata*, *Crisidia cornuta*, *Alcyonidium gelatinosum*, *Pedicellina cernua*, *Barentsia gracilis*.

Asterias rubens, *Henricia sanguinolenta*, *Ophiothrix fragilis*, *Ophiocomina nigra*, *Amphiura chiajei*, *Amphipholis squamata*, *Echinus esculentus*, *Psammechinus miliaris*, *Cucumaria saxicola*, *Cancer pagurus*, *Athanas nitescens*, *Leander serratus*, *Hyas araneus*, *Pilumnus hirtellus*, *Porcellana longicornis*, *P. platycheles*, *Galathea squamifera*, *Portunus depurator*, *Balanus crenatus*, *Verruca stroemia*, *Nebalia bipes*, *Idotea* spp., *Pycnogonum littorale*, *Nymphon gracile*.

Acmaea virginea, *Patina pellucida*, *Nassarius reticulatus*, *Trivia monacha*, *Lacuna* spp., *Lamellaria* spp., *Rissoa* spp., *Monodonta lineata*, *Goniodoris nodosa*, *Ancula cristata*, *Archidoris britannica*, *Aeolidia papillosa*, *Elysia viridis*, *Aplysia punctata*, *Anomia ephippium*, *Chlamys varius*, *Tonicella rubra*, *Lepidochitona cinerea*.

Clavelina lepadiformis, *Dendrodoa grossularia*, *Morchellium argus*, *Aplidium proliferum*.

Centronotus gunnellus, *Cottus scorpius*, *Lepadogaster* spp., *Liparis montagui*, *Nerophis lumbriciformis*, *Onos mustelus*, *O. tricirratus*, *Gobius* spp., *Spinachia vulgaris*, *Agonus cataphractus*, and many species mentioned above under *Fucus* zone.

ESTUARIES: Anemones and jellyfish.

Nereis diversicolor, Oligochaete worms.

Gammarus spp., *Sphaeroma* spp., *Crangon vulgaris*, *Carcinus maenas*, *Littorina littorea*, *L. saxatilis*, *Cantharidus striatus*, *Hydrobia* spp., *Haminoea* spp., *Patella vulgata*, *Nucella lapillus*, *Leucophytia bidentata*, *Phytia myosotis*, *Cardium edule*, *Mytilus edulis*, *Mya arenaria*, *Ostrea edulis*, *Scrobicularia plana*, *Abra tenuis*, *A. alba*, *Macoma balthica*, *Ascidiella aspersa*.

Anguilla, Blennies, *Gobius pictus*, *Pleuronectes flesus*, *Cottus bubalis*, *Liparis* spp., *Morone labrax*.

SANDY SHORE: *Peachia hastata*, *Halcampa chrysanthellum*.

Cephalothrix rufifrons, *Harmothöe lunulata*, *Sthenelais boa*, *Nephthys caeca*, *Spiophanes bombyx*, *Nerine foliosa*, *Magelona papillicornis*, *Perinereis cultrifera*, *Glycera convoluta*, *Arenicola marina*, *Lanice conchilega*, *Phascolosoma elongatum*, *Spatangus purpureus*, *Echinocardium cordatum* with *Montacuta ferruginosa*, *Leptosynapta inhaerens*.

Ensis ensis, *Venerupis* (*Paphia*) *pullastra*, *Cardium edule*, *Tellina* spp., *Lutraria lutraria*, *Montacuta ferruginosa*, *Mysella bidentata*, *Acteon tornatilis*, *Ammodytes tobianus*, *A. lanceolatus*.

Pleuronectes spp., *Trachinus draco*, *T. vipera*.

MUDDY SHORE: *Corophium volutator*.

Capitella capitata, *Scoloplos armiger*, *Scolelepis fuliginosa*, *Arenicola* spp., *Nerine foliosa*, *Pygospio elegans*, *Nereis diversicolor* and many other worms.

Tellina fabula, *Pinna fragilis*, *Scrobicularia plana*, *Cardium* spp.

BRACKISH WATER: Many small Crustacea, especially Amphipods.

Littorina littorea, *Hydrobia jenkinsi*, *Alderia modesta*, *Limapontia depressa*, *Phytia myosotis*, *Leucophytia bidentata*, *Mytilus edulis*, *Abra tenuis*.

ROCK POOLS: Sponges, e.g. *Leucosolenia* spp., *Cliona celata*.

Hydroids, e.g. *Clava*, *Sarsia* (*Syncoryne*), *Tubularia*. Anemones.

Dinophilus taeniatus, Turbellarian Worms.

Hippolyte varians, *Nebalia bipes*, Amphipods, Isopods, Mysids.

Shrimps and prawns.

Bryozoa.

Hiatella gallicana (in limestone), Nudibranchs.

Fishes.

The flooding of the pools at high water often brings in animals which are left behind when the tide recedes. These can be called *tidal migrants*.

The number of species increases in an arithmetical progression from the splash to the sublittoral zone, then decreases again in the deep sea. The sublittoral zone is capable of sustaining the largest number of individuals and species because it affords fairly uniform conditions, together with adequate protection, aeration, food supply and illumination. As the Lower shore is exposed only at the spring tides, however, it can be regarded as partly sublittoral, and is very rich in numbers and species.

Animal associations are closely correlated with the substratum. A sandy or muddy shore will favour free-living (non-fixed) forms which feed on plankton or on detritus, e.g. many Lamellibranchs, scavenging worms like *Arenicola*, burrowing Echinoderms like *Echinocardium*. Even the anemones in this region have no fixation base, but only a burrowing physa.

Passively feeding fixed forms cannot live in sand or mud, as they are unable to obtain a hold. They choose the firm rock, which becomes, therefore, the home of barnacles, hydroid zoophytes, sponges, Bryozoa, Tunicates and many tube worms.

Animals which select their food and move about in search of it are found in all the littoral areas. The number of carnivorous species is greatly in excess of the herbivorous. Amongst the plant feeders, *Patella* scrapes the surface of the rock and consumes the minute algae growing on it, then returns to its original place or "home", *Echinus* chops the weed with its five scissor-like teeth, *Aplysia* climbs the seaweed stipes and feeds on the fronds. Of the carnivorous forms, some browse on sponges (e.g. many Nudibranchs, worms), some on Hydroids and Bryozoa (e.g. *Caprella*, Pycnogonids), others on Tunicates (e.g. *Lamellaria*). Strong jaws and organs of offence and defence are not needed for this, but are important for those Carnivores which hunt for prey as active as themselves. The errant Polychaetes (e.g. *Nereis*), the majority of the Crustacea (e.g. crabs), some Gastropods (e.g. dog whelks) and the fishes belong to this group. Some develop special methods of attack, e.g. *Asterias* opens the mussel by wearing down its resistance.

A few protect themselves by boring into other animals (e.g. tube worms in sponges), or into the rock itself (e.g. *Hiatella*, *Pholas*, etc.). Protective coloration is common (*Lamellaria* on *Didemnum*, *Spinther* on *Hymeniacidon*), and certain species of spider crabs plant a garden of algae and zoophytes on their own backs, thus matching their surroundings so perfectly that when in repose they cannot be distinguished; movement alone betrays their presence.

COMMENSALS

A remarkable feature of the shore fauna is the large number of animals which live in association. In most cases there is no evidence that one of the associates gains more than any of the others, and there is therefore no question of parasitism. Such animals living together are called *commensals*. A list of common commensals is given below:

Echinocardium cordatum (heart urchin) and *Montacuta ferruginosa* (Lamellibranch). In the anterior ambulacrum.

Spatangus purpureus (purple heart urchin) and *Montacuta substriata.*

Echinus esculentus (sea urchin) and *Scalisetosus assimilis* (worm).

Antedon bifida (feather star) and *Myzostoma cirriferum* (worm).

Labidoplax digitata (Holothurian), *Harmothoe lunulata* (worm) and *Mysella bidentata* (Lamellibranch).

Leptosynapta inhaerens (Holothurian), *Harmothoe lunulata* and *Entovalva perrieri* (Lamellibranch).

Acrocnida brachiata (brittle star), *Harmothoe lunulata* (small size) and *Mysella bidentata.*

Perinereis cultrifera (worm), *Harmothoe lunulata* and *Mysella bidentata.*

Amphitrite johnstoni (tube worm), *Gattyana cirrosa* (worm) and *Harmothoe lunulata* (large size).

Amphitrite edwardsi, *Lepidasthenia argus* (worm) and *Ophiodromus flexuosus* (worm).

Marphysa sanguinea (worm) and *Harmothoe lunulata.*

Arenicola marina (worm) and *Harmothoe lunulata* (largest size).

Phascolosoma elongatum (Sipunculid worm) and *Mysella bidentata.*

Phascolosoma vulgare, *Loxosoma phascolosomatum* (Bryozoan) and *Harmothoe lunulata.*

Eupagurus bernhardus (hermit crab) and *Calliactis parasitica* (anemone), or *Hydractinia echinata* (Zoophyte), or *Ficulina ficus* (*Suberites domuncula*) (sponge), and *Nereis fucata* (worm).

Eupagurus cuanensis and *Ficulina ficus.*

Eupagurus prideauxi and *Adamsia palliata* (anemone), or *Calliactis parasitica*, or *Hydractinia echinata*, and *Nereis fucata.*

Upogebia deltaura (Decapod Crustacean) and *Lepton squamosum* (Lamellibranch).

Because of this association of different species, some zoologists speak of an *Echinocardium* habitat, an *Amphitrite* habitat, etc., meaning in each case a habitat associated with *Echinocardium* or *Amphitrite* as the typical species.

PARASITES

Some littoral animals are true parasites, living upon or within their hosts, but rarely destroying them. All degrees of parasitism are to be found, from a condition such as that of the Pycnogonids, whose larvae live in the enteron of Hydroids, to the extreme parasitism of

Sacculina and *Peltogaster* which have lost almost all their Arthropod characters and hang as chitinous bags on the abdomen of the host crab. *Bopyrus* lives under the shelter of the carapace of the prawn, *Hemioniscus* within the shell of the acorn barnacle. Some Isopods are scavengers on the bodies of fishes and may be external parasites. In the same way *Malacobdella* inhabits the mantle cavity of Lamellibranchs for food or shelter or both, *Myzostoma* crawls over the calyx of *Antedon*, and it is not easy to draw a rigid line between the parasites and the commensals. The little pea crab (*Pinnotheres*) is probably a parasite, for it takes the food of its host (a Lamellibranch) while sheltering within the shell of the latter.

BIOLOGICAL STATIONS

Courses in marine biology or facilities for biological work are obtainable at the following places round the British coasts: Plymouth, Dale Fort, Bangor, Port Erin (Isle of Man), Millport (Firth of Clyde), Cullercoats (Northumberland), Robin Hood's Bay (Yorkshire), St Andrews, Lough Ine (County Cork, Ireland).

Each of these regions has its own peculiar conditions and fauna, though apart from a few special animals most of the species are fairly general in their distribution.

Plymouth. The Laboratory of the Marine Biological Association is the largest in Great Britain, and is well situated on Citadel Hill, near Plymouth Hoe, overlooking Plymouth Sound. Marine biology courses are held in spring and autumn. The coast is high and rocky, and shore collecting grounds are not easily accessible, requiring journeys by boat or road. The chief collecting grounds are the limestone rocks below the laboratory, the shale near Mount Batten, Drake's Island, Mount Edgcumbe, Rame Head, Whitsand Bay and Wembury Bay. A famous collecting ground accessible from Plymouth is the Salcombe Estuary. A feature of the neighbourhood is the large number of river estuaries. Sandy, rocky and muddy areas occur; the best general ground is that on the reefs at Wembury Bay. Specialities of this neighbourhood are *Myriothela cocksi*, *Caryophyllia smithi*, *Phoronis*, *Lima hians*, *Pectinaria koreni*, *Alpheus macrocheles* and *Callianassa subterranea*. Many animals are exhibited alive in the aquarium. The third edition of the fauna list was published in 1957.

Dale Fort Field Centre, at the extreme south-western tip of Pembrokeshire, is predominantly a teaching station, specialising in courses in shore ecology. The deeply peninsulated coast produces rocky shores of every degree of exposure to wave action, from the Atlantic cliffs to the shelter of Milford Haven. Complicated geology gives shores of every angle of slope and of contrasting rock types. The flats exposed at the head of Dale Roads at LWST have a varied substrate, gravel and clean sand predominating, but with patches of clay and mud as well. The centre has domestic accommodation for fifty visitors.

Bangor. This area is rich in historical associations, as well as in the nature of its fauna, and the proximity of University College, and the Marine Science Laboratories, makes it an excellent centre for marine work. An Easter class is held there every year. A swift tide washes the muddy and rocky shores of the Menai Straits. On the Anglesey side are good grounds at Church Island and Penmon. Puffin Island, off the north-east coast, had one of the first biological stations in Great Britain, founded by Professor Herdman and the Liverpool Marine Biology Committee in 1877, but abandoned in 1892, when the first station at Port Erin was built. *Priapulus caudatus*, *Cerianthus lloydi*, large *Nereis virens*, *Chaetopterus variopedatus* and *Saccoglossus* sp. are found.

Port Erin, Isle of Man. The Manx shores have been worked for many years. Professor Edward Forbes collected in the island from 1832, and may be regarded as the pioneer of marine biology in Great Britain. Professor Herdman and the L.M.B.C. founded a small station on the shore of Port Erin Bay in 1892, and this building was later replaced by a much larger laboratory near the breakwater. The feature of the area is that varied collecting grounds are accessible on foot, in Port Erin Bay, Port St Mary Bay, Calf Sound, Scarlett Point, Castletown Bay and Fleshwick Bay. Sandy and muddy shores, loose rocks, rocky ledges, reefs, crevices and rock pools are all within easy reach. The rocks are mainly Manx slates, but Carboniferous limestone and volcanic dykes occur on the east coast near Port St Mary. *Echinus esculentus* is common and grows to a very large size. *Peachia hastata* occurs in the sand in Port Erin Bay. Worms, Zoophytes, Decapod Crustacea and Molluscs are well represented. A fauna list was published in 1937.

Millport. The Scottish Marine Biological Association's Station is situated on the island of Cumbrae in the middle of the Firth of Clyde. The sheltered waters of the Clyde estuary form an admirable site for plant and animal life in the sea, and there is a rich flora and fauna. The sea water is remarkable for its purity, and there are good shore grounds all round the island. On the mainland there are extensive sands. For deeper water work a depth of 62 fathoms can be reached within half-an-hour from the laboratory. Classes are held during the Easter vacation for students.

Cullercoats. The Dove Marine Laboratory is built on a quay just above sea-level at Cullercoats, Northumberland. It is connected with the University of Newcastle (Armstrong College, Newcastle-upon-Tyne). The coast is rocky and exposed, and was a famous collecting ground during the nineteenth century, when Alder and Hancock, Norman, the two Bradys and others worked there. Loose stone, sand, mud, estuarine and salt-marsh areas also occur in the vicinity. Species found in the neighbourhood include: *Cordylophora lacustris*, *Pennatula phosphorea*, *Alderia modesta* and many other Nudibranchs, *Fusus antiquus*, *Paragnathia formica*, *Nephrops norvegicus*, *Lithodes maia*, *Balanus hameri*, *Nereis virens*, *Priapulus caudatus*, *Hormathia digitata* and *Solaster endeca*.

Robin Hood's Bay. At Robin Hood's Bay, 5 miles south of Whitby, in Yorkshire, is a small station belonging to the University of Leeds. The semicircular bay is bounded on the landward side by high crumbling cliffs of boulder clay and shale. Two freshwater streams enter the bay. The shore shelves gradually, exposing, at the spring tides, a complete algal zonation from *Enteromorpha* to *Laminaria*. The substratum is blue Lias and shale, alternating with thin bands of limestone, and sandstone and ironstone also occur. The cracks and fissures in this friable material afford hiding places for many species of worms, while Molluscs such as *Hiatella* and *Barnea* bore in it. On the north exposed portion of the bay are rock pools among the large boulders, where Nudibranchs are found. Similar pools between the flat ironstone rocks of the southern extremity abound in Mysids and other small Crustacea.

Lough Ine, near Baltimore, County Cork, is one of the richest collecting grounds, and was first investigated by Rowland Southern in 1916 and later by Professor Renouf of University College, Cork, who built and equipped a laboratory there. Lough Ine is a landlocked bay communicating with the open sea by means of rapids, over a rocky sill about 6 feet high, through which a fierce tide runs. *Inside* the lough collecting is best on the neap tides, since the lough drains over the sill for 3½ hours after the beginning of the slack flood tide outside. *Outside* the lough the spring tides afford the greatest range of exposed shore. Further, deep water close to the shore within the lough causes fiord conditions, and the warm currents bring many southern species to the Irish shore. Inhabitants of this area are *Paracentrotus lividus*, *Chaetopterus variopedatus*, *Parerythropodium hibernicum*, *Corynactis viridis*, *Caryophyllia smithi*, *Anomia ephippium*, *Phoronis hippocrepia*, *Drepanophorus* sp. and *Crania anomala*.

St Andrews. The Gatty Marine Laboratory at St Andrews, where Professor McIntosh compiled his memoir on the Nemerteans and Annelids, is well equipped for teaching and research. The foreshore slopes gently and the rock is soft calcareous sandstone. There are sandy bays and an estuary within walking distance of the laboratory. The station possesses a Diesel launch, equipped with fishing and collecting gear for the exploration of the waters of the bay, and there is a dinghy with outboard motor. Advanced courses in Marine Biology are held during the Easter vacation.

Although a laboratory and a library are very useful adjuncts to a good collecting ground, they are not all-important, and there are many parts of the coast where the fauna may be even richer than in the places above mentioned. Thus the South Wales coast, the west coast of Scotland and the coast of Cornwall are all very good, and from such centres as Tenby, Oban and Falmouth much good faunistic work has been done.

While each region has its own peculiarities, these are constantly changing, and no one can predict definitely what the fauna of a certain coast will be. Some animals disappear gradually or suddenly, others appear and establish themselves. We know little of the conditions causing these spasmodic migrations. For example, the beautiful gregarious anemone *Corynactis viridis* was recorded by Gravely in the pools on the south of the Isle of Man in 1906. Later it disappeared. In 1935 it was found in 35 fathoms off the Calf of Man, and in 1936 in the rock pools of Calf Sound, having re-established itself there. In 1937 it was so abundant that hundreds of specimens in orange, pink, green and magenta colour varieties were found.

PROTOZOA*

(Plate II, fig. 2)

AUTHORITIES

HERON-ALLEN, E. & EARLAND, A. 1930. "The Foraminifera of the Plymouth District." *J. R. micr. Soc.* L, pt. 1, pp. 46–84; pt. 2, pp. 161–99.

JOHNSTONE, J., SCOTT, A. & CHADWICK, H. C. 1924. *Marine Plankton.* University Press of Liverpool, Ltd. (The illustrations were printed in a handbook entitled *Microscopic Life in the Sea*, Port Erin Marine Biological Station, Isle of Man, 1924, and subsequent dates.)

LANKESTER, E. RAY. 1903, 1909. *A Treatise on Zoology*, pt. 1, fascicles 1 and 2. Various authors: Foraminifera, J. J. Lister; Radiolaria, F. W. Gamble; Infusoria, S. J. Hickson.

LEBOUR, M. V. 1926. *The Dinoflagellates of Northern Seas.* Plymouth.

The Protozoa are non-cellular animals moving by protoplasmic extrusions of the naked body (Rhizopods), by cilia (Infusoria) or by flagella (Flagellates). Some are fixed, and many are parasitic. There may be a skeleton or "house", either secreted or formed of agglutinated material such as sand, sponge spicules, etc.

Many Protozoa occur on the shore, but most are too small to be included here. Reference may be made to the following:

RHIZOPODA

LOBOSA

Marine Amoebae may be met with in slide mounts of other small forms such as Coelenterates, Bryozoa, etc.

FORAMINIFERA

Scrapings of sand from a sea beach reveal the shells of many members of this order, e.g. *Miliolina*, *Lagena*, *Rotalia*, *Polystomella*, etc.

Gromia oviformis. Lankester (Lister), 1903, p. 49. Attached to weed or stone may be found smooth, rounded or ovate balls on

* In general, characters used in the Keys are not repeated in descriptions of species.

short stalks, about 2 mm. in diameter and the colour of mud. The test is composed of a flexible chitinous secretion and has only one aperture for the extrusion of pseudopodia.

Haliphysema tumanowiczi (Pl. II, fig. 2). Lankester (Lister), 1903, p. 85. Attached to zoophytes and the Bryozoan *Crisia*. Small, shiny, yellow, club-shaped heads, fixed by basal disks to the animal colony, and about 3–4 mm. in height. When examined microscopically, the club-shaped portion exhibits projecting sponge spicules, but these are merely odd spicules incorporated into the skeletal case of a small gregarious Foraminiferan.

RADIOLARIA

There are few representatives of this order around our coasts. The plankton has some floating forms, notably *Acanthometron pellucidum* (*Marine Plankton*, pl. 2).

MASTIGOPHORA

FLAGELLATA

Green and brown patches on a sandy shore are often due to aggregations of Gymnodinian flagellates. Others are found in the plankton, such as the species of *Ceratium*; and the phosphorescent *Noctiluca scintillans* (*Marine Plankton*, pl. 2).

INFUSORIA

Ciliates and Suctorians are numerous, though small. The former sometimes swarm amongst seaweed, some are parasitic within the bodies of worms, Coelenterates, Molluscs, etc., others are attached by stalks to zoophytes. The Suctorians are fixed to algae, zoophytes, etc. A small triangular one on a stalk is a species of *Acineta*, while a round-headed form with much-branched nucleus is *Ephelota* (Lankester (Hickson), 1903, p. 424).

PORIFERA

(Plate II, figs. 3–5)

AUTHORITIES

BOWERBANK, J. S. 1864–82. *A Monograph of the British Spongiadae.* Ray Soc. Publ. vols. I–IV, 1864, 1866, 1874, 1882.

BURTON, M. 1930. "Additions to the sponge fauna at Plymouth." *J. Mar. biol. Ass. U.K.* XVI, 489–507.

LENDENFELD, ROBERT VON. 1889. *A Monograph of the Horny Sponges.* London, Roy. Soc.

MINCHIN, E. A. 1900. "The Porifera", in *A Treatise on Zoology*, ed. Ray Lankester, pt. II.

—— 1904. "The characters and synonymy of the British species of sponges of the genus *Leucosolenia.*" *Proc. zool. Soc. Lond.* II, 349–96.

MONTAGU, G. 1818. "An essay on sponges, with descriptions of all the species that have been discovered on the coast of Great Britain." *Mem. Werner Soc.* II, 67–122.

Sponges are fixed, irregularly branched forms which attach themselves by a flattened base, or by a stalk, or more rarely by a rooting tuft of long spicules.

The main cavity of the body is the enteron (gut cavity or flagellate chamber), through which a continuous stream of water carrying small animals and plants is passed. The water enters the sponge through minute *pores* (*ostia*) in the walls, and after going through the enteron (gastric cavity) is forced out through large apertures, the *oscula.* In the more highly organised sponges the water passages may be elaborated by the formation of canals in the soft tissues of the sponge mass, but the digestive portion is always a chamber with flagellate cells in its walls.

Sponges reproduce both asexually and sexually. Budding with or without separation of the buds is common. Eggs usually develop from isolated germ cells lying in the non-cellular gelatinous region between the ectoderm and the endoderm; they are fertilised within the body of the parent colony and frequently undergo their early development there.

Sense organs are absent and there is very little differentiation of the cells. The endoderm cells are collared and flagellate; the current caused by the flagella can be demonstrated in the following way: Slice off a piece of crumb of bread sponge (*Halichondria panicea*) from its base on a rock, and place it in a deep vessel in good sea water. Inject a small quantity of aqueous methylene blue solution beneath the sponge, by means of a cannula fitted with a rubber bulb. The blue liquid will pass through the canals of the sponge and will issue by the oscula, giving the colony the appearance of a group of smoking volcanoes.

Nearly all sponges make skeletal supports called *spicules* (Pl. II, fig. 4) or secrete a horny substance called *spongin* (Pl. II, fig. 5); a few do both. The spicules may be calcareous or siliceous; the former are soluble in acid, the latter are not. A knowledge of the spicules, their composition and arrangement is essential in classification, and often for the identification of the species, since the form of the sponge is so variable under different conditions that external characters are unreliable. Many siliceous sponges possess two grades of spicules, large ones or *megascleres* and small ones or *microscleres*. Calcareous sponges have megascleres only.

Megascleres may be one-, three-, four- or many-rayed, these types being known as *monaxon*, *triaxon*, *tetraxon* and *polyaxon* respectively. The chief types of monaxon spicules are as follows:

Rhabdus: pointed at both ends.

Style: pointed at one end, blunt at the other.

Tylostyle: pointed at one end, enlarged to form a knob at the other. Some tylostyles are spiny.

Tylote: knobbed at both ends.

The triaxon or triradiate spicule may have equal or unequal rays.

A tetraxon or tetraradiate spicule with three short rays and one long ray (as in *Pachymatisma*) is called an *orthotriaene*. The polyaxon spicule of the same sponge is known as a *sterraster*.

The chief types of microsclere found in littoral sponges are the *rod*, the *six-rayed triaxon*, the *sigma*, which is nearly a crescent, and the *chela*, which is a short rod with an umbrella-like formation at one or each end.

Few of the British littoral sponges have a horny non-spicular skeleton, though in one family, the Haploscleridae, a varying amount of spongin is present in addition to the spicules. In some the spongin encloses the spicules (intraspicular condition), in others the spicule has only one tip fixed in the spongin and the other end is free (extraspicular condition).

The colour of sponges is often brilliant. Pure white is rare, but yellow, orange, scarlet, green, lilac and purplish black are common in the littoral forms. Sponges are very variable both in colour and form. Only the common and easily recognisable ones are included here. Hand sections or teased preparations are useful for determining the pattern and arrangement of the skeleton. Isolated spicules can be prepared by boiling a piece of the sponge in caustic potash. Nudibranchs, limpets and winkles browse on sponges. The richest littoral zone for sponges is at low water neap tide.

KEY TO LITTORAL SPONGES

1. Spicules and spongin absent. Sponge slimy. **24**
   Spicules present. **2**
   Skeleton horny, reinforced by foreign matter, broken spicules, sand grains, etc. **Keratosa. 25**
2. Spicules soluble in hydrochloric acid. **Calcarea. 3**
   Spicules insoluble in hydrochloric acid. **10**
3. Sponge a simple vase-shaped structure, with thin wall perforated by pores leading to a central flagellate chamber. Osculum apical. **Homocoela. Leucosoleniidae. 4**
   Sponge may be vase-shaped, but the wall is thickened, and the pores lead to flagellate chambers excavated in the wall. Central chamber non-digestive (non-flagellate). **Heterocoela. 7**
4. One osculum to each sponge tube. **5**
   One osculum to a group of branched, anastomosing decumbent tubes. Sponge coloured. **Leucosolenia coriacea**
   One osculum to a group of anastomosing erect tubes. Sponge pure white. **L. clathrus**
5. Arborescent, with an osculum at the end of each branch. **6**
   Tubes rarely branched. **L. variabilis**
6. Tubes short, cylindrical, in dense masses on weed. 8 mm. high. **L. botryoides**
   Tubes long, delicate, much branched, forming compact, crowded colonies on weed. 20 mm. high. **L. complicata**
7. Radial flagellate chambers projecting through the outer wall of the sponge (i.e. no dermal cortex). **Sycettidae. 8**
   Radial chambers not projecting because covered by a dermal cortex. **Grantiidae. 9**
8. Sponge erect, vase-shaped, cylindrical, usually with a fringe of tall spicules surrounding the circular apical osculum. Dull yellow. 30 mm. high. **Sycon ciliatum**
9. Sponge erect, compressed, vase-shaped, with flattened apical osculum. Pale yellow. 50 mm. high. **Grantia compressa**
   Sponge incrusting, snow white. Oscula numerous. 50 mm. in diameter. **Leuconia nivea**

10. Megascleres three-, four- or many-rayed. **Tetraxonida.** 11
Megascleres needle-shaped. **Monaxonida.** 13
11. Mega- and micro-scleres, amongst the former spherical or ovate spheroids (sterrasters). **Geodiidae.** 12
12. Sponge massive, incrusting, usually elephant grey in colour. Surface smooth, with a hard rind, the oscula prominent, arranged in lines, or scattered. **Pachymatisma johnstoni**
13. Skeleton of spicules only. 14
Skeleton of spicules reinforced by spongin. **Haploscleridae.** 22
14. Sponge boring into shells or limestone rock. **Clionidae (Cliona celata)**
Sponge not of boring habit. 15
15. Microscleres present. Incrusting. **Desmacidonidae.** 16
No microscleres. 17
16. Sponge buff, yellow or orange. **Myxilla incrustans**
Sponge dark red. **Microciona atrasanguinea**
17. Surface quite smooth, oscula inconspicuous. Shape variable. **Suberitidae.** 18
Surface thrown into elevations and depressions or with tag-like processes. Always incrusting. 19
18. Fig-shaped, or on a shell containing a hermit crab. Yellow. **Suberites domuncula (Ficulina ficus)**
Thin, filmy, incrusting and of Reckitt's blue colour. **Terpios fugax**
19. Sponge low and incrusting. **Axinellidae.** 20
Sponge mound-shaped, with tag-like projections bearing the oscula. Bright yellow. **Polymastiidae (Polymastia mammillaris)**
20. Oscula prominent, on volcano-like elevations, yellow, orange or green. **Halichondria (panicea)**
Oscula not prominent. **Hymeniacidon.** 21
21. Colour blood red, scarlet or orange. **H. perleve (sanguinea)**
22. Sponge erect, with many long finger-shaped branches and lateral oscula. Yellow or green. 15 cm. Spicules embedded in spongin. **Haliclona**
Sponge incrusting. 23
23. Brown or lilac. Spicules arranged in rectangular formation, fastened by their tips with spongin. **Adocia (cinerea)**
Brick red. Spicules projecting irregularly from spongin. **Ophlitaspongia seriata**
24. Sponge a slimy, yellowish brown film. **Halisarca (dujardini)**
Sponge lobulate, brown. **Oscarella lobularis**
25. Surface of sponge corrugated, the tips of the elevations stiffened by incorporated sand grains. **Spongelia (fragilis)**

CALCAREA

In these sponges there is a spicular skeleton composed of calcium carbonate, and the sponge body is relatively simple, each branch or tube of the sponge ending usually in a single osculum or exhalent aperture. The ostia (pores or inhalent apertures) are in the walls, and lead in the more primitive sponges (Homocoela) to a central flagellate chamber or enteron, in the more advanced members (Heterocoela) to chambers in the wall discharging into a central non-flagellate or paragastric cavity.

HOMOCOELA

LEUCOSOLENIIDAE.

Usually erect sponges, with monaxon, triaxon and sometimes tetraxon spicules.

Leucosolenia. Simple sponges formed of tubes with apical oscula, incrusting, bushy or arborescent.

L. botryoides (Pl. II, fig. 3*a*). B. 1866, p. 28; 1874, pl. 3 (*variabilis*); Minchin, 1904, p. 387. Arborescent, with short cylindrical thin-walled branches. Oscula large. Spicules with three equal arms. Minchin says *botryoides* means a bunch of grapes, but a "bunch of bananas" would be a better comparison. Sponge tougher and stouter than other species. Grows always on *Fucus* and zoophytes, the masses crowded together round the stems of the weed. White. About 8 mm. high but sometimes more. Very common.

L. clathrus. Minchin in Lankester's *Zoology*, fig. 7, p. 6 (as *Clathrina clathrus*). This beautiful sponge, which is usually of a pure white colour, consists of an incrusting base from which arise delicate tubes anastomosing and converging towards a small number of apical oscula. It spreads over areas of about 4 sq. cm. and attains a height of about 1 cm., but is strongly contractile, so that the height is extremely variable. On stones in littoral zone. Common locally.

L. coriacea. B. 1866, p. 34; 1874, pl. 3. A sessile, incrusting form, with much branched twisted tubes which anastomose, so that one osculum serves for a group of tubes. Spicules regularly triradiate, not massed round the oscula. Grows on rocks and is of a yellow, brown, red or bluish grey colour. About 15 mm. high. Common.

L. complicata. Minchin, 1904, p. 362. Tubes tall, arborescent. Colonies compact, crowded and much branched, with an osculum at the end of each branch. A delicate, fragile sponge. White. 20 mm. high. In rock pools, on weed, etc. Very common.

L. variabilis. Minchin, 1904, p. 373. Both spreading and bushy forms occur. Colony a close network of tubes ending in large oscula in spreading form, oscular tubes long and slender in bushy form, and rarely branched. 20 mm. high, on rocks and algae. *L. botryoides*, *complicata* and *variabilis* are possibly one species.

HETEROCOELA

SYCETTIDAE.

Flagellate chambers elongated, radially arranged in the walls of the sponge round a central non-flagellate cavity which discharges by an apical osculum. The chambers in the wall project on the outer surface and are not covered by a dermal cortex.

Sycon ciliatum (Pl. II, fig. 3*b*). B. 1866, p. 19, as *Grantia ciliata;* 1874, pl. 2. An erect, vase-shaped, cylindrical sponge, with rough surface, and usually with an edging of tall, fine, straight spicules surrounding the osculum and looking like giant cilia. The lower portion is usually narrowed to form a short pedicel for fixation to a weed or stone. Spicules of two kinds, monaxonid and triaxonid. Cream or yellow in colour. 30 mm. in height or more. Common. Some authors distinguish two species, *S. coronatum* and *S. ciliatum.*

GRANTIIDAE.

Resembling the Sycettidae in the arrangement of the flagellate chambers, but differing in that there is a dermal cortex covering the projecting ends of the chambers, so that the surface is smooth.

Grantia. Wall of sponge of regular thickness.

G. compressa. B. 1866, p. 17; 1874, pl. 1. One of the commonest of the littoral sponges. The sponge body is flat, with a large compressed elongated apical osculum, and short pedicel for attachment to weed. The shape varies; it may be oval, broadly ovate or irregular. Spicules triradiate and monoradiate, the latter curved and with one dilated end, like a hockey stick. Colour pale yellow. 50 mm. or more high. On rock or weed.

Leuconia. Wall of sponge very irregular in shape and thickness. Canals in wall sinuous.

L. nivea. B. 1866, p. 36; 1874, pl. 5. Sponge smooth and pure white, forming an incrusting mass reaching 5 cm in diameter, on stones, rocks, on the shore or in caves. May be low, raised into crests, or conical. Oscula numerous. Spicules: large and medium triradiate, minute dagger-shaped and monaxons.

Other species of *Leuconia* occur locally.

DEMOSPONGIAE

The remaining British sponges are grouped in this rather odd collection, for the Hexactinellids are not represented in this country. The canal system reaches a high degree of complexity. The skeleton is siliceous or horny, or a mixture of both; only two families are devoid of any skeleton. The Monaxonida have one-rayed spicules, the Tetraxonida four-rayed spicules, though other types may be present as well, e.g. monaxons, polyaxons. There is every gradation to the horny sponge without spicules. Some, especially littoral forms, have a skeleton composed of monaxonid spicules arranged in rectangles, the angles being glued together with the horny substance (spongin). In others, the spongin flows round the spicules (intraspiculate), or the spicules may project out of the spongin (extraspiculate). Some members of this group possess minute spicules (microscleres) as well as the ordinary ones (megascleres).

MONAXONIDA

Only monaxonid spicules are present.

CLIONIDAE.

Boring sponges which make round holes in limestone, dead shells, etc.

Cliona celata. B. 1866, p. 212, as *Hymeniacidon celata*; 1874, pl. 38. This species appears in two forms, the massive and incrusting. Only the latter, which is the boring form, is found on the shore. The sponge body is simple, with smooth surface, and yellow or dull orange in colour. It bores into limestone rocks and shells, the surface of which may be riddled with round holes of various sizes. Out of each hole projects a lobule of the sponge about 2 mm. high. The commonest shells attacked are oyster, scallop and *Venus*.

SUBERITIDAE.

Incrusting or erect and very fleshy. The megascleres are rod-like and there are no microscleres.

Suberites (*Ficulina*). Sponge body massive, sometimes low and incrusting, sometimes elevated and rounded, with a stout stem and spreading base. The same species varies greatly in different habitats. The surface is smooth, the oscula few and inconspicuous.

S. domuncula (*Ficulina ficus*). B. 1866, p. 200 as *Hymeniacidon suberea*; 1874, pl. 36. Neither name expresses accurately the habit of this sponge, since it varies according to its habitat. When growing on a shell which is the home of a hermit crab (*domuncula*) it completely encloses the shell and afterwards continues the spiral growth, becoming massive and regular in form. When living alone it may assume a fig-shaped form (*ficus*). In either case it is rather compact and fleshy, smooth and yellow or orange in colour. 18 cm. long in shallow water, but smaller (7–10 cm.) on the shore.

Terpios fugax. One of the most brilliantly coloured British sponges, though insignificant in size. It forms incrustations about 1–2 cm. in diameter on stones and rocks between tide marks, and is of a vivid blue colour—the same shade as Reckitt's blue. The incrustation often appears level with the surface of the stone, so that the sponge looks like a blue lichen, but on scraping is found to occupy a concavity in the stone. It is rarely, however, more than a few millimetres in thickness. Oscula are inconspicuous. Monaxonid spicules are abundant.

AXINELLIDAE.

Sponge erect, or spreading. Skeleton usually with plumose arrangement of spicules. Monaxon spicules tylostyles, none spiny. No microscleres.

Halichondria panicea (Pl. II, fig. 3*c*). B. 1866, p. 229; 1874, pls. 39 and 40. Crumb of bread sponge. Attached to the undersides of rocks and on weed, in yellow, green and orange varieties, but not scarlet or brick red. Form very variable, but usually resembles a range of miniature volcanoes, since the large rounded oscula are elevated on conical protuberances. The surface is smooth, but when dried shows a reticulate appearance. 1–2 cm. high. Very common.

Hymeniacidon perleve (*sanguinea*). B. 1866, p. 168; 1874, pl. 32. Sponge incrusting, forming large masses on rock; of fairly regular thickness. Surface may be almost smooth, or ridged and furrowed. Oscula small, numerous. Orange, deep blood red and scarlet. A small oval Polychaete worm, *Spinther miniaceus*, browses on this sponge and is coloured exactly like it. Colony may be 15 cm. long and more than 2 cm. thick. Mid-tide level and below.

HAPLOSCLERIDAE.

The skeleton is reticulate, with fibres, and consists of monaxonid megascleres but no microscleres (except in *Ophlitaspongia*), reinforced by spongin. If microscleres are present they are sigmata, never chelae.

Haliclona oculata (Pl. II, fig. 3*d*). B. 1866, p. 361; 1874, pl. 66. Sponge erect, fixed by a narrow base from which arises a stem branching into many long, finger-shaped branches. Oscula occur *laterally* and are numerous. The sponge is fleshy, of a yellow or greenish colour, and varies in height from 10 to 30 cm. Hand sections show numerous small monaxon spicules, completely embedded in spongin, but no microscleres. On rocks, weed.

Adocia cinerea (=*Reniera*) (Pl. II, fig. 5). B. 1866, p. 274; 1874, pl. 48 as *Isodictya*. A brown or lilac-coloured sponge with smooth surface, incrusting but often showing lobular growth. Hand sections reveal a characteristic rectangular arrangement of the monaxon spicules, whose tips are held in place by isolated secretions of spongin. The spongin does not enclose the spicules as in *Haliclona*. Up to 2 cm. high.

Ophlitaspongia seriata. B. 1866, p. 376 as *Chalina*; 1874, pl. 65. Sponge incrusting, thick and spongy. Surface rather flat but finely papillose (hispid) with numerous, fairly regularly spaced oscula. Skeleton of monaxonid spicules with one end embedded in spongin and the other end free and projecting (extraspiculate), not intraspiculate as in *Haliclona*. Microscleres also present but rare. Colour brick red. It has a superficial resemblance to *Hymeni-*

acidon, but has a smoother surface, more regular oscula and a much duller red colour. (Sometimes placed in Desmacidonidae because of the microscleres.)

DESMACIDONIDAE.

Spicules frequently spiny and hooked. Mega- and micro-scleres present Some of the microscleres are chelae.

Myxilla. Megascleres are straight, spiny styles and tylostyles; microscleres are chelae and sigmata.

M. incrustans. B. 1866, p. 249 as *Halichondria*; 1874, pl. 44. A massive incrusting sponge, with wrinkled surface and scattered oscula. Buff, yellow or orange. Often in deep water, but occurring also at extreme low-tide mark, together with other species of the genus.

Microciona. With spiny spicules. Microscleres curved.

M. atrasanguinea. B. 1866, p. 138; 1874, pl. 24. Incrusting. Surface thin, finely irregular, with small oscula. Spicules spiny and smooth, a few curved, but no chelae or sigmata present. Dark red colour.

POLYMASTIIDAE.

No microscleres. Megascleres massed in a cortex.

Polymastia mammillaris (Pl. II, fig. 3*e*). B. 1866, p. 71; 1874, pl. 12. Bright yellow, forming massive rounded growths on rocks at extreme low-water mark or in deeper water. The surface of the mass is often prolonged into numerous tag-like processes, each of which bears a terminal osculum. The spicules are long and straight and some have a knobbed end. To 6 cm. diameter, 1 cm. high.

TETRAXONIDA

GEODIIDAE.

Both mega- and micro-scleres present, the former tetraxons with one very long arm, or polyaxons of spherical shape, the latter of at least two kinds—small rod-like spiny ones and six-rayed triaxons.

Pachymatisma johnstoni (Pl. II, fig. 3*f*). B. 1866, p. 51; 1874, pl. 8. Forming large incrusting masses of irregular shape on rocks and in caves. The base is the broadest portion of the sponge, and from it arises a conical, undulating or flattened projection of considerable thickness. Surface smooth but very rigid and crustlike, owing to the well-defined cortex loaded with spicules. Oscula large and conspicuous and usually arranged in lines on the surface on the summits of the undulations. Colour a dull elephant grey though lighter specimens occur, shallow-water specimens reddish. Beneath the crust the sponge is yellow or white. To about 10 cm. long, 2–3 cm. high.

The order Myxospongida is no longer used; *Halisarca* is regarded as an aberrant member of the Keratosa, and *Oscarella* as related to the Tetraxonida.

Halisarca dujardini. B. 1866, p. 224; 1874, pl. 38 as *Hymeniacidon*. Smooth, gelatinous or slimy incrustations on rocks, stones, weeds and shells of Molluscs and crabs. Deep dull yellow or orange colour. Extremely thin, hardly 2 mm. thick. Common.

Oscarella lobularis (Pl. II, fig. 3*g*). As the specific name implies, this sponge is lobulated. It is smooth and firmer than *Halisarca*. About 1 cm. in height. Yellow or pinkish-brown.

KERATOSA

No spicules. Skeleton of horny fibres.

SPONGELIIDAE.

Solid spongin fibres, reinforced by foreign matter. No intrinsic spicules.

Spongelia fragilis (*Dysidea*). B. 1864, pl. 14; 1866, p. 381; 1874, pl. 69; 1882, p. 188; L. p. 662, pl. 37, fig. 10. One of the few British horny sponges. Irregular, incrusting when young, erect and massive when mature. Surface covered with regularly arranged cones, oscula small and inconspicuous. Colour variable, often dull yellow. Skeleton of reticulating horny fibres within which are incorporated all kinds of foreign materials, varying with the situation in which the sponge is found, but typically containing large numbers of sand grains packed closely together within the fibres, and stiffening the apices of the cones. Spicules may also occur within the spongin fibres, but these are foreign and have not been made by the *Spongelia*. Common on most shores. 1–2 cm. high.

EXPLANATION OF PLATE II. PROTOZOA AND PORIFERA

Fig. 2. *Haliphysema tumanowiczii*. Two specimens growing on *Crisia*. × 6.

Fig. 3. Common littoral sponges. *a*, *Leucosolenia botryoides*, a homocoelous calcareous type whose central cavity is the enteron. × 5. *b*, *Sycon ciliatum*. Heterocoela, with gastric chambers in the wall, the central chamber not gastric in function. × 2. *c*, *Halichondria panicea*. An incrusting form, with complex canal system. × $\frac{2}{3}$. *d*, *Haliclona oculata*. Erect, dichotomously branched stems, with a lateral row of oscula on each. × 1. *e*, *Polymastia mammillaris*. Incrusting, with short processes scattered over the surface. × 1. *f*, *Pachymatisma johnstoni*. Incrusting, with sunk oscula on the summit of the elephant grey surface. × $\frac{1}{2}$. *g*, *Oscarella lobularis*. Soft nodular outgrowths on an incrusting base. × $\frac{2}{3}$.

Fig. 4. Types of spicules in littoral sponges. *a*, *b*, *c*, calcareous types from *Grantia compressa*. *d–j*, siliceous megascleres. *k–o*, siliceous microscleres (highly magnified). *a*, *b*, *d*, *e*, *f* are monaxon, *c* and *h* triaxon, *g* and *i* tetraxon, *j* polyaxon; *d*, rhabdus; *e*, style; *f*, tylostyle (*Suberites*); *g*, orthotriaene (*Pachymatisma*); *h*, triaxon; *i*, tetraxon (*Tethya*); *j*, young and old sterrasters (*Pachymatisma*); *k*, six-rayed triaxon (*Pachymatisma*); *l*, rod (*Pachymatisma*); *m*, chela (*Myxilla*); *n* and *o*, sigma (*Myxilla*). The organic axis of the spicules has not been figured.

Fig. 5. *Adocia cinerea* (*Reniera*). Part of a hand section to show a skeletal combination of monaxonid spicules and spongin. The meshes vary from square to irregularly rectangular, pentagonal and hexagonal. × 25.

br. branch; ch. flagellate chamber; d. basal disk; inc. incrusting base; os. osculum; pap. papilla; s.s. incorporated sponge spicule; spic. spicule; spon. spongin.

II. PROTOZOA AND PORIFERA

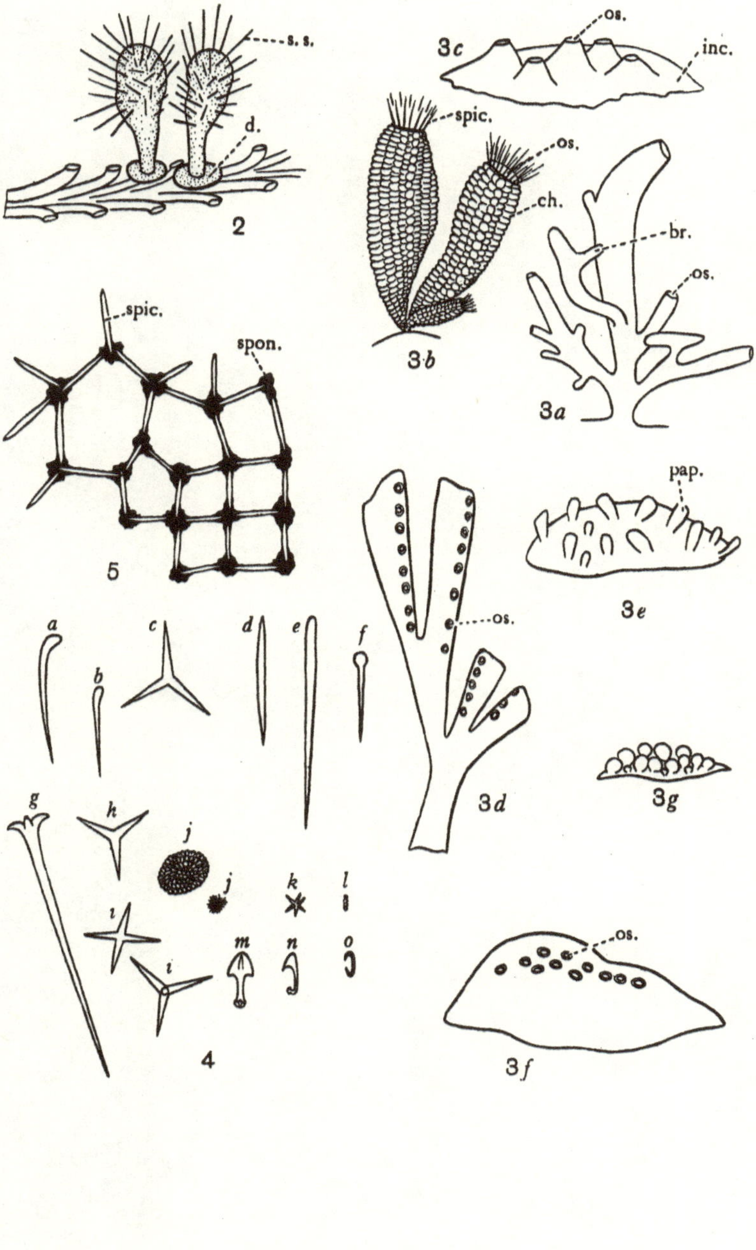

(*facing p.* 26)

COELENTERATA

(Plates III, IV)

AUTHORITIES

ALLMAN, G. T. 1871. *A Monograph of the Gymnoblastic or Tubularian Hydroids.* Ray Soc. Publ.

GOSSE, P. H. 1860. *A History of the British Sea Anemones and Corals.* Van Voorst.

HINCKS, T. 1868. *A History of British Hydroid Zoophytes.* 2 vols. Gurney and Jackson.

STEPHENSON, T. A. 1928. *British Sea Anemones,* vol. I. Ray Soc. Publ. Also *J. Mar. biol. Ass. U.K.* XIII, 1925, 880–90.

—— 1935. *British Sea Anemones,* vol. II. Ray Soc. Publ.

Popularly known as zoophytes, jellyfish, corals, anemones and comb-jellies. The only cavity in the body is the gut cavity or *coelenteron.* The symmetry is usually radial, though in many a bilateral symmetry is superimposed upon this (e.g. in sea anemones, Ctenophores, etc.).

Many Coelenterates are colonial, the individual being known as a polyp or *zooid.* This is usually a flower-like structure, possessing a mouth surrounded by one or more circlets of *tentacles.* The only communication between the gut and the exterior is the mouth.

Some Coelenterates have a free-swimming stage, the *medusa* or jelly-fish stage. The medusa is sexual; in structure it is a modified zooid. Many Coelenterates are permanently fixed, others temporarily. A few are free. Asexual reproduction is common. All the members of the phylum are aquatic; most are marine.

Class I. HYDROZOA

(Pl. III, figs. 6–11)

The zooid is simple, with one or more circlets of tentacles, which are never feathered or pinnate. The mouth opens directly to the enteron; there is no invagination of the ectoderm to form a stomodaeum. The coelenteron is also simple, and its walls are rarely thrown into folds.

There are two main forms of zooids: (*a*) the nutritive zooids or *hydranths,* usually asexual, derived from the eggs, and (*b*) the sexual zooids or *medusae*, formed from buds. The former are fixed, the latter typically free-swimming. The eggs or sperms of the medusa are discharged directly to the exterior (i.e. not into the enteron). Typically the hydroid and medusoid stages alternate, but in some species one or other may be suppressed. The hydroid stage is usually colonial, and frequently secretes a horny or calcareous exoskeleton or *perisarc*. The individuals of a colony may become modified for special purposes and the colony may be di-, tri- or poly-morphic, having zooids of two, three or many different kinds. The medusoid stage is solitary, rarely exhibiting budding. It possesses sense organs at the bases of its tentacles; these may be eye-spots (*ocelli*) and balancing organs (*statocysts*). The *gonads* are borne either on the *radial canals* or on the mouth stalk (*manubrium*). Many Hydroids are known as zoophytes, because of their resemblance to plants.

In the Gymnoblastic Hydroids the perisarc consists of rooting processes (*hydrorhiza*) and upright stems (*hydrocaulus*) which may be simple or branched. The perisarc never encloses either the zooid (*hydranth*) or the reproductive body (*gonophore*), these being naked. The gonozooids show every stage of reduction and degradation from free medusae (*Sarsia*), through simplified fixed medusoids (*Tubularia*), to non-medusiform *sporosacs* (*Clava*). The free medusa is without statocysts, but usually has ocelli.

In the Calyptoblastic Hydroids the perisarc forms a protective covering, not only for the stems and hydrorhizae, but also for the hydranths and the medusa buds, as the *hydrotheca* and *gonotheca* respectively. Medusa buds are usually borne on reduced hydranths called *blastostyles*. In some families medusae are freed, in others the medusa stage is wholly or partly suppressed and the genital products develop on reduced medusae or directly on the blastostyles within the gonothecae. In one family (Plumulariidae) specially modified protective structures known as *nematophores* are present.

Class II. SCYPHOZOA

(Pl. III, fig. 12)

The prominent stage is the sexual medusoid or jelly-fish, the hydroid stage (*scyphula* or scyphistoma) being small and giving rise to young medusae (*ephyrae*) by a peculiar method of transverse division called strobilation. The coelenteron is always ridged, and the free edges of the ridges bear *gastric filaments*. The gonads are endodermal and discharge into the enteron, their products passing out through the mouth. The sense organs of the medusa are formed from modified tentacles. These open *tentaculocysts* or rhopalia bear *statoliths* and sometimes *ocelli* and *olfactory pits*. The medusa may be free or temporarily fixed.

Class III. ANTHOZOA
(Pl. IV)

The zooid is relatively large. Its mouth does not open directly into the enteron as in the Hydrozoa, but into a tubular invagination of the ectoderm, called the *stomodaeum* or actinopharynx. This is seldom round, but is compressed laterally to an oval shape, with a ciliated groove at one end or at both ends. The single groove is called the *sulcus* (ventral groove or siphonoglyphe of some authors). If two grooves are present, the second or so-called dorsal groove is the *sulculus*.

The stomodaeum is fastened to the body wall by radial partitions or *mesenteries*. At the level of the enteron these become free, so that they project inwards from the body wall, but they are usually attached at their lower ends to the basal disk. The mesenteries bear muscles and gonads, both being arranged in a definite order in each subclass. The gonads are endodermal and discharge into the coelenteron through the mouth cavity. The edge of the mesentery may be thickened to form a *mesenteric filament*. There is no medusoid phase.

Subclass Alcyonaria. The zooid has eight, and never more than eight tentacles, which are hollow and pinnate. A ciliated sulcus is present. The eight mesenteries are fixed (complete) in the stomodeal region but are free (incomplete) below. The muscle groups (*banners*) of the mesenteries are always situated on the sulcar aspect of the mesenteries. Each mesentery has a mesenteric filament; the asulcar (sulcular) mesenteries are longer than the others and do not bear gonads. The zooids may be dimorphic and the two types are then known as *autozooids* and *siphonozooids*. The latter pump fresh water through the canal system of the colony.

A calcareous skeleton is present and usually consists of spicules which are typically warted. In addition, a horny skeleton, the *sclerobase*, may form an axial rod, or such a rod may be formed of fused spicules. The soft tissue, composed of mesogloea (with strands of denser mesogloea, at one time thought to be tubular, and spicules), in which the zooids are embedded, is called the coenenchyme or *scleroderm*. Nearly all the Alcyonaria are colonial, budding occurring by stolon formation or from the fleshy surface of the colony. Usually fixed. The group includes the sea-fans, sea-pens and others.

Subclass ZOANTHARIA. The zooid has six or more than eight hollow tentacles, which are usually simple, never pinnate. The mesenteries vary in number and the arrangement is never the same as in the Alcyonaria, since the muscles of the sulcar and asulcar mesenteries face away from the sulcar and sulcular grooves of the stomodaeum. If a skeleton is present it is never spicular, but may be calcareous or horny. Both simple and colonial forms occur, and they may be fixed permanently or temporarily or, more rarely, are free. The group includes the anemones and corals.

Anemones. The term "anemone" is used popularly to describe any large zooid in which there is no skeleton. Zoologically it means members of the subclass Zoantharia, and it is doubtful if all belong to the same order. The Zoantharian, unlike the more uniform Alcyonarian zooid, varies in internal structure. In a typical anemone, the zooid is fixed to the substratum by an adhesive *base*, from which rises a cylindrical *column*, surmounted by a flower-like *disk*. The disk is separated from the column by the *margin*, usually with a sphincter muscle. On the margin there may be a circular *fossa* or ditch, bounded by an upward, collar-like prolongation of the column known as the *parapet*. The disk consists of (*a*) the peripheral zone of tentacles, (*b*) a flat area, the *peristome*, (*c*) the mouth perforation in the centre. The tentacles are simple and vary considerably in number. There may be five, six, or multiples of these numbers, but except during development in a very few anemones, there are never eight. In some species the number may reach several hundreds in a full-grown specimen. New tentacles are added by the formation of circlets outside the earlier ones, so that the first formed or primary tentacles are in the inner ring, nearest to the mouth. The row of secondary tentacles alternates with the primaries, and the tertiary row usually has double the number. Thus, a common formula would be 6, 6, 12, etc. The primary tentacles are the largest. The peristome is usually marked with radial lines representing the attachment of the mesenteries; the areas between the lines are known as *radii*. The mouth is an elongated slit in the centre of the disk. It functions for the intake of food and for the ejection of waste matter and genital products (eggs, sperms or young anemones). At each end of the mouth there is in most anemones a ciliated food groove or *siphonoglyphe*. A few have such a groove at one end only. In the Zoantharia neither internally nor externally is there any distinction between the two grooves, which are known as the *sulcus* and *sulculus*. The tentacles opposite the two grooves are called *directive tentacles*.

Anemones are divided into the Cerianthidea and the Actiniaria. The Cerianthidea, or tube-forming anemones, differ from the Actiniaria in the arrangement of the mesenteries, in the absence of muscles on the mesenteries and of a sulcar groove on the mouth.

Members of the Actiniaria do not make a permanent tube, though many secrete mucus and incorporate sand or stones in it to form a temporary protection. Some are not even fixed, but burrow in sand. Such anemones have no adhesive base, and the body may be divided into regions. The disk and the part near it are often transparent and are called the *capitulum*, the intermediate thick opaque region is the *scapus*, and the lower, often bulbous, burrowing end is the *physa* (Edwardsiidae, Halcampidae).

In the fixed anemones, the column may be imperforate or perforate. In the latter case long stinging threads called *acontia* can be protruded (see Sagartiidae) through apertures called *cinclides*. In many anemones the column bears suckers, which may be elevated on warts (*verrucae*).

Corals. The true corals or Madreporaria have very few British littoral representatives and those only small solitary ones. The living coral consists of soft and hard parts. The hard part or *corallum* is the ectodermal chair in which the soft part or anemone (zooid) sits. The *base* of the corallum is usually fixed to the substratum, and from the base arises the wall or *theca*, sometimes reinforced by an epitheca. The theca may exhibit externally parallel ridges or *costae* which are often continuous, within the cup, with radiating calcareous lamellae, the *septa*. Transverse partitions joining the septa are called *pali*. The septa do not reach the centre of the corallum, this region being occupied by a trabeculate calcareous *columella*.

Class IV. CTENOPHORA

The comb-jellies. These are solitary, pelagic, transparent, gelatinous forms in which the radial symmetry has largely given place to a bilateral arrangement. Swimming is effected by eight meridional rows of *costae* (swimming plates) formed of fused cilia (the *combs*). The mouth is at the lower end of the spherical or ovate body, and it is compressed laterally in what is named the stomodeal (*sagittal*) plane. The aboral pole bears a sensory plate (*statocyst*) and two or more excretory pores communicating with the enteron. The stomodaeum leads to a chamber compressed in a plane at right angles to the stomodeal plane; this chamber is the *infundibulum*, and the plane through it is the infundibular or *transverse plane*. A pair of tentacles, retractile into sheaths in the gelatinous substance of the body, lie in the infundibular plane. They are used for catching prey and not for swimming.

Ctenophores are hermaphrodite, and the genital products escape by the mouth. There is no Hydroid stage. They differ from all other Coelenterates in retaining cilia as organs of locomotion throughout life.

KEY TO COELENTERATES

Zooids without stomodaeum, gastric ridges or filaments. Often colonial. Typically with alternate Hydroid and Medusoid stages. Sex cells discharged directly to the exterior. Zoophytes. **Hydrozoa**, p. 32

Medusae with gastric ridges and filaments. Sex cells discharged into coelenteron. Jellyfish. **Scyphozoa**, p. 38

Zooids with stomodaeum and internal mesenteries. Mouth laterally compressed. Sex cells discharged into coelenteron. Anemones and corals. **Anthozoa**, p. 39

Pelagic forms with ciliated swimming combs. Comb-jellies. **Ctenophora**, p. 46

HYDROZOA

1. Zooids of Hydroid stage not retractile into cups (hydrothecae). No permanent gonothecae. **(Anthomedusae) Gymnoblastea. 2**
   Zooids retractile into hydrothecae. Permanent gonothecae. **(Leptomedusae) Calyptoblastea. 6**
2. Tentacles knobbed (capitate). **5**
   Tentacles filiform. **3**

3. Tentacles scattered on hydranth. No free medusae. Sporosacs crowded. **Clavidae**, p. 33
Tentacles in whorls or circlets. **4**

4. A single circlet of tentacles. Free medusae, or sporosacs. Colony much branched. **Bougainvilliidae**, p. 33
Two circlets of tentacles. Zooids gregarious, sometimes colonial, usually single. **Tubulariidae**, p. 34

5. Colonial. Zooids small. **Corynidae**, p. 34
Solitary. Zooids to 5 cm. high. **Myriothelidae**, p. 34

6. Zooids arranged uniserially on the branches. Nematophores present. **Plumulariidae**, p. 37
Zooids arranged biserially. **Sertulariidae**, p. 36
Zooids pedunculate, neither uniserial nor biserial. Hydrotheca bell-shaped. Hypostome trumpet-shaped. **Campanulariidae**, p. 35

GYMNOBLASTEA (ANTHOMEDUSAE)

CLAVIDAE.

Clava. Zooids arising singly from the hydrorhiza.

C. squamata (Pl. III, fig. 7). A. p. 243, pl. 1; H. p. 4. About 2·5 cm. high, yellowish red in colour, with paler tentacles about twenty in number. The hydrorhizal tubes form a dense mass. On *Fucus*.

C. multicornis. A. p. 246, pl. 2; H. p. 2. Smaller (6 mm. high), pale brown to pink in colour. Its hydrorhizal tubes form an open network so that the colony does not become crowded on the weed or stone.

BOUGAINVILLIIDAE.

Hydrorhizae well developed, the hydrocaulus (upright portion of the colony) branched or unbranched. Some colonies produce medusae (*Bougainvillia*), others sporosacs (*Bimeria*).

Bougainvillia ramosa. A. p. 311, pl. 9; H. p. 109. Much branched. 2–7 cm. high. Pale pink, with yellow perisarc. Blastostyles stalked, numerous, just below hydranth. Each blastostyle produces one free medusa. On rock, Hydroids, shells, etc. Lower shore to deeper water.

Garveia nutans. A. p. 295, pl. 12; H. p. 102. Branches regularly alternate. 1–2·5 cm. high. Orange or red. Hydranth bending its head downwards (*nutans* = nodding). Sporosacs numerous, on short stalks at base of colony. On stones and weeds. Lower shore.

CORYNIDAE.

The tentacles, which are usually scattered on the hydranth, bear nematocysts on their knobbed ends only. *Coryne* has regular annulations of the perisarc and bears sporosacs, *Sarsia* has faint, irregularly arranged annulations and bears medusae.

Coryne muscoides (Pl. III, fig. 6). A. p. 268, pl. 4; H. p. 41 (*C. vaginata*). Colony tall, branched, 3–10 cm. high. Rose coloured. In rock pools.

Sarsia eximia. A. p. 282, pl. 5; H. p. 50. as *Syncoryne*. Colony prolific, irregularly branched. 5–10 cm. high, red or pink. Lower shore.

MYRIOTHELIDAE.

This family has one British member, a solitary zooid of relatively large size.

Myriothela cocksi (Pl. III, fig. 8). A. p. 382; H. p. 77, pl. 12; both as *M. phrygia*. The hydranth is shaped like the spadix of an arum lily. 5 cm. high. White, pink or brown. Attached at the base by rooting processes, above which are fixed numerous sporosacs on short branches. Tentacles very numerous, covering three-fourths of the body, small and capitate. On stones. Lower shore to shallow water.

TUBULARIIDAE.

Hydranth with two circlets of filiform tentacles, oral and aboral in position. Gonophores with fixed sporosacs (*Tubularia*) or free medusae (*Hybocodon*).

Tubularia. Hydrorhizae are filiform and adherent, stems large, upright, simple or branched. Oral tentacles short and stand erect round the conical hypostome; aboral tentacles long, and, like the bunches of sporosacs, drooping. The medusoid stage is not set free but discharges the genital products through an opening in the sporosac. Hydranth dies in autumn.

T. indivisa. A. p. 400, pl. 20; H. p. 115. Rarely branched. To 7 cm. or more in height. Pale pink to scarlet, with brown perisarc. On rocks, etc. Lower shore.

T. larynx. A. p. 406, pl. 21; H. p. 118. Stems branched. 2·5–6 cm. high. Rose coloured with pale yellow perisarc. On rocks, etc. Lower shore to deep water.

Hybocodon prolifer. A. p. 422. Solitary or in small groups, to 5 cm. high. Orange red, often lined with red. Gonophores on short stalks, densely clustered, producing medusae with one tentacle and five red radial stripes. In pools. Lower shore.

CALYPTOBLASTEA (LEPTOMEDUSAE)

CAMPANULARIIDAE.

Zooids stalked, with large trumpet-shaped hypostomes. Branching variable. Hydrothecae bell-shaped. Sporosacs in some species, free medusae in others, both housed in gonothecae. The family includes some of the most abundant littoral zoophytes.

1. Colony producing free medusae. **2**
   Medusae abortive. Fixed sporosacs. **3**
2. Stems simple or only slightly branched. Gonothecae usually directly on stolon. Medusae bell-shaped. **Clytia**
   Stems much branched. Gonothecae axillary. Medusae flattened. **Obelia**
3. Gonothecae with tapering narrow neck, or if truncated, ringed. **Campanularia**
   Gonothecae truncated, smooth. **Laomedea**

Campanularia verticillata. H. p. 167, pl. 32. Much branched, branches erect in whorls resembling the plant called horsetail. 7 cm. or more high. White. Hydrotheca with denticulate margin, its stalk ringed. Gonotheca on main stem with short stalk. Intertidal.

Clytia johnstoni. H. p. 143, pl. 24. Stems long, transparent, simple or slightly branched. 3 mm. high. Hydrothecae with denticulate rims. Gonothecae ringed, usually on creeping stolon. Free globular medusae. On red weed. Intertidal.

Obelia. This well-known genus has a branched, creeping stolon, erect, much branched, jointed or ringed stems, and gonothecae, bearing free medusae, in the axils of the stalked hydranths.

O. geniculata (Pl. III, fig. 9). H. p. 149, pl. 25. Stem zigzag, jointed at the base of each branch, branches alternate, with a kind of shelf or shoulder at each bend. Branches mainly simple. 3 cm. high, white. Gonothecae shaped like the Grecian vase called the amphora. Common in summer. On weeds, etc. Intertidal.

O. dichotoma. H. p. 156, pl. 28. Slender, with straight stems not broadened to form a shelf at the joints. Branches often compound. 3 cm. high. Horn coloured. Often on other zoophytes.

Laomedea. Resembles *Campanularia* in the somewhat irregular compound branching. The gonothecae are, however, truncated and lack the long neck characteristic of those of *Campanularia.*

L. flexuosa. H. p. 168, pl. 33, as *Campanularia.* Stem slender, flexible, simply pinnate or irregularly branched. 2·5 cm. high. Yellow. Ringed. Hydrothecae with smooth rims, on long ringed pedicels. Gonothecae large, in axils of branches, containing sporosacs. Common in littoral zone from spring onwards.

SERTULARIIDAE.

Hydrothecae sessile, inserted in the stems and branches on both sides, either alternately or opposite. Gonozooids degenerate and fixed to the colony, the ova developing within the gonotheça or in a gelatinous capsule (acrocyst) attached to it. No free medusa.

1. Hydrothecae opposite, gonothecae smooth. 2
   Hydrothecae alternate, gonothecae ringed. **Sertularella**
2. Stem appearing serrated. Hydrothecae curved outwards, deep set. **Dynamena**
   Stem hair-like. Hydrothecae straight and very small. **Sertularia**

Dynamena. Stems much branched. Hydrothecae opposite, curved. Gonothecae smooth.

D. pumila. H. p. 260, pl. 53, as *Sertularia.* Stems crowded on creeping stolons, straight or gently curved. Branches opposite, forming pinnate fronds. Stems and branches jointed. 1–4 cm. high. Horn coloured. Hydrothecae opposite, short, tubular, with sinuous rims, but not denticulate. Gonothecae ovate, almost sessile, with short tubular apertures. Ova extruded from female gonozooid in a gelatinous investment (acrocyst). Grows abundantly on *Fucus*, etc., and is the commonest littoral zoophyte.

Sertularia. Stems fine and hair-like. Hydrothecae opposite, straight. Gonothecae smooth.

S. operculata. Sea hair. H. p. 263, pl. 54. Stems long, very slender (hair-like), slightly flexuous, branches compound, alternate. Very prolific, growing to a height of 10–20 cm., in dense tufts. Pale yellow to reddish in colour. Hydrothecal rim produced into a fine point laterally with a small denticulation on each side of it.

Gonothecae large, pear-shaped, scattered over the stems, with plain circular apertures and lids, resembling the capsules of a moss plant. On *Fucus* and *Laminaria digitata*. *S. argentea* is the sea fern of commerce.

Sertularella. Stem branched, jointed. Hydrothecae alternate, with a toothed orifice, supplied with folds forming an operculum. Gonothecae transversely ringed.

S. polyzonias. H. p. 236, pl. 46. Stems irregularly branched and jointed but not ringed. Hydrothecae with wide four-toothed apertures. Gonothecae large, ovate, irregularly ringed. Hydranths with about twenty tentacles. 4–6 cm. high. On shells, seaweeds, etc.

S. fusiformis. H. p. 243, pl. 47. Stems slender, not much branched, jointed, ringed below the calycles. Hydrothecae with narrowed four-toothed apertures. Gonothecae ringed, with toothed apertures. 2·5 cm. high. Between tide marks. North and west coasts.

PLUMULARIIDAE.

The hydrothecae are sessile and are arranged on one side of the axis only. The colony is protected by nematophores, modified zooids, in minute hydrothecae, each zooid being shaped like a long flexible tentacle, bearing nematocysts and often adhesive cells.

1. Nematophores confined to the hydranths. 2
   Nematophores numerous, not confined to the hydranths, but on axils of stems, etc. **Plumularia**
2. One nematophore to each hydranth. **Kirchenpaueria**
   Several nematophores to each hydranth. Gonothecae in enclosed basket-shaped structures (corbulae). **Aglaophenia**

Plumularia. Shoots plumose, simple or branched, jointed, delicate. Nematophores of several kinds, on stems and branches. Two species are littoral, *P. setacea* and *P. halecioides*, the first regularly plumose, the second less regular in its branching.

P. setacea (Pl. III, fig. 10). H. p. 296, pl. 66. Stolons give rise to a forest of small plumes. Fronds very delicate, simply pinnate, regularly jointed, the pinnae alternate, one to each internode. Branches of alternately long and short joints, the former bearing one hydrotheca each. To 4 cm. high. Horn colour. Nematophores elongate, usually three in connection with each hydranth, and one on the short segment without a hydranth. Gonothecae in the axils of the branches, tall with smooth tubular neck. On weed, zoophytes, from littoral zone to shallow water.

P. halecioides. H. p. 306, pl. 67. Stem compound and often irregularly branched. Pinnae alternate, short, rarely bearing more than three or four zooids. 3 cm. high. Pale horn colour. Hydrothecae distant, separated by two or three joints. Nematophores minute, two to each hydranth. Gonothecae large, ovate, ribbed or ringed transversely, with a broad truncated top. On stones, etc., near low-water mark, often on other hydroids.

Kirchenpaueria pinnata. H. p. 295, pl. 65, as *Plumularia.* This large hydroid has numerous, tall, white or pale yellow fronds springing from a basal stolon. The pinnae are alternate. Height 5–15 cm. Each joint bears a zooid in a broad smooth-rimmed hydrotheca, beneath which lies a single minute nematophore. Gonothecae, shaped like the wrinkled pods of a plant, are found in double rows along the main stem. On shells, stones and weed. Low water to deep water.

Aglaophenia pluma. H. p. 286, pl. 63. The podded coralline. Pinnae alternate, simple, one to each internode on the stem, giving a regularly branched arrangement. 3–10 cm. high. Dark brown in colour. Hydrothecal rim toothed. Three nematophores to each zooid. On weeds, shells and rocks. Lower shore. South and west coasts.

SCYPHOZOA

Many members of this class are pelagic and are seen on the shore only after being washed up in a dying condition after a storm. Some are estuarine. Three inhabit the littoral zone as adults, and in each of these temporary fixation of the funnel-shaped umbrella of the medusoid is brought about by means of an aboral stalked sucker.

1. Tentacles of umbrella in eight groups with one or more rows in each. **Depastridae**
   Tentacles of umbrella in eight large pedunculate bunches. **Lucernariidae. 2**
2. Marginal anchors (adhesive glands) between the bunches of tentacles. **Haliclystus**
   No marginal anchors. Gonads extending along arms. **Lucernariopsis**

No marginal anchors. Gonads in four gastro-genital pouches between arms. **Craterolophus**

Depastrum cyathiforme. Umbrella with eight shallow lobes, and four interradial horseshoe-shaped gonads. 1·5 cm. high. Usually, of a dirty brown colour. Lower shore to shallow water, on weed and stones.

Haliclystus auricula (Pl. III, fig. 12). Umbrella with deeply incised bays between the bunches of tentacles. Four gonads situated interradially, forking at the bases of the lobes to give eight bands which extend as far as the tentacle groups. 2·5 cm. high. Pale green, rarely red. Often on *Zostera.*

Lucernariopsis campanulata. Similar to *Haliclystus* in shape, but lacking the marginal anchors. 2·5–4 cm. high. Typically on red weed and coloured deep red, though pale and even green specimens are found. Lower shore. Southern coasts.

Craterolophus convolvulus. A northern species. Isle of Man, etc.

ANTHOZOA

1. Tentacles eight, always pinnate. Eight complete mesenteries. **Alcyonaria.** 2
   Tentacles never eight, usually numerous, simple. **Zoantharia.** 3
2. Colony fixed, without skeletal axial rod but with spicules scattered in the fleshy part of the colony. **Alcyonacea**
3. Calcareous exoskeleton present (corals). **Madreporaria**
   No skeleton (anemones). 4
4. Anemone secreting a tube. **Cerianthidea**
   Anemone not making a tube for itself. **Actiniaria**

ALCYONARIA

ALCYONACEA

Alcyonium digitatum. Dead men's fingers. Fleshy lobes bearing the beautiful flower-like zooids (Pl. IV, fig. 13), and fixed by a base, at first incrusting, but later becoming stalk-like and sterile. Zooids completely retractile, expanding in dull light. Colour white, pink, orange or yellow. A large colony may measure 15–20 cm. in height. On rocky coasts. Often abundant.

ZOANTHARIA

CERIANTHIDEA

CERIANTHIDAE.

Solitary anemones secreting tubes of felted nematocysts.

Cerianthus. Column elongated, cylindrical, bulbous at perforated lower end. Surface smooth. No parapet or fossa. Disk funnel-shaped, bearing two groups of tentacles, one peripheral and the other stomodeal, not strongly contractile. The tube is loose and non-adherent, usually much longer than the anemone's body. It is closed at the lower end and is usually dirty owing to the agglutination of foreign particles.

C. lloydi. Gosse, 1860, p. 268. A rich chestnut brown anemone living in muddy or gravelly sand, on the surface of which the disk and tentacles are expanded. Tentacles about sixty-four in number. Expands to 15 cm. or more in length but contracts strongly when dug out of its burrow. Disk about 1 cm. in diameter, and the expanded crown 4 cm. Local. Menai Straits, Channel Islands, Plymouth.

ACTINIARIA

The sea anemones (Pl. IV, fig. 14) are usually simple zooids of relatively large size. A few are colonial. Some secrete a temporary tube, or incorporate sand and stones with a slimy secretion to form a house. The tentacles and mesenteries are numerous, usually in multiples of six.

1. Anemones living in sand, without definite fixation disk but with rounded lower end adapted for burrowing (physa). 2
   Fixed anemones with adhesive muscular basal disk. 4
2. Column elongate, divided into regions. 3
   Column thick, pear-shaped, not clearly divided into regions. **Ilyanthidae**, p. 41
3. Tentacles in three or four cycles, long, tapering, more than twelve in number. **Edwardsiidae**, p. 41
   Tentacles few, short, blunt, in two cycles. **Halcampidae**, p. 41
4. Column thick, usually with suckers, and perforated for the extrusion of stinging threads (acontia). **Sagartiidae**, p 44
   Column smooth, perforated. Tentacles numerous, short, plumose. **Metridiidae**, p. 43
   Column not perforated for acontia. 5

5. Tentacles simple, neither knobbed nor drooping, in cycles. **Actiniidae**, p. 42

Tentacles knobbed. Zooids semi-transparent, small. Gregarious. **Corallinomorphidae**, p. 45

EDWARDSIIDAE.

Small elongated anemones. Tentacles slender, in three to four cycles, more than twelve in number. Not fixed by a basal disk, but either burrowing in sand or inhabiting crevices, holes in rocks or old burrows of Molluscs such as *Hiatella* (*Saxicava*). Body divided into three regions, capitulum, scapus and physa.

Milne-Edwardsia. Column long, cylindrical, slender, characterised by the delicate transparency of the capitulum and the thick opacity of the scapus. Physa bulbous, thin-walled. Column with suckers, imperforate. Tentacles in several rows, more than twelve.

Milne-Edwardsia carnea. S. 1935, p. 35; Gosse, 1860, p. 259, as *Edwardsia*. A small anemone which when extended does not exceed 3 cm. in height. Capitulum flesh-coloured, mouth scarlet. Tentacles twenty-eight, in three or four rows, usually clear crimson or pink and spotted, long, delicate, pointed, translucent. South and west coasts, in the cavities of eroded rocks and in caves at extreme low water. Local.

HALCAMPIDAE.

Elongated or vermiform anemones living in sand.

Halcampa. Physa, scapus and capitulum present, the first with perforations. Tentacles eight to twelve.

H. chrysanthellum. S. 1935, p. 69; Gosse, 1860, p. 247. Tentacles twelve, rather short, blunt and stout, in two rows, of equal length. Mouth creamy white or buff, disk with brown markings. 2–5 cm. high. In sand or mud, under stones and in pools, from shore to shallow water. Local.

ILYANTHIDAE.

Sand burrowers with retractile head (capitulum), thick column often provided with small suckers, and a rounded physa. The regions are not clearly marked.

Peachia hastata (Pl. IV, fig. 15). S. 1928, pls. 2 and 4; 1935, p. 96; Gosse, 1860, p. 235. Peach coloured (though its name is derived from that of Mr Peach), to 10 cm. long and 2·5 cm. in diameter. The capitulum bears twelve small tentacles in a single circlet, and a large compressed mouth whose sulcar end is prolonged into a thick-lobed lip called the conchula. The column is inflated and thick and is covered with small suckers. The larvae are parasitic in medusae. At extreme low-water mark, buried in sand, mud or gravel.

ACTINIIDAE.

A large family. There is a large pedal disk or base for attachment, the column is smooth or warted but not perforate, and the disk is usually separated from the column by a ridge and a ditch (parapet and fossa). Bead-like tubercles are common on the parapet. Tentacles numerous, arranged in cycles.

1. Column smooth-walled. **2**
   Column warted. **3**
2. Vividly coloured marginal tubercles present. **Actinia**
   No tubercles. Margin of disk crenate. Tentacles scarcely retractile, green or brown, shot with violet. **Anemonia**
3. Warts scattered. Column broad and short. Tentacles short, thick, banded transversely. **Tealia**
   Warts in longitudinal lines. **4**
4. Large and small warts in alternate lines or longitudinal bands on the column. **Bunodactis**
   Warts equal in size, coloured or spotted. **Anthopleura**

Actinia equina. Beadlet anemone. S. 1935, p. 113; Gosse, 1860, p. 175, as *A. mesembryanthemum*. Base broad, wider than the column, which is itself thick and short. Twenty-four bright blue marginal tubercles (acrorhagi) present round the edge of the disk. Tentacles in six circlets, rather long, 192 altogether (6, 6, 12, 24, 48, 96). Highly contractile. Diameter of base 10 cm.; diameter of column 2·5 cm.; disk diameter 7 cm. Colour variable, brown, red, green or grey, sometimes spotted. Viviparous. Common on rocks.

Anemonia sulcata. Opelet anemone. S. 1935, p. 124; Gosse, 1860, p. 160, as *Anthea cereus*. Base broader than column, which is smooth and satin-like on the surface. Tentacles smooth, long, not fully retractile, up to 100 or 200 in number. 5 cm. high, tentacles to 10 cm. Oviparous. Colour variable. Sometimes brown, but usually the tentacles are tipped with lilac and green, like shot silk. Common in intertidal zone in rock crevices on west and south coasts, rare on the east coast.

Tealia felina. Dahlia wartlet. S. 1935, p. 139; Gosse, 1860, p. 209, as *T. crassicornis.* Base much broader than column, the latter very short, studded with grey adhesive warts to which shells and stones are attached. Tentacles about eighty in number (10, 10, 20, etc.), short, thick, conical, retractile, banded transversely. Column 5 cm. high, 7 cm. diameter; disk 12 cm. diameter. Oviparous. Colour variable, crimson, dull green, olive, grey and white may all be present in the same specimen. Intertidal pools to deep water.

Bunodactis verrucosa. Gem anemone. S. 1935, p. 156; Gosse, 1860, p. 190, as *Bunodes gemmacea.* Base broader than column, which is pillar-like and marked by six or seven vertical bands on which lie white rounded warts. Between these bands are dark bands of smaller, darker coloured warts. The disk is flat with distinct radii. Tentacles about forty-eight (6, 6, 12, 24), rather long, completely retractile, a mark like the letter B at the base of each tentacle. Viviparous. Column to 5 cm. high, 2·5 cm. diameter. Colour usually rose pink with white warts. On west and south coasts.

Anthopleura. The warts are alike in size and colour.

A. thallia. Glaucous pimplet. S. 1935, p. 162; Gosse, 1860, p. 195, as *Bunodes.* Warts in about thirty-six vertical rows, strongly adhesive. Tentacles forty-eight to eighty in four rows, with scattered white spots, irregularly arranged, retractile. Column 5 cm. high, 3 cm. in diameter. Disk 5 cm. in diameter. Colour olive green (*thallia* = olive shoot), grey or blue, with darker warts. On rocks. West and south coasts.

A. balli. Red-specked pimplet. S. 1935, p. 167; Gosse, 1860, p. 198, as *Bunodes.* Warts spotted with crimson, in forty-eight vertical rows. Tentacles ninety-six. in five rows (6, 6, 12, 24, 48), pellucid, with white flecks, not strongly retractile. Column 2 cm. high, 2·5 cm. in diameter, and 2·5 cm. across the disk. Colour red, crimson and yellow. Oviparous. Intertidal to deep water. South and west coasts.

METRIDIIDAE.

Metridium. Cinclides scattered, small. Disk lobed in older specimens, tentacles very numerous, pointed.

M. senile. S. 1935, p. 214; Gosse, 1860, p. 12 as *Actinoloba dianthus.* Base large, irregular, column with parapet, disk frilled or lobed. 10 cm. high, 6 cm. in diameter. Colour brown, orange, yellow, pink, cream or pure white. At extreme low water, in pools and in shallow water.

SAGARTIIDAE.

Contractile anemones with a broad pedal disk. Body wall perforated by cinclides for acontia. No parapet. Suckers usually present but not on prominent warts. Disk completely retractile.

Tentacles very numerous, more than 200 (about 500) in six to eight rows, the outermost rows small and tubercle-like. **Cereus**
Tentacles in three to five rows, fewer than 200. **Sagartia**

Cereus pedunculatus. Daisy anemone. S. 1935, p. 363; Gosse, 1860, p. 27, as *Sagartia bellis*. Vase or trumpet-shaped, disk broad and flat, broader than base or column, with distinct radii and numerous short tentacles in six to eight cycles (6, 6, 12, 24, 48, 96, etc.), the first (innermost row) 6 mm. long, the outermost row mere tubercles. Column with cinclides and suckers, former dark dots, latter grey spots. Height to 7 cm., diameter 3–5 cm. Colour variable, flesh colour to brown, cream, yellow, etc. Throat white. Viviparous. In crevices in tide pools, in mud, attached to small stones, spreading the disk on the surface of the mud.

Sagartia. This genus contains a number of very variable species whose colour varieties have often been considered specific. Five species are now recognised, of which four are littoral. Column with cinclides and usually with suckers.

1. Suckers present on column. 2
   Suckers absent. 3
2. Acontia freely extruded. Tentacles rather irregular in five cycles, often with seven or more in the first row. Brightly coloured. Reproduce by laceration. **S. elegans**
   Acontia not freely extruded. Column often with adherent sand or gravel. Tentacles in four to six cycles and six in the first cycle, barred transversely and marked with a letter B at the base. Viviparous. **S. troglodytes**
3. Cinclides scattered. White or yellow disk. Tentacles fewer than 100. **S. sphyrodeta**
   Cinclides forming a zone in a single series below margin of disk. Buff or brown. Tentacles more than 100. **S. anguicoma**

Sagartia elegans. S. 1935, p. 306; Gosse, 1860, pp. 41–67, as *S. miniata*, etc. A very variable species. Flower-like, usually brightly coloured, and emitting its acontia freely. Tentacles in four to five cycles, slender but not very long. Common in rock pools and in crevices at extreme low-tide level to shallow water. Height 4–5 cm., diameter of disk 4·5 cm.

S. troglodytes. S. 1935, p. 324. Base very broad, irregular. Column low in contraction but capable of considerable expansion. Upper part of column with numerous grey suckers to which foreign matter becomes attached. Cinclides among the suckers and in the lower part of the column which is free from suckers. Disk narrower than the base as a rule, dull, speckled. Yellowish grey with darker rim. Markings variable, often green. Tentacles in six cycles (6, 6, 12, etc.), about 190, blunt, slender, translucent, with B-shaped markings at base. Diameter of base 3·5 cm., disk 2 cm. Height 4 cm. Viviparous. Common, shore to shallow water. Buried in sand or mud, attached to stones, shells, etc., or in holes in rock, crevices in pools, etc. Occasionally estuarine.

S. (Actinothoe) sphyrodeta. S. 1935, p. 342. Small delicate species. Base well developed. Column smooth, with cinclides but no suckers. Greyish white. Disk white. Tentacles white with a grey zone at base. Tentacles irregular in five cycles, not very long. Base diameter 1 cm. Height 3 cm. Disk diameter 2 cm. Acontia emitted readily. Reproduced by longitudinal fission and by eggs. Shore to shallow water. On overhanging rocks or in caves.

S. (Actinothoe) anguicoma. S. 1935, p. 355. Base broad. Column varying from low pillar to tall graceful one. Cinclides near top and bottom of column. Acontia emitted reluctantly. Disk light brown with brown throat. Tentacles in five cycles (12, 12, 24, etc.), slender, tapering and very long when expanded, transparent, almost colourless, may be longitudinally striped. Base diameter 4 cm., disk and tentacles 5–6 cm., height 6–7 cm. Littoral to shallow water. Often buried in sand and mud, on stones or on empty shells, etc.

CORALLINOMORPHIDAE.

It is doubtful whether this family belongs to the Actiniaria, but the single species has been included here for convenience. Tentacles are all alike, in a double circlet, knobbed at the tips, retractile. Small anemones of brilliant colours.

Corynactis viridis. Gosse, 1860, p. 289, pl. 9, figs. 1–5. Base expanded, adherent to next anemone so that a group looks like a colony. Column tall, with parapet, smooth, without suckers or perforations. Tentacles in four rows (16, 24, 32, 32 = 104). Radii of disk well marked. 2 cm. high, 1 cm. broad. Gregarious anemones, of pellucid emerald green, purple, orange and rose pink shades. Under overhanging ledges and in rock pools. Local.

MADREPORARIA

Zooid resembles an anemone. Often colonial. Secreted calcareous exoskeleton or corallum always present.

TURBINOLIDAE.

Mostly solitary corals, attached or embedded in sand. Septa radial, imperforate.

Caryophyllia smithi. Devonshire cup coral. Gosse, 1860, p. 310, pl. 10, figs. 12–13. Corallum with expanded base, rounded or elliptical. Cup wall is a true theca, with epitheca below. Five cycles of septa connected by pali. Spongy calcareous columella. Zooid translucent, with about fifty tentacles in three rows. Tentacles knobbed. Pale pink, with brown girdle round the white throat. Gregarious. 2 cm. in diameter. Often on shells. Littoral to deep water. South and west coasts.

CTENOPHORA

The comb-jellies are pelagic and are found swimming in the open sea near the coast in August and September. Occasionally they may get washed into rock pools and their access to the sea cut off temporarily.

Pleurobrachia pileus. Body ovate or spherical, eight ribs (costae), bearing the glistening yellow combs, and a pair of long retractile tentacles issuing from the sides. The species reaches 12–15 mm. in length.

III. COELENTERATA

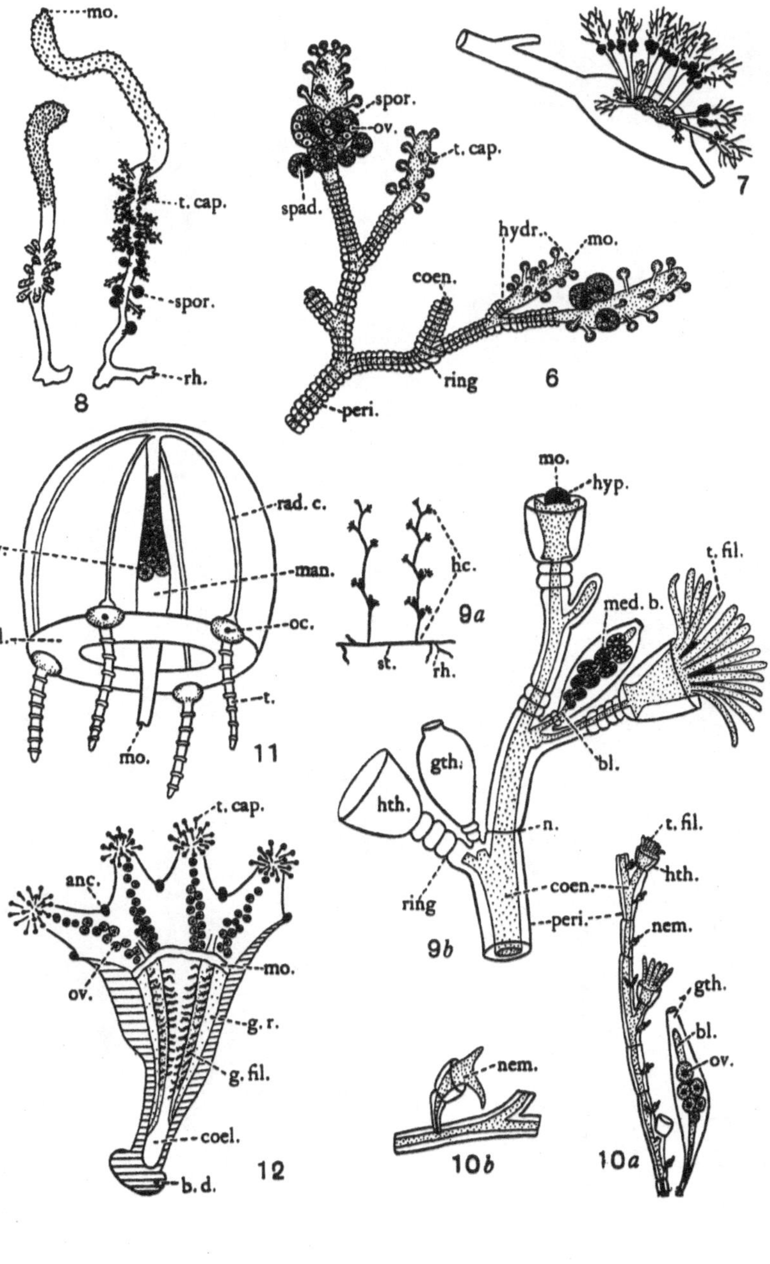

(*facing p.* 47)

EXPLANATION OF PLATE III

Figs. 6–11, *Hydrozoa*. Fig. 12, *Scyphozoa*

Fig. 6. *Coryne muscoides*. Gymnoblastea. Part of the hydrocaulus. ×12.

Fig. 7. *Clava squamata*. A small colony growing on *Fucus*. Note the filiform tentacles arranged in whorls above the bunches of sporosacs. The encrusting base of rhizoids is compact. ×1½.

Fig. 8. *Myriothela cocksi*. Two zooids, one contracted, the other expanded and bearing sporosacs. ×1.

Fig. 9. A calyptoblastic hydroid. *a*, A small portion of a colony. ×3. *b*, Detailed structure. The left side shows the skeletal parts only, the right both the living and skeletal parts. The hypostomial area has been shown in black. ×12.

Fig. 10. *Plumularia setacea*. *a*, A small portion of the colony. ×15. *b*, A nematophore. ×50.

Fig. 11. *Sarsia*, the medusa. Gymnoblastea. Female specimen. ×8.

Fig. 12. *Haliclystus auricula*. Scyphozoa. A female specimen cut in half longitudinally and flattened out. The cut edge is indicated by parallel lines. The relation between the gonads and the coelenteron is not shown. Note the bands of tissue strapping down the mouth radially. ×4½.

anc.	marginal anchor	n.	node
b.d.	basal disc	nem.	nematophore
bl.	blastostyle	oc.	ocellus
coel.	coelenteron	ov.	ovum
coen.	coenosarc	peri.	perisarc
g.fil.	gastric filament	rad.c.	radial canal
g.r.	gastric ridge	rh.	hydrorhiza
g.th.	gonotheca	ring	ring on perisarc
hc.	hydrocaulus	spad.	spadix
hth.	hydrotheca	spor.	sporosac
hydr.	hydranth	st.	stolon
hyp.	hypostome	t.	tentacle
man.	manubrium	t.cap.	capitate tentacle
med.b.	medusa bud	t.fil.	filiform tentacle
mo.	mouth	vel.	velum

EXPLANATION OF PLATE IV. ANTHOZOA

Fig. 13. *Alcyonium digitatum.* Alcyonaria. *a*, Female zooid. $\times 8$. *b*, Spicules. $\times 25$. The zooid is figured in optical section and the flat surface of one side mesentery is shown by dots. The remaining mesenteries are indicated by their edges only.

Fig. 14. Scheme to show the structure of a Zoantharian zooid with basal disk. A little more than half of an anemone possessing twelve tentacles. A portion of the column wall is shown, perforated by a cinclis, through which an acontium protrudes. Cut edges are cross-hatched, with the exception of the edges of the stomodaeum.

Fig. 15. *Peachia hastata.* Lateral view of an expanded specimen with mouth open to show the lip or conchula. In the sand anemones the base is inflated during burrowing and there is no basal disk. $\times \frac{3}{4}$.

ac.	acontium
b.d.	basal disc
cap.	capitulum
ci.	cinclis
coel.	coelenteron
col.	column
con.	conchula
d.	disk
fos.	fossa
gon.	gonad
mar.	margin
mar.sph.	marginal sphincter
mes.c.	fixed (complete) mesentery
mes.d.	the two dorsal mesenteries
mes.f.	free mesentery
mes.fil.	mesenteric filament
mo.	mouth
or.sph.	oral sphincter
ov.	ovum
par.	parapet
phy.	physa
r.	radius of the peristome
sc.	scapus
si.	siphonoglyphe
std.	stomodaeum
su.	sucker
t.dir.	directive primary tentacle
t.p.	primary tentacle
t.pin.	pinnate tentacle
t.s.	secondary tentacle
ver.	verruca (wart)

IV. COELENTERATA

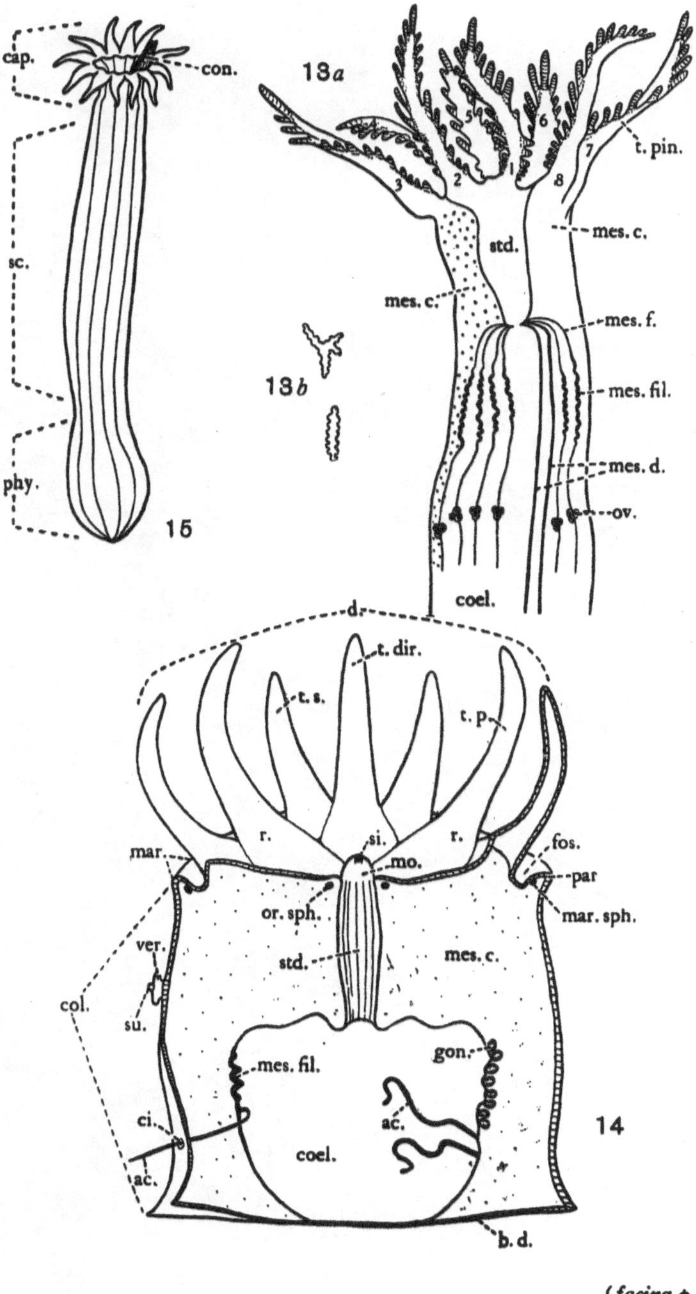

(*facing p.* 48)

PLATYHELMINTHES

(Plate V, figs. 16–18)

AUTHORITIES

GAMBLE, F. W. 1893. "Contributions to a knowledge of British marine Turbellaria." *Quart. J. Micr. Sci.* xxxiv, 433–528.

GIARD, A. 1886. "Sur un Rhabdocoele nouveau parasite, et nidulant." *C.R. Acad. Sci., Paris*, CIII, 499–503.

VON GRAFF, L. 1882. *Monographie der Turbellarien. I. Rhabdocoelida*, pp. 1–141. Leipzig.

—— 1905. Turbellaria. I. Acoela. In *Das Tierreich*, pp. 1–85.

—— 1913. Turbellaria. II. Rhabdocoelida. In *Das Tierreich*, pp. 1–484.

The flatworms have a depressed body, ventral mouth and a prostomial region anterior to the mouth containing the brain. The gonads and excretory organs open on the ventral surface. The flatworms are usually hermaphrodite, often with elaborate accessory organs. The sense organs are simple and consist of eyes and statocysts. No anus. Some are free-living, others are ecto- or endo-parasites, like the flukes and tapeworms. They inhabit fresh and salt water, rarely the land.

Class I. TURBELLARIA

Usually free-living, with ciliated epidermis. Body flattened, oval or leaf-shaped. Usually less than 3 cm. long. Carnivorous.

Class II. TREMATODA

Ecto- and endo-parasitic forms.

Class III. CESTODA

Endoparasitic tapeworms.

The parasitic forms have been omitted.

A large number of Turbellaria inhabit the littoral zone, but many are so small that they are not included here. They were described by Gamble (1893). The class is divided into four orders: Acoela without gut, Rhabdocoelida with gut simple and unbranched, Tricladida with gut shaped like an inverted tuning fork, and Polycladida with much-branched gut.

The **Acoela** rarely exceed 2 mm. in length, but any one species may be present in large numbers. For example, *Convoluta convoluta* and *C. flavibacillum* may be obtained by washing weeds from rock pools, or from sand around the base of *Corallina*. *C. convoluta* (Gamble, 1893, p. 445 as *C. paradoxa*) is about 2 mm. long, greenish brown, with red eyes. *C. flavibacillum* (Gamble, p. 448) is yellow with reddish eyes.

On stones between tide marks occurs a yellowish flask-shaped capsule about 3 mm. long, with a short neck. This is the encysted stage of a **Rhabdocoele,** *Fecampia erythrocephala*, Giard, p. 499, and the adult dies within the capsule after depositing rose-coloured eggs. This adult is a flatworm 10 mm. in length and is parasitic in young shore crabs (*Carcinus*) or hermit crabs (*Eupagurus bernhardus*). Although the capsule may be abundant on a certain portion of the coast, the adult parasite may never have been found there.

The commonest **Triclad** is found, often in great abundance, under stones in places where a trickle of fresh water runs down a cliff or beach. *Procerodes ulvae* (Pl. V, fig. 16) (Gamble, p. 493 as *Gunda*) is a grey-banded or streaked worm, with truncate anterior margin, short neck, and broadened posterior end. Two eyes, length 3 to 7 mm. Egg capsules reddish brown, 1 mm. in diameter, in groups on stones.

The **Polyclads** are the largest Turbellaria.

Leptoplana tremellaris (Pl. V, fig. 17). Gamble, p. 498. Delicate, ovate, about 25 mm. long, pale brown in colour, often with white patches, and groups of black eye-spots over the brain and antero-lateral portions of the body. No tentacles. Under stones, on sponges or on weed. Can swim by means of tremulous, sinuous movements of the sides of the body. Common.

Oligocladus sanguinolentus (Pl. V, fig. 18). Gamble, p. 509. Body delicate, elliptical, semi-transparent, creamy white, with splashes of carmine and brown. Two long pointed tentacles on the anterior margin, and numerous eyes. Length about 10 mm., breadth 4 mm.

Stylostomum ellipse. Gamble, p. 511 as *variabile*. Elliptical, white or yellowish white, with red, orange, brown or black gut, three pairs of cephalic eyes only, and a tentacular group of eyes at the base of the small tentacles. 5–9 mm.

Cycloporus papillosus. Gamble, p. 504. Elliptical, yellowish white, with dark blotches and spots. Surface usually papillose. Two small tentacles. Four groups of numerous small eyes. 10–14 mm. On Tunicates and sponges.

Prostheceraeus vittatus. Gamble, p. 504. Large, 33 × 15 mm., brightly coloured, yellow, with longitudinal dark stripes. Two large tentacles. Two small cephalic eye groups. Under boulders or in *Laminaria* holdfasts. South and west coasts.

NEMERTINI

(Plate V, figs. 19, 20)

AUTHORITIES

BÜRGER, O. 1895. "Die Nemertinen des Golfes von Neapel," *Fauna u. Flora Neapel*, XXII, 1–743.

—— 1904. Nemertini in *Das Tierreich*, pp. 1–151.

MCINTOSH, W. C. 1873, 1874. *A monograph of the British Annelids*, vol. I, part I. The Nemerteans. Ray Soc. Publ. pp. 1–213.

RICHES, T. H. 1893. "A list of the Nemertines of Plymouth Sound." *J. Mar. biol. Ass. U.K.* III, 1–29.

Like the free-living flatworms, the Nemertine worms are ciliated, unsegmented worms, but differ from them in shape and in the possession of blood vessels and an anus. They are elongated, highly contractile worms, either round or flattened in section, and vary from a few millimetres to several metres in length. There is a protrusible, anteriorly placed "proboscis", which is, however, not connected with the gut. It may be armed with a stylet. Eyes are often present, and the head may be flattened and bear a pair of "head slits", whose function is obscure. Nearly all are marine and carnivorous. Their soft, velvety bodies and habit of breaking up or of contracting violently make them difficult to preserve. Alive, their colours are often brilliant—purple, scarlet, chocolate brown, cream, bright yellow, green and pink being common, either singly or in combination.

The identification of Nemertines involves a study of their histology, which is difficult for the collector. The following key, therefore, is based on features which may be seen in the living animal:

1. Free-living. Body elongated, worm-like. — 2
   Parasitic within the mantle cavity of Lamellibranchs. Body broad with posterior sucker. — **Malacobdella**
2. Eyes present. — 3
   Eyes absent. — 6
3. Four eyes arranged in a square or oblong. — **Tetrastemma**
   Morè than four eyes. — 4
4. Colour pale pink, with two rose-pink ovate spots on head. Body flat. Cerebral and marginal eye groups. — **Amphiporus**
   Colour pronounced—red, green or purplish-black. — 5

5. Proboscis armed with a stylet but shorter than the flattened body. **Eunemertes**
Proboscis unarmed and very long. Body rounded or slightly flattened. **Lineus**
6. Head slits present, though shallow. **7**
No head slits. Pale yellow or pink, with reddish head. **Cephalothrix**
7. Brick red with white proboscis. **Micrura**
Red or chocolate coloured, with white circular bands and longitudinal stripes. **Tubulanus**

PROTONEMERTINI

TUBULANIDAE.

Tubulanus (*Carinella*). Body elongated, tapering posteriorly. Anterior end rounded and blunt, broader than the rest of the body.

T. annulatus (Pl. V, fig. 20). M. 1874, p. 203, as *Carinella*; B. 1904, p. 14. No eyes. Head paler than the body, which is rounded and of a rich red or chocolate brown colour, banded at regular intervals with narrow white circular bands, as well as striped longitudinally with white in most specimens. It is one of the most beautiful Nemertines. It secretes a tube of mucus under stones or in rock crevices, from lower shore to deep water. 16–90 cm. long, but only about 2 mm. broad.

MESONEMERTINI

CEPHALOTHRICIDAE.

Cephalothrix rufifrons = *C. bioculata* Oersted. M. 1874, p. 208, as *C. linearis*; B. 1895, p. 539; 1904, p. 18. Head nearly cylindrical, slightly tapering. No eyes or head slits in adult. Body elongated, pale yellow or flesh coloured with reddish anterior end. Under stones or in sand, often gregarious. 8 cm. long, but highly contractile, often alternately dilated and contracted.

C. linearis. B. 1895, p. 538. Eyes in larva, head has two red pigment specks in adult but no eyes. Among *Corallina*, in sand and muddy sand.

METANEMERTINI

AMPHIPORIDAE.

Amphiporus lactifloreus. Pink ribbon worm. M. 1874, p. 156; B. 1904, p. 38. Proboscis armed with stylet, large, as long as body. Eyes numerous, not arranged in a square, but in two groups on each side, one group over brain, the other marginal. Body rather short, flattened ventrally. White, rose or grey. Head spatulate with two rose pink oval spots over the brain ganglia. Head slits obliquely transverse. Common. Under stones on lower shore. 2·5–7 cm. long, 6 mm. broad.

TETRASTEMMIDAE.

Tetrastemma flavida. B. 1895, p. 583; 1904, p. 64 (*Prostoma*). Eyes four, arranged in a square or oblong. Head bluntly rounded; not wider than the rest of the body. Body flattened. Head slits oblique. Pink, with pale margins and opaque patches on the head. 3–4 cm. long. Under stones, sluggish. Common. (McIntosh's *T. flavida* is *Nemertopsis flavida* and is related to *Eunemertes*.)

Many other species occur.

Eunemertes (*Emplectonema*). Proboscis rather small, with stylet. Body elongated. Eyes numerous. Snout broad.

E. gracilis. M. 1874, p. 176, as *Nemertes*; B. 1904, p. 22. Long, greenish, slender worm, to 20 cm. or more in length and only 2 mm. broad. Flattened and pale beneath. Eyes in two or three groups on each side. No marked head slits. Under stones.

E. neesi. M. 1874, p. 178, as *Nemertes*; B. 1904, p. 22. Streaked with purplish brown, iridescent. Flat and pale beneath. Head snake-like. Eyes numerous in two dense clusters, not always visible owing to the streaky markings on the head. 10–40 cm. long, 6 mm. broad. Head slits shallow. Often gregarious. Under stones.

MALACOBDELLIDAE.

Malacobdella. A broad leech-like Nemertine living in the mantle cavity of certain bivalve Molluscs and acquiring a posterior sucker for attachment.

M. grossa. B. 1904, p. 74. This peculiar Nemertine resembles a Trematode in its general shape, and a leech because of its posterior sucker. It is found, living singly, in the mantle cavity of *Zirfaea crispata*, *Cyprina islandica* and other Lamellibranchs. Eyes are absent. The dorsal proboscis is nearly as long as the gut when withdrawn and does not open to the exterior directly, but into the anterior end of the gut. 30 mm. long, 15 mm. broad.

HETERONEMERTINI

LINEIDAE.

No stylets.

Lineus. Deep head slits. Spatulate head. Eyes numerous on side of head. Body elongated, rounded or slightly flattened, tapering posteriorly. *L. ruber* and *L. gesserensis* are red, the former very slender, the latter twice the breadth of the former. *L. longissimus*, the giant Nemertine, looks like a velvety bootlace. Eggs deposited in a gelatinous rope, white or dirty white.

L. longissimus. M. 1874, p. 181, as *L. marinus*; B. 1904, p. 93. This is the largest British Nemertine, attaining a length of 5 m. or more, but capable of contracting violently. A large specimen will fill both hands. 3–8 mm. broad. Colour dark brown to black, velvety, iridescent. Tip of snout pale, often with longitudinal pale bands and stripes, and slightly bilobed. Eyes numerous but difficult to observe owing to the dark colour of the body. Feeds on worms (Annelids). Commoner in the north than in the south.

L. ruber. M. 1874, p. 188, as *L. sanguineus*; B. 1904, p. 101. Eyes fairly regularly arranged, three or four in a row on each side. Reddish brown. 10–20 cm. long, 1 mm. broad when dead. Feeds on Annelids. Common.

L. gesserensis (Pl. V, fig. 19). M. 1874, p. 185. 10 cm. long, 2 mm. broad; not so slender as *L. ruber*. Eyes three to six on each side. Reddish brown and olive varieties, colour very variable. Common in north.

Micrura aurantiaca. M. 1874, p. 201. Body not very long. Distinct tail, head plainly marked, snout truncated. No eyes. A white patch at tip of snout. Body rounded dorsally, flattened ventrally. 7–10 cm. long. Brick red, proboscis white.

ANNELIDA

(Plates V, VI, VII, figs. 21–32)

AUTHORITIES

BEDDARD, F. 1889. "Notes on Marine Oligochaeta of Plymouth." *J. Mar. biol. Ass. U.K.* I.

Cambridge Natural History. 1896. Vol. II. *Worms, Rotifers, and Polyzoa. Archiannelida, Polychaeta and Myzostomaria*, by W. Blaxland Benham.

FAUVEL, P. 1923. *Faune de France*. V. *Polychètes Errantes*.

—— 1927. *Faune de France*. XVI. *Polychètes Sedentaires*.

McINTOSH, W. C. *A monograph of the British Annelids*, Polychaeta. 1900: vol. I, pt. 2. 1908 and 1910: vol. II, pts 1 and 2. 1915: vol. III, pts 1 and 2. 1922 and 1923: vol. IV, pts 1 and 2.

Annelida are typically ringed worms, i.e. they exhibit marked metamerism, the body being divided into rings or segments forming a longitudinal series. The phylum contains three classes.

I. The ARCHIANNELIDA, or primitive worms.

II. The CHAETOPODA, or bristle worms. The largest class.

III. The HIRUDINEA, or leeches, usually ectoparasites, with large posterior suckers for attachment to the host (not dealt with here).

ARCHIANNELIDA

This class contains a number of aberrant worms of doubtful relationship, belonging to three families. It is possible that some are primitive and others degenerate.

Dinophilus taeniatus. *C.N.H.* II, 242. A small red or orange worm, about 2 mm. long. It has no tentacles, bristles (chaetae) or parapodia (bristle-bearing feet), but its body has five indistinct segments and there are ten metameric girdles of cilia on the body and two on the head. The head end is broadened and bears a pair of eyes. Abundant in rock pools at equinoctial spring tides.

Protodrilus flavocapitatus (Pl. V, fig. 21). Fauvel, 1927, p. 424. Under stones where fresh water runs into the sea, and often associated with the flat worm *Procerodes ulvae*, occurs a pale yellow or reddish worm 8–10 mm. long and 0·2 mm. wide, bearing a pair of rather long tentacles and two eyes on the head. The tail has two leaf-like lobes and the body has ciliated segmental bands, but neither parapodia nor chaetae.

Nerilla antennata. Fauvel, 1927, p. 434, fig. 146; M. 1908, p. 147, amongst the Syllidae. *Nerilla* is more advanced than the other Archiannelids. The segmentation is distinct and each segment carries a pair of simple parapodia with two bundles of fine chaetae. The head bears three filiform tentacles, a pair of cirri and a pair of fleshy palps. There are four eyes. The tail end has a pair of caudal cirri and there are incomplete ciliated girdles on the body. The worm resembles somewhat a member of the Polychaet family Syllidae. It is transparent, almost colourless, 1–2 mm. long and 0·2 mm. wide. It is found amongst algae and corallines.

CHAETOPODA

The Chaetopods or bristle worms form the largest class of the Annelids and are mostly marine worms, though a few inhabit fresh water, or, like the earthworm, are terrestrial.

The body is always distinctly ringed, and typically each segment bears a pair of muscular locomotor feet or swimming paddles, the *parapodia.* A parapodium carries bristles or chaetae, and often one or more tactile organs (*parapodial cirri*). Gills (*branchiae*) may be present, either on the parapodia or the head. The head may be supplied with tentacles, eyes and statocysts, and the mouth may be protrusible as an offensive proboscis.

Chaetopods include Polychaetes and Oligochaetes.

The **Polychaeta** are marine.

The **Oligochaeta** contain fresh water and terrestrial forms, without well-defined parapodia or distinct head. As the name implies, the chaetae are few in number. About ten species of Oligochaetes are estuarine, occurring under stones where fresh water is running into the sea. The commonest species is *Clitellio arenarius.* This worm is often found in large numbers, the worms entangling themselves in a coiled mass in the mud under stones. They are about 3 cm. long, less than 1 mm. wide, red or red and orange in colour, and somewhat resembling the terrestrial Brandling worm in appearance (Beddard, F., 1889, pp. 69–71).

POLYCHAETA

This very large subclass contains both free-living and sedentary members, the latter inhabiting tubes. Although there are transitional forms between these two, it is convenient to retain the old classification into Errant and Sedentary Polychaetes.

The ERRANT POLYCHAETES are active carnivorous worms, capable of creeping and often swimming. The segments are similar, except at the extremities, the anterior end forming a distinct head and the posterior a tail or *pygidium*. The head consists of a *prostomium* or head proper and a *peristomium* or buccal segment. The prostomium may bear *antennae* dorsally, fleshy *palps* laterally and one or more pairs of *eyes*. The peristomium, which may be longer than the head, has the mouth aperture ventral, and laterally may bear several pairs of peristomial tentacles (*tentacular cirri*). The peristomium is really a body segment but rarely has parapodia. The mouth aperture opens to a protrusible pharynx or "*proboscis*" usually furnished with chitinous jaws. The remaining body segments are similar. Each bears laterally a pair of fleshy muscular lobes, the *parapodia* (creeping or swimming feet). A typical parapodium consists of a dorsal portion or *notopodium*, and a ventral portion or *neuropodium*. In some Errant Polychaetes the neuropodium may be a looking-glass image of the notopodium, in others the two may differ in one or more particulars. A typical notopodium (e.g. *Nereis*) is lobed, bears a tentacle-like *dorsal cirrus* dorsally, and between the lobes a bundle of chaetae fixed in a chaeta sac stiffened by a stout bristle called the *aciculum*. The neuropodium has lobes, a bundle of chaetae, an aciculum and a *ventral cirrus*. When both notopodium and neuropodium are present, the parapodium is described as *biramose*; if one portion be absent, it is said to be *uniramose*. Branchiae, when present, are borne on the body. They may consist of branched filamentous projections from the notopodia (e.g. *Eunice*) or flat plates called *elytra* (e.g. *Aphrodite*) attached subcentrally. Other worms develop flat, basally placed scales used as paddles for swimming. The pygidium or anal segment bears no parapodia but usually has a pair of *anal cirri* (*urites*).

The SEDENTARY POLYCHAETES are typically tube-inhabiting forms, though not all are permanently fixed in their tubes.

Some, like *Amphiglena mediterranea*, can leave their tubes, and have eyes on their tails. The sedentary forms are in many respects degenerate. The head is small, and is often devoid of eyes or tentacles. In some it becomes greatly modified for tube-building purposes. There is usually no protrusible proboscis, and jaws are absent. In many the body is divided into regions, called head, thorax and abdomen, the last two varying in the shape and size of the segments and in the nature of the parapodia. These parapodia, being no longer required for walking or swimming, become reduced or modified, their lobes and cirri tend to disappear, and their chaetae, especially those of the neuropodia, assume the form of small hooks (*uncini* or crotchets) for holding on to the tube. Branchiae, when present, are situated on the head, where they form a crown or circlet, or on the body. Sedentary Polychaetes feed on vegetable matter, or on debris.

The number of species is so large that only the commonest are included here. A list of the known species will be found in Fauvel.

Head distinct, body segments all alike, pharynx protrusible and often armed with jaws. Carnivorous. **Errantia**

Head small, body often divided into regions, no jaws but sometimes a protrusible pharynx. Debris feeders. Usually tubicolous. **Sedentaria**, p. 77

ERRANTIA

1. Elytra on certain parapodia, others with cirri. **Aphroditidae**, p. 61

   No elytra. 2

2. Prostomium reduced. Body covered with papillae, slender and elongated. **Sphaerodoridae**, p. 74

   Prostomium reduced. One to five antennae. Setae simple, bifurcated. Cirri filiform. No jaws. Short, broad, ovate worms. **Amphinomidae**, p. 64

   Prostomium well developed. 3

3. Proboscis armed with chitinous jaws. 4

   Proboscis unarmed, or with concealed jaws. 7

4. Antennae of even number, two or four. 5

   Odd number of antennae. (None in *Lumbriconereis*, p. 80.) 6

5. Head large. Two antennae, two palps, four eyes, several pairs of tentacular cirri. Proboscis with two large lateral teeth. **Nereidae**, p. 71

   Head small. Four minute antennae, no palps. Proboscis with four or fewer jaws. **Glyceridae**, p. 75

6. Three antennae two simple palps, uniramose parapodia without branchiae but with large dorsal tentacle-like cirrus. **Syllidae**, p. 66

One to seven antennae, two palps, uniramose parapodia often with branchiae. Complex jaws of three pieces. **Eunicidae**, p. 76

7. Parapodia biramose, with small branchiae. Jaws concealed within the papillose proboscis. **Nephthydidae**, p. 73

Parapodia uniramose, with foliaceous lobes folded on the back. **Phyllodocidae**, p. 65

Parapodia with reduced notopodia, dorsal cirri present but no branchiae. **Hesionidae**, p. 70

SEDENTARIA

1. Body segments all similar to one another, but may differ in length. **2**

   Body divided into two or more regions. **5**

2. Tentaculiform palps present on prostomium. **3**

   No palps on prostomium. **4**

3. Two large palps. Dorsal cirriform branchiae on all body segments. **Spionidae, p.** 78

   Two palps and two bundles of branchiae on head, protected by a palisade of chaetae. **Chlorhaemidae, p.** 82

4. One or more pairs of filiform branchiae on anterior segments, simple filiform branchiae on others. **Cirratulidae, p.** 81

   Prostomium surrounded by branchial lobes. Ventral chaetae long uncini, numerous. Pygidium plain. **Oweniidae, p.** 85

   Prostomium obliquely truncate. Pygidium usually bearing cirri. Ventral chaetae long uncini. **Maldanidae, p.** 84

5. Tentacular crown on the head. **10**

   No tentacular crown on the head. **6**

6. Branchiae on some of the body segments. **7**

   No branchiae on body, but coelomic vesicles may be present. **8**

7. Proboscis rosette-shaped. Parapodia well developed, branchiae simple. Caudal cirri present. **Ariciidae, p.** 77

   Proboscis globular. Parapodia reduced, branchiae arborescent, confined to certain segments. No caudal cirri. **Arenicolidae, p.** 83

   Filiform branchiae on anterior segments, retractile tentacles on prostomium. **Ampharetidae, p.** 87

8. Palps present. **9**

   No palps. Earthworm-like forms with pouched coelomic vesicles. **Capitellidae, p.** 82

9. Small, slender worms with acorn-shaped prostomium and very long papillose palps. **Magelonidae, p.** 80

   Large, grotesque worms, with small transverse prostomium, two grooved palps and much modified body segments. **Chaetopteridae, p.** 80

10. Tube rigid. 11
    Tube variable, usually of mucus, impregnated with sand or mud. 13
11. Tube calcareous, with firm operculum. **Serpulidae**, p. 94
    Tube encrusted with sand. 12
12. Tube open at both ends, short, conical, free, and carried about by the worm. **Amphictenidae**, p. 86
    Tube hard, usually encrusting. Worm with a reflexed achaetous tail. **Sabellariidae**, p. 85
13. Tentacular crown of simple filaments. Two or three pairs of red branchiae on thoracic segments (except in *Polycirrus*). **Terebellidae**, p. 88
    Tentacular crown of feather-like filaments. No thoracic branchiae. **Sabellidae**, p. 91

ERRANTIA

APHRODITIDAE.

A large family, all the members having flat scales or elytra folded over the back. In some of the deeper water representatives, however (e.g. *Aphrodite*, the sea mouse), the elytra are hidden by a felting of soft hairs. The elytra are probably expanded dorsal cirri and lie on every other segment, thus alternating with normal dorsal cirri. They are attached subcentrally and not terminally. The elytra break off readily, and for classificatory purposes it is necessary to count them at once on obtaining the specimen, and also to notice whether the tail portion is uncovered. The body is usually short and broad, and may be oval or oblong. The prostomium is small, hidden by the first few segments, and bilobed. There are from one to three antennae, two palps and four eyes, also two pairs of tentacular cirri. The pharynx is cylindrical and papillose and bears in most genera four chitinous jaws. Parapodia biramose, with simple chaetae and dorsal and ventral cirri. Two anal cirri. The smaller members of the family are often called the Polynoid worms, because they resemble the worm *Polynoe*. See p. 8 for Commensals.

1. Body oval to vermiform, but relatively short and broad. Prostomium bilobed. 2
   Body elongated with numerous segments (50–200). 6
2. Twelve pairs of elytra. **Lepidonotus**
   Fifteen pairs of elytra. 3
   Eighteen pairs of elytra. **Halosydna**
3. Body incompletely covered with elytra, tail region free. **Lagisca**
   Elytra covering body posteriorly. 4

4. Elytra transparent, colourless. Chaetae with lateral spiny pouches. On *Echinus* from shallow water. **Scalisetosus**
Elytra opaque, coloured. Chaetae without lateral spiny pouches. 5

5. Some ventral chaetae bifid at tip. Dorsal chaetae short and thick. **Harmothoe**
Ventral chaetae pointed, thicker than capillary dorsal chaetae. **Gattyana**

6. Fifteen pairs of elytra, posterior half without elytra. **Polynoe**
Elytra very numerous. 7

7. Elytra large, imbricating. Colour pale. **Sthenelais**
Elytra small, not meeting mid-dorsally. Transverse bands of colour on body. **Lepidasthenia**

Lepidonotus. Body short, linear, of uniform width, with twenty-six chaetigerous segments.

1. Elytra reniform, granular, and fringed along the posterior margin, overlapping mesially. **L. squamatus**
Elytra orbicular, slightly granular, not fringed and not overlapping. **L. clava**

L. squamatus. F. 1923, p. 45; M. 1900, p. 274, pl. 25. Elytra covering the whole body. 15–30 mm. long. Very variable in colour—pale or clear yellow, or brownish with a mosaic pattern on the elytra. Under stones. A broad, stiff species.

L. clava. F. 1923, p. 46; M. 1900, p. 280, pl. 26. A southern and western species. Elytra leaving a part of the back and a portion of the tail uncovered, marbled with brown and white specks. 25–30 mm. long and 8 mm. broad. Under stones.

Harmothoe. Body short, linear or tapering posteriorly, with about forty chaetigerous segments and bilobed head. Dorsal cirri and anal cirri rather long. Many species.

1. Elytra fringed and furnished with horny papillae. 2
Elytra not fringed. **H. lunulata**

2. Elytra with papillae of different sizes—small horny ones and a few large pear-shaped ones. **H. impar**
Elytra with short fringe and small uniform papillae. **H. imbricata**

H. imbricata. F. 1923, p. 55; M. 1900, p. 314, pl. 26. Body oblong or oval with thirty-seven chaetigerous segments. 30–50 mm. Colour variable, greyish blue or brown, with a light fleck on the centre of each elytron, occasionally banded or edged with white.

Under stones and often in worm tubes, littoral to shallow water.

H. impar. F. 1923, p. 59; M. 1900, p. 353, pl. 26, as *Evarne*. Body very fragile, oval. Thirty-five to forty chaetigerous segments, anal cirri very long. 12–25 mm. Brownish green, ornamented with bands and flecks or with marbled elytra. Phosphorescent. Under stones and among weed.

H. lunulata. F. 1923, p. 70; M. 1900, p. 342. Body elongated, linear, thirty-six to thirty-nine chaetigerous segments. 10–35 mm. Colour variable. Head red, elytra brown, marked with rings, crescents or spots in dark on a lighter ground. Phosphorescent. Under stones, often commensal with tubicolous annelids, or in *Laminaria* holdfasts.

Scalisetosus assimilis. F. 1923, p. 74; M. 1900, p. 377. This delicate, elongated worm is found crawling amongst the spines of the sea urchin (*Echinus esculentus*) taken in shallow water. There are about forty chaetigerous segments and fifteen pairs of thin, rounded, transparent elytra not covering the body mesially but extending over the tail. 18–20 mm. Almost transparent but usually with a brown or yellow dorsal band.

Lagisca extenuata (Pl. VI, fig. 24). F. 1923, p. 76; M. 1900, p. 298, pl. 26, as *L. floccosa*. Body elongated, forty to fifty chaetigerous segments, attenuated posteriorly with a "tail" of eight to fifteen segments not covered by elytra. Two long anal cirri. Fifteen pairs of elytra, first pair rounded, others reniform or oval, imbricating, covered with conical papillae, a few larger papillae on posterior border and a fringe on the edge. 30–40 mm. long, 8–10 mm. wide. Colour variable, elytra marbled with brown and grey or red; usually with a central clear spot. Under stones. Commonest Polynoid on the south coast.

Halosydna gelatinosa. F. 1923, p. 84; M. 1900, p. 384, pl. 25. As the specific name implies, this worm is semitransparent. About forty-three chaetigerous segments. The four eyes are larger than in the other genera. Long dorsal cirri. Elytra on the whole length of the body, large, suborbicular, soft, without a fringe, but often irregular on the edge, and covered with fine chitinous papillae. Two long anal cirri. 5–9 cm. Body yellow or brown, elytra gelatinous, whitish or brown, almost transparent. Under stones on lower shore.

Sthenelais boa. F. 1923, p. 110; M. 1900, p. 408, pl. 26. This giant member of the family inhabits sandy shores, or places where sandy pockets lie under or between rocks. It burrows rapidly. 150 to 200 chaetigerous segments, convex dorsally, without dorsal cirri but with a richly ciliated sickle-shaped branchia on the elytrophore and on each segment not bearing an elytron. The elytra extend along the whole body and overlap on the back, are at first round and then reniform; about 156 pairs. Two long anal cirri. 10–20 cm., or even more. Colour variable, but usually homogeneous grey, yellow or brown, rarely banded transversely with darker shades.

Polynoe. Body elongated, vermiform, with numerous segments, usually from fifty to 100. Dorsal cirri resembling antennae, ventral cirri short. Fifteen pairs of elytra, all in the anterior half of the body, the posterior half being naked. Two urites (anal cirri).

P. scolopendrina. F. 1923, p. 80; M. 1900, p. 389. A handsome, elongated, iridescent and phosphorescent worm. 30–120 mm. Variable in colour, reddish anteriorly. Naked segments with a median transverse line and a pair of brown dots. Elytra marbled with brown, encircled with a clear area. Often commensal with other worms.

Gattyana cirrosa. F. 1923, p. 49; M. 1900, p. 285, pl. 25. Median antenna twice the length of lateral antennae. Elytra large, imbricating, fringed, opaque, with white, grey, brown or black spots. Chaetae spiny. 50 mm. long, 10 mm. wide. Often in tubes of *Amphitrite*, *Chaetopterus*, etc.

Lepidasthenia argus. F. 1923, p. 87. Elongated, with about 200 segments. Elytra orbicular, smooth, translucent, but with brown and white patches posteriorly. Notopodium reduced to aciculum and a few fine chaetae. Ventral chaetae spiny near the tip and bifid. 20 cm. Pale yellow, with brown transverse bands. Commensal in tubes of *Amphitrite* on south coasts.

AMPHINOMIDAE.

A small family, whose sole littoral representative is one of the most perfect examples of protective resemblance in the animal kingdom.

Spinther miniaceus. F. 1923, p. 140; M. 1900, p. 232, pl. 24. A short, elliptical worm, about 9 mm. long, whose flattened under-surface is attached to sponges, particularly to the vermilion or crimson coloured *Hymeniacidon sanguinea*. It resembles its host so exactly in colour that detection is difficult. Twelve to twenty-four chaetigerous segments, the notopodia with membranous leaf-like lamellae in place of cirri and the neuropodia with hooked chaetae. A pair of short, stout anal cirri. Proboscis unarmed. 9 mm. long, 1 mm. thick, 5 mm. wide. Colour varies according to the colour of the sponge on which it lives, e.g. on green varieties of *Halichondria panicea* it will be green.

PHYLLODOCIDAE.

Body vermiform, usually long and thin, with very numerous segments. The prostomium is often conical, oval or heart-shaped, and bears two black eyes and four or five antennae. The protrusible proboscis is unarmed. The first few segments bear tentacular cirri (one to three), and on the remainder the uniramose parapodia bear dorsal and ventral foliaceous cirri. Two anal cirri (urites). The members of the family are often brilliantly coloured, blue and green being common. In *Phyllodoce* and *Eulalia*, the body is long and thin, with four pairs of tentacular cirri and uniramose parapodia. *Phyllodoce* has four antennae and a heart-shaped prostomium, *Eulalia* five antennae and a broadly ovate prostomium.

Phyllodoce. Three species are common on the shore.

1. Colour predominantly greenish blue. Urites filiform. 2
   Colour predominantly brown, with very slender green or yellow body. Urites cylindrical. **P. maculata**
2. Prostomium broader than long. Foliaceous cirri oval. **P. laminosa**
   Prostomium longer than broad. **P. lamelligera**

P. laminosa (Pl. VI, fig. 23*a*). F. 1923, p. 145. Proboscis cylindrical with a crown of sixteen to twenty rounded papillae and about fourteen whorls of papillae on the anterior portion. Basal part with small conical papillae. 15–75 cm. Iridescent blue with transverse bands of green or brown. Ventral surface pinkish. Dorsal cirri sea green or olive. Under stones and in crevices. Some claim that this is a variety of *P. lamelligera*.

P. lamelligera. Large green paddle worm. F. 1923, p. 147; M. 1908, p. 76, pl. 47. Proboscis pear-shaped with a crown of sixteen papillae, ridged anterior portion and six longitudinal rows of eight to sixteen papillae at the base (fixed portion). Dorsal foliaceous cirri lanceolate. 6–60 cm. long, 6 mm. broad. Colour a beautiful metallic iridescent blue. Cirri green or yellowish brown. Under large stones on lower shore. Eggs in green, gelatinous, pear-shaped masses on stalks, fastened to weed, shells, etc.

P. maculata. Small brown paddle worm (Pl. VI, fig. 25). F. 1923, p. 152; M. 1908, p. 89, pl. 45. Prostomium heart-shaped. Proboscis relatively short, cylindrical, crowned with papillae, the anterior portion ridged, the posterior or basal part with small conical papillae, in seven to twelve rows. Dorsal foliaceous cirri obtusely oval. 3–10 cm. long, 2 mm. broad. Body yellow or green with three brown spots or transverse bars on each segment. Dorsal cirri brown. Ova green. Under stones in sandy areas.

Eulalia viridis. F. 1923, p. 160; M. 1908, p. 55, pl. 44. Prostomium rounded in front, broad, with two large black eyes, two pairs of lateral antennae and a median antenna. Very long cylindrical proboscis, with a crown of fifteen to thirty rounded papillae and the sides covered with small round papillae. Dorsal foliaceous cirri lanceolate. 5–15 cm. long, 2–3 mm. wide. Colour uniform emerald or grass green, rarely with black lines and spots. Under stones and amongst *Laminaria.* Often found crawling over vertical rock surfaces, amongst sponges, ascidians, and anemones, when the tide has receded. The egg capsule is a green gelatinous pear-shaped body, the stalk pendent on *Fucus*, etc., measuring about 1 cm. long and 8 mm. broad.

Eulalia sanguinea. F. 1923, p. 166; M. 1908, p. 66, as *Eumida.* Body rather short, with from sixty to 140 segments. Prostomium heart-shaped. Sides of proboscis smooth. Dorsal cirri heart-shaped, ventral cirri oval. 30–60 mm. Brown or reddish. Among stones and in *Laminaria* holdfasts.

SYLLIDAE.

A large family containing small slender worms rarely exceeding 5 cm. in length, with elongated, delicate bodies, and relatively long filamentous dorsal cirri. Prostomium distinct, usually rounded, with four eyes, three antennae, and two rounded and often flattened palps. One or two pairs of tentacular cirri. Proboscis protrusible, armed with jaws and chitinous pieces. Two anal cirri. Many reproduce asexually by stolons or by strobilation (transverse budding).

1. No ventral cirri. **8**
   Ventral cirri present. **2**
2. Palps free, dorsal cirri jointed, proboscis armed with a single large tooth on anterior part of pharynx. **Syllis**
   Palps united. Cirri smooth or indistinctly articulated. **3**
3. Palps united at base only. **4**
   Palps united throughout their length. **6**
4. Proboscis sinuous. A large tooth and a circlet of teeth. Occipital wings on the prostomium and very long dorsal cirri. **Pterosyllis**
   Proboscis straight. **5**
5. Proboscis armed with one large anterior tooth. Two pairs of tentacular cirri. Compound chaetae like short billhooks. **Eusyllis**
   Proboscis armed with a semi-circle of great recurved teeth. **Odontosyllis**
6. One pair of tentacular cirri. **7**
   Two pairs of tentacular cirri. Parapodial cirri fusiform, elongated. **Grubea**
7. Tentacular cirri small. Dorsal and ventral cirri cylindrical and very small. **Exogone**
   Tentacular cirri well developed. Dorsal cirri pointed. **Sphaerosyllis**
8. Dorsal cirri filiform or cylindrical. **Autolytus**
   Dorsal cirri flattened, foliaceous, body banded with red and white. **Myrianida**

Syllis. All the body and head processes simple, filiform, made up of a series of bead-like joints. Reproduction by stolons.

S. (*Typosyllis*) ***krohni*** (Pl. VI, fig. 26). F. 1923, p. 259; M. 1908, p. 192, pl. 49. Cirri alternately long and short, curled over the back. Three antennae equal. 15–30 mm. long. Transverse stripes of violet brown on anterior segments. Cirri tipped with white opaque points. Among algae and under stones.

*S. **prolifera***. F. 1923, p. 261; M. 1908, p. 166, pl. 51, as *Pionosyllis hyalina*. Cirri alternately long and short. Median antenna longest. 10–25 mm. long. Colour variable, brown, grey or red. Often pigmented anteriorly, cirri spotted. Among algae.

Odontosyllis. Body fragile. Two pairs of tentacular cirri. A rounded swelling on the peristomial segment overlaps the prostomium. Anterior border of proboscis bears a transverse row of backwardly directed teeth.

O. ctenostoma. F. 1923, p. 277; M. 1908, p. 182. Proboscis with six large recurved teeth. Dorsal cirri long, cylindrical, ridged but not jointed, alternately long and short, about $1\frac{1}{4}$ times the width of the worm. Ventral cirri flattened. Chaetae become oar-like for swimming when worm is mature. Phosphorescent. 10–20 mm. long. Green or yellowish green with transverse band of greyish granules on each segment. Among algae under stones.

Pterosyllis. The generic name is derived from the pair of wing-like outgrowths which extend backwards from the prostomium. Body short, often fewer than twenty segments.

P. formosa. F. 1923, p. 280; M. 1908, p. 225, pl. 48, as *Amblyosyllis lineata.* Thirteen chaetigerous segments and about sixteen segments altogether. Median antenna very long, lateral ones less than half its length. All the cirri extremely long. 10–15 mm. long. Variable in colour, often banded with brown and cream on body, cirri colourless. Among algae, shells, sponges, etc.

Eusyllis. Body fragile. Antennae and cirri faintly ringed.

E. blomstrandi. F. 1923, p. 293; M. 1908, p. 173, pl. 59, as *E. tubifex.* About fifty chaetigerous segments. Median antenna longer than lateral. Dorsal cirri alternately long and short. Phosphorescent. 6–10 mm. long. Orange or yellow, with brown tips to cirri. Common. Often among Hydroids and Bryozoa, or making mucous tubes on *Laminaria.*

Grubea. Characterised by the complete fusion of the palps and the fusiform (radish-shaped) non-articulated cirri. They are small worms, 2–3 mm. long, found in shells, *Laminaria* holdfasts, etc.

1.	Dorsal cirri truncated.	**G. pusilla**
	Dorsal cirri pointed at tips.	**2**
2.	Tentacular cirri shorter than prostomium and palps.	**G. clavata**
	Tentacular cirri longer than prostomium and palps.	**G. limbata**

G. clavata. F. 1923, p. 296; M. 1910, p. 235. Twenty-five to thirty-five chaetigerous segments. Three subequal antennae. Fused palps as long as prostomium. Dorsal cirri with bulbous bases, drawn out to a point. At maturity the male develops long natatory chaetae. Eggs attached dorsally to body of female by a transparent membrane, three to five per segment. 2–3 mm. long. Colourless except for transverse bands of brown pigment. Among algae and *Laminaria* holdfasts.

G. limbata. F. 1923, p. 298. Twenty-seven to thirty chaetigerous segments. Tentacular segment (peristomium) visible externally but partly hidden by a projection of the first chaetigerous segment. Other features as in *G. clavata.* Some authors regard it as a variety of that species. 2–3 mm. long. Colourless. Among algae, Ascidians and Serpulids. Southern species.

G. pusilla. F. 1923, p. 299. Twenty-eight to thirty-five chaetigerous segments. Dorsal cirri ovate, truncate. 1–2·5 mm. long. Colourless, but with brown proboscis, yellow dorsal cirri and carmine eggs. In *Laminaria* holdfasts, shells, etc.

Sphaerosyllis. Very small worms. Proboscis rectangular, armed with one large tooth and with a terminal crown of papillae. Antennae, tentacular cirri and dorsal cirri bulbous at base, pointed at tip.

S. erinaceus. F. 1923, p. 302; M. 1923, p. 450. Body covered with adhesive papillae, and encrusted with sand. Twenty to twenty-two chaetigerous segments. Eggs carried on the back of the female. Male with natatory chaetae at maturity. 1–2 mm. Colourless. Among algae, *Laminaria* holdfasts, etc.

Exogone. Very small worms. Four to six eyes. Proboscis armed with one tooth, rectangular.

E. gemmifera. F. 1923, p. 305; M. 1908, p. 151. Twenty-four to thirty-three chaetigerous segments. Median antenna longer than lateral ones, the set almost in a straight line. Eggs fixed to ventral surface of body. 2–4 mm. long. Colourless or yellowish. Among algae, Bryozoa and Ascidians.

Autolytus. This genus is well known for two reasons. Its habit of budding by stolons, though common amongst the Syllidae, occurs in the pelagic phase, and in this condition it is easily caught in the tow-net and can be preserved as a slide mount. Secondly, the mature female, which is also pelagic, carries her eggs as a large mass pendent on the ventral side of the body, and was given the generic name *Sacconereis*, before her identity with *Autolytus* was understood. Many species.

Proboscis sinuous with a crown of teeth. Palps very small, soldered together. Two pairs of tentacular cirri. Sexually dimorphic. The adult male (*Polybostrichus*) and adult female (*Sacconereis*) lose their proboscides.

1. Proboscis very long, convoluted within the body. **A. longeferiens**
   Proboscis sinuous, forming an S. **2**
2. Proboscis with ten equal teeth. White or yellow ova. **A. prolifer**
   Proboscis with about thirty equal teeth. Body grey. **A. ehbiensis**

A. prolifer. F. 1923, p. 311; M. 1908, p. 215, pl. 48. About sixty chaetigerous segments. First dorsal cirri very long—at least three times as long as the others, which are equal. The pelagic male (*Polybostrichus*) has large fused palps, the first few segments small and sexual, the remainder broad with natatory chaetae. The pelagic female (*Sacconereis*) has one large bag of eggs carried about the middle of the body. 5–15 mm. White or yellow. Among old shells, *Laminaria* holdfasts, etc.

A. ehbiensis. F. 1923, p. 314; M. 1910, p. 243. Thirty-eight to fifty-five chaetigerous segments. Proboscis only slightly sinuous, crowned with about thirty little equal teeth. 4–7 mm. Uniform grey. Among *Laminaria* holdfasts, etc.

A. longeferiens. F. 1923, p. 319; M. 1910, p. 245. Fifty-one to eighty-eight chaetigerous segments. Proboscis exceedingly long and coiled within the body. First dorsal cirrus longer than the others, the remainder alternately long and short. 10–20 mm. Orange, with three longitudinal reddish bands anteriorly. Eggs grey. Among *Laminaria* holdfasts. Southern species.

Myrianida. Body linear. Palps united, small. Proboscis sinuous, crowned with a circlet of teeth. Asexual development by stolons, forming long chains.

M. pinnigera. F. 1923, p. 322; M. 1908, p. 229. A very beautiful worm. White with transverse bars of carmine, orange or yellow. About sixty-six chaetigerous segments on the stock individual. Stolons forming a long chain of fifteen to thirty individuals. 15–32 mm., for stock individual.

HESIONIDAE.

Small worms. Body short and cylindrical, or elongated and resembling that of a Syllid worm. Prostomium simple or bilobed, with four eyes, usually only two antennae, and two two-jointed palps. Proboscis usually without jaws. Tentacular cirri present. Long dorsal cirri more or less distinctly jointed. Dorsal ramus of parapodia reduced, often without chaetae, ventral ramus well developed, with compound chaetae, and short ventral cirri. Two urites.

1. Six pairs of tentacular cirri, on first three fused segments. Body marked with transverse bars, dark. **Castalia**
Eight pairs of tentacular cirri, on four fused segments. Segments not marked by transverse bars. **Kefersteinia**

Kefersteinia. Body elongated, prostomium short. Parapodium with long, jointed, dorsal cirrus and an aciculum, but no chaetae in the notopodium; compound chaetae and a small ventral cirrus in the neuropodium. Two long urites.

K. cirrata. F. 1923, p. 238; M. 1908, p. 12, pl. 46, as *Castalia fusca.* Very fragile, easily breaks into fragments. Fifty to sixty-five chaetigerous segments. 20–75 mm. long. Yellow, brown, red or purple, with white cirri and well-marked dorsal blood vessel. Under stones and in *Laminaria* holdfasts.

Castalia. Similar to *Kefersteinia* except that there are six pairs of long filiform tentacular cirri and the proboscis bears two horny ridges.

C. punctata. F. 1923, p. 240; M. 1908, p. 121, pl. 46. An active, restless worm. 10–25 mm. long, 3–4 mm. wide. Yellow, with dark, transverse bars across the segments and a dark spot at the base of each foot. Eyes red. Among shells, in *Laminaria* holdfasts, as well as in deeper water.

NEREIDAE.

This well-defined family contains the worm *Nereis* which is used as the type of the Errant Polychaete worms.

The Nereids are active, rather large, bulky worms, with numerous body segments, four eyes, two antennae, two ovoid fleshy palps terminating in a button-like segment, and four pairs of tentacular cirri. Proboscis protrusible, with two denticulate jaws and usually chitinous denticles on the sides of the proboscis (paragnaths). Parapodia biramose, except on the first two segments. Dorsal and ventral cirri present, notopodium bilobed, neuropodium bilobed. Chaetae compound. Two anal cirri. The muscles of the body wall are strongly developed. Sexual dimorphism in the breeding season is in many accompanied by the assumption of pelagic habits and modifications of portions of the body and their parapodia, which become adapted for swimming. This phase is called the heteronereis.

1. Paragnaths of proboscis in groups. Tentacular cirri rarely reaching beyond the fifth body segment. **Nereis**

Paragnaths of proboscis in transverse lines. Tentacular cirri reaching to fifth chaetigerous segment. **Perinereis**
Paragnaths of proboscis feebly developed, very small. Tentacular cirri very long, reaching to tenth chaetigerous segment at least. **Platynereis**

Nereis. Paragnaths of proboscis black, conical, disposed in distinct groups, but not in transverse lines.

1. A large foliaceous lobe on each pale green parapodium. **N. virens**
   Parapodia normal. 2
2. Anterior notopodia trilobed. 3
   Notopodia bilobed. 4
3. Cirri shorter than the lobes. **N. diversicolor**
   Dorsal cirrus longer than notopodial lobe. **N. irrorata**
4. Noto- and neuropodia subequal. **N. pelagica**
   Notopodia longer than neuropodia. 5
5. Paragnaths reduced to two groups. Body long. **N. longissima**
   Paragnaths numerous, small. **N. fucata**

Nereis pelagica. F. 1923, p. 336; M. 1910, p. 267, pl. 52. Eighty to 100 chaetigerous segments. Antennae usually shorter than palps, which are elongated and subcylindrical. 6–12 cm. long. Colour greenish, reddish, yellow or bronze with metallic reflections. Dorsal vessel shows as a red line along the back. Common among weeds, under old limpet or mussel shells, under layers of slaty rocks, etc. Heteronereis in male and female.

N. diversicolor. F. 1923, p. 344; M. 1910, p. 312, pl. 52. Ninety to 120 chaetigerous segments. Antennae much shorter than the large fleshy palps. Large parapodia with long chaetae. No heteronereis phase. 6–12 cm. Colour very variable, green, yellow, red and orange, with distinct red line down the back. In mud and sand, often estuarine.

N. fucata (Pl. V, fig. 22). F. 1923, p. 344; M. 1910, p. 336, pl. 52. Ninety to 120 chaetigerous segments. Antennae and palps of about equal length. Parapodia long, dorsal cirri longer than lobes. Heteronereis in both sexes. 10–20 cm. Yellow to brick red with a white longitudinal band on each side of the dorsal red vessel. Palps and parapodia white. In Gastropod shells inhabited by the hermit crab *Eupagurus bernhardus*, etc., but young individuals often found free on the shore. Pelagic when mature.

N. virens. F. 1923, p. 348; M. 1910, p. 330, pl. 53. This handsome worm somewhat resembles a *Phyllodoce* in that the parapodia develop foliaceous outgrowths over the back. It is, however, much stouter in build, and of a paler green colour. The foliaceous outgrowth is due to the enlargement of the dorsal lobe of the notopodium, the dorsal cirrus being correspondingly small. Heteronereis in both sexes. 20–30 cm. or more. Iridescent, blue and purple on the back, parapodia green, bordered with yellow. Sometimes used as bait. In muddy sand, under rocks. Lines its burrows with mucus. A northern form. On some coasts it may be no larger and thicker than *Phyllodoce*, but on the north Welsh coast it reaches 40–50 cm., and is as thick as a man's finger.

N. irrorata. F. 1923, p. 340. About 140 chaetigerous segments. Heteronereis in both sexes. 30 cm. long. Brick red, white and grey laterally. In sand, south coast.

N. longissima. F. 1923, p. 351. More than 200 chaetigerous segments. Tentacular cirri short. Parapodial cirri shorter than the pointed lobes. Heteronereis in both sexes. 50 cm. Red, with iridescent blue on the back. In sand.

Perinereis. This genus has been split off from *Nereis* on very slender grounds. The paragnaths of the proboscis are arranged in transverse lines, not in groups. Heteronereis in both sexes.

P. cultrifera (Pl. VI, fig. 23*c*). F. 1923, p. 352; M. 1910, p. 280, pl. 52, as *Nereis*. Two curved, transversely placed paragnaths, shaped like a pair of eyebrows. Between them three small ones arranged in a triangle. Noto- and neuro-podia subequal. 10–25 cm. Greenish bronze with reddish parapodia and dorsal vessel. Under stones, among weed and in mud.

Platynereis. Paragnaths of proboscis very small.

P. dumerili. F. 1923, p. 359; M. 1910, p. 302, pl. 52, as *Nereis*. Seventy to ninety segments. A small worm. Tentacular cirri very long, reaching back to tenth or even fifteenth segment. 2–6 cm. Colour variable, green, yellow, rose red, violet or purplish red. Usually in a mucous or membranous tube.

Nephthydidae.

A small homogeneous family of large stout worms. Body linear, broad, with numerous short segments, flattened ventrally, convex dorsally. Prostomium minute with small eyes often hidden under the first, greatly reduced, chaetigerous segment. Four short antennae.

Proboscis eversible, bearing bilobed papillae terminally and softer papillae posteriorly. Two horny jaws completely concealed. Parapodia biramose with flat lobes, without dorsal cirri but with ventral cirri and a gill between the noto- and neuro-podia. Chaetae simple. One anal cirrus. In sand and mud at all depths. Worms active, creep and swim well. Carnivorous. Used as bait.

Nephthys. Proboscis cylindrical or ovate, bearing twelve to twenty-four (twenty-two in British species) longitudinal rows of soft papillae, and terminated by two vertical lips, each furnished with ten bifurcate papillae. A pair of concealed jaws. Chaetae arranged in fan formation on the large lobes of the parapodia, one aciculum in each ramus. Branchiae cirriform, recurved.

1. All the chaetae straight and smooth. Branchial cirrus small. 2
   Some of the chaetae bent (geniculate) and toothed. Posterior branchial cirri almost as long as the branchiae. **N. cirrosa**
2. Proboscis with five to six transverse rows ot papillae but no median papilla. Chaetae long, silky. **N. caeca**
   Proboscis with two to four transverse rows of papillae and a dorsal median papilla. Chaetae shorter and darker. **N. hombergii**

N. caeca. F. 1923, p. 365; M. 1908, p. 8, pl. 43. Ninety to 150 chaetigerous segments. Parapodial lobes large, rounded, simple, occasionally bilobed. 15–25 cm. Greyish yellow and red, iridescent. Chaetae long, yellow, branchiae red. In sand, often with *Arenicola.*

N. hombergi. F. 1923, p. 367; M. 1908, p. 17, pl. 43. About 130 chaetigerous segments (90–200). Bristles shorter than in *N. caeca* and darker. 10–20 cm. Greyish red or pearly white, iridescent, branchiae red. Common. Used as bait, and known as the Catworm.

N. cirrosa. F. 1923, p. 369; M. 1908, p. 36, pl. 57. Body of ninety to ninety-five segments. Proboscis with a large dorsal median papilla. 6–10 cm. Whitish yellow, chaetae yellow, branchiae red.

SPHAERODORIDAE.

Body short and broad, or elongated, cylindrical. Prostomium not very distinct, covered with papillae, of which two well-developed ones look like antennae. Two or four eyes under the skin. Buccal segment (peristomium) achaetous with a pair of large papillae resembling tentacular cirri. Proboscis cylindrical, unarmed. Skin covered with little papillae, and bearing laterally a certain number of large spherical capsules. Parapodia uniramose, covered with papillae. Chaetae simple or compound. Pygidium bearing papillae and an unpaired cirrus. On rocky shores among shells and calcareous algae.

Ephesia. With four eyes.

E. gracilis. F. 1923, p. 377; M. 1915, p. 47, pl. 88. An elongated worm from 1–6 cm. in length whose body is entirely covered with irregular papillae, masking the 100–120 segments. Large lateral papillae form a line of button-like excrescences along each side of the body, and every papilla has a terminal knob.

GLYCERIDAE (Pl. VI, fig. 27).

Body elongated, rounded, usually tapering at extremities, with numerous segments often ringed two or three times. Body wall firm and muscular, often with a thick cuticle. Prostomium long, conical, ringed, terminated by four small antennae arranged in a cross. Proboscis eversible, cylindrical, furnished with jaws and papillae. Parapodia small, biramose, except anteriorly. Branchiae compound, simple or absent. Dorsal chaetae simple, ventral compound. Two anal cirri.

Glycera. Body segments with two to three rings. Head with eight or more rings, resembling in shape the pointed anterior end of a fly maggot. Proboscis pear-shaped. Two first chaetigerous segments uniramose, without dorsal cirrus. All the others biramose, with globular dorsal cirrus, two-lobed notopodium, one- or two-lobed neuropodium and a ventral cirrus.

1. Branchiae present. 2
   No branchiae. 3
2. Branchiae not retractile, finger-shaped. **G. convoluta**
   Branchiae retractile, globular. **G. gigantea**
3. Prostomium with eight rings. Anal cirri long, filiform. **G. capitata**
   Prostomium with more than eight rings. Anal cirri short, broad. **G. lapidum**

G. capitata. F. 1923, p. 385. 140–170 setigerous segments. Notopodial lobe smaller than neuropodial lobe, the latter with two bundles of chaetae. 30–60 mm. Greyish white or yellow. Very similar to *G. lapidum*.

G. convoluta. F. 1923, p. 383. Body transparent, much attenuated posteriorly. 120–180 biannulate segments. Prostomium conical with fourteen to sixteen rings. Two long, filiform anal cirri. 60–100 mm. Red, rose-coloured or pearly, active. Rolls up in a spiral when touched.

G. gigantea. F. 1923, p. 387; M. 1910, p. 482 as *G. siphonostoma.* Body long, 300–400 bi-annulate segments. 35 cm. Rose to grey. In sand.

G. lapidum. F. 1923, p. 386; M. 1910, p. 477, pl. 55. Body with 60–170 segments each with three rings. Prostomium small, conical, with eleven rings. 25–80 mm. White, pearly or greyish, transparent, with gut showing darkly through the skin.

EUNICIDAE.

Body vermiform, usually elongated. Prostomium distinct, with or without antennae and palps. Two first segments without parapodia or chaetae. Branchiae often present on dorsal cirrus. Usually no ventral cirrus. Chaetae simple or compound. Eyes two, four or none. Pygidium with two or four anal cirri (urites). Very muscular body wall. Proboscis armed with ventral jaw of one piece and dorsal jaw often of two or four series of paired pieces, more or less denticulate. Frequently with a membranous tube. An odd collection of worms.

1.	Antennae present.	2
	No antennae, palps or ventral cirri.	**Lumbriconereis**
2.	Antennae, palps, dorsal and ventral cirri.	3
	One antenna, no palps.	**Nematonereis**
3.	Five antennae, branchiae present.	4
	Three antennae, no branchiae.	**Lysidice**
4.	One pair of tentacular cirri.	**Eunice**
	No tentacular cirri.	**Marphysa**

Eunice. Body very long. One pair of tentacular cirri. Branchiae simple or pectinate, usually the latter. Four anal cirri.

E. harassi (Pl. VI, fig. 23*b*). F. 1923, p. 399; M. 1910, p. 425, pl. 54, as *E. fasciata.* Median antenna longer than lateral antennae, with six to ten cylindrical joints. Tentacular cirri barely as long as the peristomial segments. Branchiae commencing on the third or fourth chaetigerous segment, at first simple, then pectinate. Two long urites, filiform, and usually two very short ones. 15–20 cm. Red to brown, with red branchiae.

Marphysa. Differs from *Eunice* in absence of tentacular cirri. Branchiae simple or branched.

M. sanguinea. F. 1923, p. 408; M. 1910, p. 442, pl. 54. Fat, reddish worm with distinct dorsal vessel, iridescent. Up to 300 chaetigerous segments. Branchiae with four to seven filaments arising in a bunch. To 30 or more cm. Rather fragile and breaks up readily. In cracks and fissures, burrowing in sand or mud. Used as bait in the Channel Islands. *M. belli* has pectinate branchiae.

Lysidice. Head broad and flattened, bilobed. No tentacular cirri. Simple capillary, pectinate and compound chaetae. Four urites.

L. ninetta. F. 1923, p. 411; M. 1910, p. 456, pl. 55, as *L. punctata.* Body cylindrical. Fifty to 200 segments. 5–15 cm. Iridescent, white, red, orange and violet. In rock fissures.

Lumbriconereis. Prostomium conical. No branchiae. Four urites. Body long, cylindrical, of uniform thickness, earthworm-like. No eyes. Parapodia small.

L. latreilli. F. 1923, p. 431; M. 1910, p. 376, pl. 54. Prostomium a short blunt cone. 5–15 cm. Red or brown with metallic reflections. Under stones between tide marks.

Nematonereis unicornis. F. 1923, p. 412; M. 1910, p. 453. Slender, 300–400 chaetigerous segments, the first two achaetous. No tentacular cirri. 20 cm. Pale pink anteriorly, orange or green posteriorly, iridescent. In sand.

SEDENTARIA

ARICIIDAE.

This family forms a link between the active, carnivorous errant forms and the more passively feeding sedentary forms. Body vermiform, with numerous segments, divided into a flattened and enlarged thoracic region, and a longer rounded abdominal region. Prostomium conical, cylindrical or globular, without appendages but sometimes with two small eyes. Proboscis unarmed. Short simple dorsal branchiae. Chaetae all simple. Pygidium with two or four urites. Usually found burrowing in sand and mud.

Scoloplos armiger. F. 1927, p. 20; M. 1910, p. 510, pl. 56. This reddish, orange or brown worm is from 5–12 cm. long and 2 mm. broad. With 200 chaetigerous segments. Prostomium conical. Eyes concealed. Peristomium apodous and achaetous. Thorax of from twelve to twenty fat segments with branchiae, dorsal cirri and rather short chaetae directed laterally. In the abdominal region the cirri are shorter, but the chaetae much longer and directed dorsalwards. Two filiform urites. Proboscis rosette-like, adapted for feeding on the organic matter in the muddy sand in which the worm usually lives. Eggs in pale green or brown pear-shaped masses, 2 × 1 cm. moored to sand by thin stalks 5 cm. long.

SPIONIDAE.

Small, slender worms. Body not divided into regions. The worms inhabit clean sand, and may form tubes or penetrate into crevices. No antennae, but prostomium may be prolonged at the corners to form frontal horns. Usually four eyes are present and two very long tentacle-like palps (tentacular cirri). The proboscis is feebly developed and unarmed. Both dorsal and ventral cirri are lamelliform, and some of the parapodia may bear branchiae. The chaetae are usually simple and hooked.

1. Prostomium with frontal horns. **2**
   Prostomium without frontal horns. **3**
2. Two anal cirri. No branchiae but ciliated transverse bands on segments. **Spiophanes**
   Six to eight anal cirri. Branchiae present. **Scolelepis**
3. Fifth chaetigerous segment modified for prehension, with large dorsal chaetae. **Polydora**
   Fifth chaetigerous segment not different from the rest. **4**
4. Branchiae on all segments after the first. Anal region cup-shaped. **Nerine**
   Branchiae on middle segments only. Anal region with four short foot-like processes. **Pygospio**

Scolelepis. Peristomium with two long coiled tentacles (palps). Branchiae on all the chaetigerous segments except the first. Anus surrounded by cirri.

S. fuliginosa. F. 1927, p. 28; M. 1915, p. 160. 5–6 cm. long and about 2 mm. in diameter, with about 100 segments. Its colour is reddish, with darker pigment anteriorly and often with a white spot on the prostomium. The pygidium bears six to eight foliaceous cirri. The worm lives in sand.

Nerine. Rag worms. Prostomium rounded or conical and may bear a small median tentacle posteriorly. Two long tentacular cirri which are frequently carried folded along the back and concealed between the branchiae. Branchiae from second chaetigerous segment to the tail. Eggs pelagic, with reticulate shell.

N. foliosa. F. 1927, p. 34; M. 1915, p. 142, pl. 90. A green or orange worm with red branchiae. 10–16 cm. long, rather fragile. The hooked chaetae have a single point encircled by the hood. In sand.

N. cirratulus. F. 1927, p. 36; M. 1915, p. 148, pl. 90. Colour greenish blue. 5–8 cm. long. The hooked chaetae have two points. Red branchiae. In sand.

Spiophanes. Prostomium triangular, prolonged antero-laterally into a pair of frontal horns, constricted behind the four small eyes, then pointed posteriorly. Two rather stout, curling tentacular cirri.

S. bombyx (Pl. VII, fig. 29). F. 1927, p. 41; M. 1915, p. 182, pl. 93. This worm is common in sand, where it constructs casings of sand. It can be collected by sieving. The body is filiform with between 100 and 200 segments. Length 5–6 cm., breadth 1·5 mm. Anterior region bright red, then brownish green, finally reddish owing to the gut showing through.

Pygospio. Prostomium bluntly bifid anteriorly, pointed posteriorly, bearing four or more small eyes. Two long tentacular cirri. Dorsal and ventral chaetae capillary. Sexual dimorphism.

P. elegans. F. 1927, p. 46; M. 1915, p. 189, pl. 93. Thin and small, with from fifty to sixty chaetigerous segments. The middle segments of the body bear branchiae, seven to nine pairs in the female, twenty to twenty-eight pairs in the male; 10–15 mm. Yellow or green with brown intestine and red dorsal blood vessel. Inhabiting tubes made of fine sand agglutinated with mucus, in fissures of rocks. Gregarious.

Polydora. Rock-boring worms. Burrow U-shaped, lined by mucus. Worms fragile, semi-transparent. Usually four small eyes, and two long coiled tentacular cirri. Fifth segment strongly modified for prehension. Branchiae commence on segment 7 in *P. ciliata* and *P. flava*, and on 2 in *P. polybranchia.* Anus cup-shaped. Boring in limestone, chalk, shale, clay, oyster shells and occasionally in wooden piles.

P. ciliata. F. 1927, p. 49; M. 1915, p. 198, pl. 89. Sixty to 180 chaetigerous segments. 20–30 mm. long. Brownish yellow. Among calcareous rocks, on old shells. Branchiae seventh to thirty-eighth segment.

P. flava. F. 1927, p. 52. Elongated, 100–150 chaetigerous segments. No eyes. 45 mm. Yellow or white, brown posteriorly. In coarse sand, old shells and rock crevices.

P. polybranchia. F. 1927, p. 58. Sixty to eighty chaetigerous segments. Prostomium greatly elongated, overlapping segments 1 and 2. 15–25 mm. Greenish or yellowish red, often pigmented black. Inhabiting sand, old shells, and tubes in rocks.

Magelonidae.

Body filiform with two distinct regions. Prostomium enlarged, oval or acorn-shaped, with thinned margins and strong muscular central region strengthened by chitinous pieces. Neither antennae nor eyes. Two long papillose tentacles. Dorsal and ventral cirri lamelliform on biramose parapodia. No branchiae. Chaetae simple. Anal cirri present.

Magelona. Palps nearly as long as the body, vascular, with closely arranged papillae. Blood pink. Eight segments form the thorax, then a single segment differing from all the others, and an abdomen comprising all the rest.

M. papillicornis (Pl. VII, fig. 30). F. 1927, p. 64; M. 1915, p. 223, pl. 90. About 150 segments. Thorax with fine capillary chaetae, ninth segment with well-developed dorsal lamellae and a large fan of flattened fine chaetae terminating in spatulate tips, as well as fine chaetae. In the abdominal region the noto- and neuropodia are large, oval, with short hooded and hooked chaetae. 5–17 cm. Colour rose, to grey and green posteriorly, but variable. Does not make tubes but burrows in sand.

Chaetopteridae.

This family is represented by a peculiar tubicolous worm.

Chaetopterus variopedatus. F. 1927, p. 77; M. 1915, p. 120, pl. 89. It occurs on south and west coasts from extreme low water to 15 fathoms. The tube is tough and membranous, often with agglutinated sand and debris, and is wide and open at both ends. It is usually firmly fixed in a rock crevice. The worm is large, attaining a length of 25 cm. and a breadth of 1 cm. The soft body is divided into three dissimilar regions. The anterior region consists of head and thorax. The prostomium is very small and bears two eyes, and a pair of filiform palps, each with a ciliated longitudinal groove.

The thorax is composed of nine flattened segments bearing chaetae on simple parapodia, and a median ventral food groove leading to the mouth. The middle region is the most modified and consists of five segments (11–16). The anterior segment bears a pair of elongated lateral "wings" with ciliated food groove communicating with the median groove above mentioned. The twelfth segment adheres to the tube by means of a dorsal and ventral sucker, while the dorsal portions of segments 13, 14 and 15 have been modified as "fans" whose function is to cause a current of water to wash over the animal by their movement. The third and posterior region is the least modified and consists of about thirty similar segments with soft, biramose parapodia, the notopodium of one lobe, the neuropodium of two lobes. The tube is white, opaque, parchment-like and cylindrical in section. The worm is phosphorescent. It is creamy white and yellow in colour, and sometimes green. Being extremely fragile, it is difficult to extract it whole from its tube. The break usually occurs between the wings and the sucker, and the halves are each capable of regenerating the portion lost.

CIRRATULIDAE.

Body thick, cylindrical, tapering at the end, with numerous short segments. Prostomium small, but distinct, without antennae, with or without eyes. Peristomial segment achaetous, with three rings. Proboscis soft, unarmed. One pair of large cirri, recalling those of the Spionidae, in some genera, or numerous tentacular filaments inserted on the dorsal face of one of the anterior segments. Parapodial branchiae dorsal, simple, forming long, contractile filaments and giving the name Red Threads to these worms. No dorsal or ventral cirri. Chaetae simple, on small biramose parapodia. Pygidium usually lobed, but without anal cirri. Burrowers in mud, throwing off the filaments when disturbed. Eggs in yellow masses.

Cirratulus. Tentacular cirri and branchiae numerous.

C. tentaculatus. F. 1927, p. 91, as *Audouinia*; M. 1915, p. 242, pl. 91. No eyes in adult. Segments very short, very numerous, 300 or more. Branchiae commence at the first chaetigerous segment and extend all down the body, except at the extreme posterior end. Tentacular filaments in two distinct groups on segments 6–7 as a rule (i.e. behind the first branchial filament). 15–20 cm. Yellowish orange, brown or bronze green, with red branchiae and tentacular filaments, the latter paler and more curled than the former. Under stones in sand and muddy sand. Common.

C. cirratus. F. 1927, p. 94; M. 1915, p. 249, pl. 91. Prostomium broad, with four to eight pairs of small black eyes. Body rounded, with about 100 segments. Branchiae commence at the first chaetigerous segment and occur on most segments of the body. Tentacular filaments in two groups of from two to eight on the first chaetigerous segment. 3–12 cm. long. Yellowish orange, red or brownish, with red and yellow filaments. Habitat as above.

CHLORHAEMIDAE.

Body cylindrical or flattened, not very long, with short, papillose segments. Prostomium and peristomium forming a short, retractile, cylindrical tube into which the eyes, the two large palps and numerous filiform branchiae can be invaginated. The chaetae of the first few body segments are usually longer than the others and are directed forwards to form a kind of cage round the head. Parapodia biramose. Dorsal chaetae capillary, ventral chaetae hooked and jointed. Blood green.

Flabelligera. Body soft, transparent, enclosed in a thick mucous covering through which the chaetae project, and in which it wanders around. Young stages on *Psammechinus miliaris*.

F. affinis. F. 1927, p. 113; M. 1915, p. 107, pl. 89. Body short, with from thirty to fifty segments. Branchiae twenty to twenty-five, green, palps yellow or orange, stomach showing red through the skin. 2–6 cm. Under stones and amongst weed, in crevices of rock, etc. On all coasts from tide marks to shallow water.

CAPITELLIDAE.

Reddish, cylindrical worms looking like earthworms. Body divided into two regions, a short anterior thoracic region and a longer thinner abdominal region. Prostomium conical, without appendages but usually with two eyes. Proboscis papillose, voluminous, unarmed. Branchiae are small coelomic pouches. Parapodia biramose, reduced to bundles of capillary chaetae or to groups of uncini or crotchets. Sometimes a caudal appendage. Burrowers in sand or mud and making mucous tubes.

1. Only capillary chaetae on thorax, only crotchets on abdomen.

Twelve thoracic segments, eleven chaetigerous. Branchiae simple or absent. **Notomastus**
Capillary chaetae and crotchets on thorax. Nine thoracic segments. No branchiae. **Capitella**

Notomastus. Peristomial segment without chaetae, remaining eleven thoracic segments with two groups of capillary chaetae on each side, without lobes. Each segment divided by a secondary ring. Abdominal segments also biannulate, the first eight to fifteen having the notopodia and neuropodia coalescent, the remainder separate, all flattened.

N. latericeus. F. 1927, p. 143; M. 1915, p. 276, pl. 92. Body very fragile, 100–150 segments. Numerous little eyes present. 15–30 cm. Red anteriorly, reddish yellow to transparent behind, where it is hidden in the sand.

Capitella. First six thoracic segments bear only capillary chaetae, the seventh capillary chaetae and crotchets, the eighth and ninth nothing but crotchets. On the abdomen crotchets only.

C. capitata. F. 1927, p. 154; M. 1915, p. 280, pl. 92. About 100 segments, less wrinkled than in *Notomastus*. Peristomial segment with chaetae. 2–10 cm. Blood red in colour. In muddy sand, under stones.

ARENICOLIDAE.

Body with numerous short segments divided by five secondary rings. Two to three distinct regions. Prostomium without antennae or palps. Proboscis soft, unarmed. Dorsal branchiae restricted to a certain number of segments, arborescent. Parapodia biramose, reduced, notopodia conical with capillary chaetae, neuropodia with crotchets set on ridges. The lobworm, *Arenicola*, is abundant on sandy and muddy shores, where its casting forms a conspicuous feature after the tide has receded. The L shaped burrow is lined with a mucous secretion which prevents its collapse when the worm is in it. The worms are fat, black or greenish black, wrinkled, with red gills. Their eggs form thin reddish layers, 20 cm. long, on the surface of the sand around the burrow.

1. Chaetae and often branchiae extending to posterior end of body. **2**
   Posterior end of body without chaetae or branchiae. Thirteen pairs of branchiae. **A. marina**

2. Fifteen to sixteen anterior chaetigerous segments without branchiae. **A. ecaudata**
   Eleven to twelve anterior chaetigerous segments without branchiae. **A. branchialis (grubii)**

A. marina. F. 1927, p. 161; M. 1915, p. 59, pl. 88. Nineteen chaetigerous segments. Thirteen pairs of branchiae. Prostomium small, trilobed. Tail long, narrowed, one-third body length. 15–20 cm. Greenish black, iridescent, branchiae bright red. Burrowing in sandy bays. Spawns in autumn.

A. ecaudata. F. 1927, p. 164; M. 1915, p. 72, pl. 88. Forty to sixty chaetigerous segments. Prostomium conical without lateral lobes. Abdomen of thirty to forty segments, all chaetigerous and branchiate. 12–25 cm. Yellowish red, brownish red or black, with metallic iridescence. Gills red. Under stones and in rock clefts. Spawns all the year round.

A. branchialis. F. 1927, p. 165, as *A. grubii*; M. 1915, p. 78, pl. 88. Thirty to forty-five chaetigerous segments. Prostomium similar to *A. ecaudata.* Abdominal segments with chaetae and branchiae except for one to three at the end, which may be without branchiae. 15–20 cm. Greenish black with metallic iridescence. Under stones and in rock clefts. Spawns in winter.

MALDANIDAE.

Body cylindrical, usually sharply truncated posteriorly. Segments few, elongated. No antennae, palps or tentacular cirri. Notopodium a rounded elevation with a few capillary chaetae, neuropodium with a row of denticulate uncini. An aciculus on some segments. No cirri or branchiae. Pygidium plain or with a circlet of cirri. Inhabiting tubes of sand.

Clymene. Head obliquely truncated, pygidium cup-shaped, with a circlet of cirri.

C. oerstedi. F. 1927, p. 173. (The *Caesicirrus neglectus* of Arwidsson, 1911.) Slender, fragile. Nineteen chaetigerous and three achaetous pre-anal segments. Pygidium with seven to ten long cirri and two to four short cirri. 10 cm. long, 1 mm. wide. Red or brown with red glandular bands around each anterior segment. Gregarious, in muddy sand.

OWENIIDAE.

Shape as in Maldanidae, but the prostomium bears a branched branchial membrane in *Owenia*, and the anal segment is small. Dorsal chaetae capillary, ventral uncinate in rows, very numerous.

Owenia fusiformis. F. 1927, p. 203; M. 1915, p. 356, pl. 92. Slender, 20–30 segments. Branchial membrane of six branched lobes. Thorax of four segments, the first achaetous, the next three short, with capillary chaetae only. First five to seven abdominal segments long, remainder short. They bear small bundles of capillary chaetae, and minute uncini arranged in transverse rows on raised bands. 10 cm. long, 3 mm. wide. Green or yellow, with red branchial membrane. Tube rigid, membranous, open at each end, covered with closely packed, overlapping sand grains. Gregarious, in sand or muddy sand. Egg masses attached by an anchoring thread, colourless, about half the diameter of those of *Phyllodoce*.

SABELLARIIDAE.

True tube builders, making strong irregular tubes of sand or gravel, often grouped in masses, and sometimes flattened on stones. Body short, cylindrical or quadrangular, divided into three regions. Prostomium indistinct, set between two large projections of the peristomium (peduncles) which bear modified chaetae (paleae) arranged in one, two or three concentric rows and forming an operculum. Two palps. Mouth unarmed, usually surrounded by filiform tentacles or by papillae. Parts of the body:

(1) Thorax: (*a*) Two short segments with reduced, uniramose parapodia. Neuropodia bear capillary chaetae, often pectinate. (*b*) Posterior thoracic region of three to four segments, with chaetae, flattened or styliform, biramose.

(2) Abdomen. Notopodium with uncini, neuropodium with capillary chaetae, biramose. Simple leaf-like branchiae on most of the segments.

(3) Post-abdomen (caudal). Unsegmented and without chaetae. Bent back on abdomen so that the anus lies close to the mouth.

Sabellaria. Operculum purple, bearing three rows of golden paleae terminally, and numerous tentacles ventrally. The two halves joined by a membrane dorsally. The tubes form dense, almost stony masses on rocks, rarely single. They may be entirely decumbent, or the lip may be raised off the substratum. The apex of the tube is crenulated.

1. A median interopercular cirrus. Each external palea with long, barbed median tooth. Strong acicular chaetae on dorsal side of peduncles of operculum. **S. spinulosa**
No interopercular cirrus. External paleae with smooth teeth. No acicular chaetae on peduncles. **S. alveolata**

S. alveolata. F. 1927, p. 207; M. 1922, p. 22, pl. 112. Thirty-two to thirty-seven segments. Ventrally the short peduncles bear numerous long tentacles. There is also a large posterior lobed lip or "building organ". Branchiae sickle-shaped, on thoracic segments and on fifteen to twenty of the abdominal segments. The coarse sandy tubes of these worms may be packed so closely on the rock that the apertures look like bee cells. Worm 30–40 mm. Body red or brownish red, purplish anteriorly, caudal region clear brown.

S. spinulosa. F. 1927, p. 208; M. 1922, p. 12, pl. 112. Thirty to forty segments. Opercular peduncles elongated. Tentacles of the peduncles shorter than in *S. alveolata*, a triangular cirriform languet between the peduncles, and a larger posterior lip or "building organ". Branchiae on thoracic segments and about thirty abdominal segments. 20–30 mm. Red or reddish brown, caudal region clear brown. Estuarine. Masses often smaller than *S. alveolata*, but the species may be mixed.

AMPHICTENIDAE.

Body short, conical, with few segments, divided into three regions, recurved posteriorly but segmented to the end. Prostomium indistinct, fused with peristomium, which forms a cephalic hood, with entire or denticulate border, numerous prehensile tentacles and two bundles of golden bristles (paleae). The paleae converge to form an operculum. Two pairs of filiform tentacular cirri. On the dorsal side, posterior to the paleae, is a collar whose border may be smooth or crenate. Thorax of three setigerous segments, uniramose, two last bearing sickle-shaped (falciform) branchiae. Dorsal chaetae capillary. Glandular areas ventrally. Abdomen with biramose parapodia, with dorsal capillary and ventral uncinate chaetae (except first). Caudal region recurved, with indistinct segments and a lobe over the anus which helps to close the open posterior end of the tube. Simple eyes, often hidden. Tube free, conical, rigid, fragile, straight or slightly curved, open at both ends, made of a single layer of fine sand grains or shells cemented together. Lying in sand with the small (posterior) end uppermost. Worm capable of locomotion in its tube.

Pectinaria. Peristomium with denticulate border, a constriction separating the abdomen from the post-abdomen.

1. Fifteen chaetigerous segments, of which twelve bear uncini. **P. koreni**
   Seventeen chaetigerous segments of which thirteen bear uncini. 2
2. Tube strongly curved. Collar crenate dorsally. **P. auricoma**
   Tube almost straight. Collar smooth dorsally. **P. belgica**

P. auricoma. F. 1927, p. 222; M. 1922, p. 43, as *Amphictene.* Cephalic region with long papillae. Ten to fifteen paleae on each side. Animal 20–40 mm. long; tube 70–80 mm. long, 9 mm. broad at mouth. Colour pale rose, with red branchiae. In sand.

P. belgica. F. 1927, p. 220; M. 1922, p. 38. 30–70 mm. Tube to 90 mm. Worm rose coloured, gills red. Rarer than *P. koreni*, with which it is often confused.

P. koreni. F. 1927, p. 221; M. 1922, p. 49, as *Lagis.* Tubes slightly curved. 20–50 mm. Tube to 80 mm. Pale rose, iridescent, branchiae red.

AMPHARETIDAE.

Thoracic region with dorsal capillary chaetae and a ventral lobe bearing uncini, abdominal region with uncini only. Prostomium with numerous filiform tentacles. Three or four pairs of filiform branchiae resembling tentacular cirri. Anterior to these a bundle of large chaetae (paleae) may be present on each side. Pygidium naked or with cirri. Tube membranous, with agglutinated sand or other foreign bodies.

Paleae and anal cirri present. **Ampharete**
No paleae or anal cirri. Two large dorsal uncini posterior to the branchiae. **Melinna**

Ampharete grubei. F. 1927, p. 227. Eight branchiae, caducous. Body short, three achaetigerous, fourteen chaetigerous segments in thorax, the latter with capillary chaetae, the last twelve with uncini. Twelve abdominal segments with uncini only, on projecting lobes, with a short dorsal cirrus. 35 mm. Male white or cream, female pink, branchiae green. Tube sandy. In sand or mud.

Melinna palmata. F. 1927, p. 239. Elongated, tapering posteriorly. Twenty-one thoracic segments, sixteen with capillary chaetae and seventeen with uncini, but none on the sixth segment. About sixty abdominal segments. 50 mm. Red or brown, branchiae green, ringed with brown. Tube sandy. In muddy sand.

TEREBELLIDAE.

Body with numerous segments, divided into two regions, thoracic and abdominal. Prostomium variable but always bearing numerous filiform tentacles, and fused with the achaetous peristomium. Eyes often present. Branchiae arborescent, more rarely filiform or pectinate on anterior body segments. No operculum. Notopodial chaetae capillary, neuropodial chaetae uncinate, on ridges. Glandular areas known as ventral shields often present on the ventral side of the thorax. Abdomen usually with uncini only. Tube mucous or membranous, encrusted with sand, mud, or the debris of shells, often fixed in the mud or on rocks, algae or hydroids. The members of this family can be readily recognised owing to the crown of long tentacles and the two or three pairs of bright red gills.

1. Branchiae three pairs (except in one species of **Amphitrite**). **2**
   Branchiae two pairs. Eyes distinct. **Nicolea**
   Neither branchiae nor eyes. **Polycirrus**
2. Dorsal chaetae denticulate **3**
   Dorsal chaetae with smooth tips. Slender worms. **4**
3. Fat worms, broadened anteriorly. No dorsal chaetae on abdominal segments. **Amphitrite**
   Short worms, less than 10 cm. long. Dorsal capillary chaetae on abdominal segments. **Terebella**
4. Uncini in two rows, back to back, avicular. Tube of coarse sand, with a fringe at upper end. **Lanice**
   Uncini in two rows, but not back to back. No fringe on tube. Eyes small, numerous. **Polymnia**

Amphitrite. Fat red worms inhabiting galleries in sand or mud, sometimes making a sand incrusted tube. Thoracic region of seventeen to twenty-five setigerous segments. Long abdominal region with numerous segments. Usually three pairs of branchiae, ramified. Deposit feeders.

1. Two pairs of branchiae. **A. gracilis**
   Three pairs of branchiae. **2**

2. Twenty-four chaetigerous thoracic segments. Up to 25 cm. **A. johnstoni**
   Nineteen chaetigerous thoracic segments. 11 cm. long. **A. groenlandica**
   Seventeen chaetigerous thoracic segments, large fat worm. 25 or more cm. long. **A. edwardsi**

A. edwardsi. F. 1927, p. 245; M. 1922, p. 114, as *A. gigantea* Body tapering, 100–150 segments. Lateral lobes present on cephalic and on second, third and fourth thoracic segments. Tube usually a simple vertical gallery. 25–30 cm. Salmon, brown or yellow. Branchiae dark red. Tentacles yellowish orange. On south and west coasts. *Lepidasthenia argus* is commensal with this species.

A. groenlandica. F. 1927, p. 250; M. 1922, p. 111, pl. 119. Ninety to 100 segments. About fourteen ventral shields. 115 mm. long, 9 mm. broad. East and west coasts to Arctic. In sand and mud.

A. johnstoni. F. 1927, p. 248; M. 1922, p. 117, pl. 113A, as *A. figulus*. Ninety to 100 segments. Thirteen to fourteen ventral glands. Uncini with five teeth. 15–26 cm. long, 8–12 mm. broad. Yellowish red or brown, branchiae red. Tentacles yellow brown. Commensal with *Gattyana cirrosa*.

A. gracilis. F. 1927, p. 252; M. 1922, p. 122, pl. 113, as *A. scylla*. Body long, gelatinous, not swollen anteriorly. 100–200 segments. Seventeen, eighteen or nineteen chaetigerous thoracic segments. Two pairs of ramified branchiae, the first pair larger than the second. Forming sinuous galleries in sand and in fissures of rocks. 6–12 cm. long. Pale red or yellowish grey, branchiae red. Tentacles pink. On all coasts.

Terebella. Branchiae arborescent or cirriform. Dorsal chaetae pectinate and often of two kinds. Uncini in a double opposed row. Thoracic and abdominal uncini of the same type.

T. lapidaria. F. 1927, p. 254. Worm short, about 160 segments. Ventral glands on thorax. 9 cm. Brownish red, branchiae bright red, tentacles yellow or orange. In rock crevices, among hydroids, etc., the thin tube encrusted with sand.

Lanice. Body almost cylindrical, long and narrowed posteriorly. Thoracic region with seventeen chaetigerous segments. Ventral glandular areas almost joined together. Tube of mucus, incrusted with coarse sand grains, and extending as narrow, occasionally branched tubes, forming a fringe projecting above the sand and enclosing the tentacular appendages. Deposit feeder.

L. conchilega. F. 1927, p. 255; M. 1922, p. 137, pl. 113A. Worm long, rather thin, soft. From 150 to 300 segments, three pairs of gills on segments 2 to 4, fourteen to seventeen pairs of ventral glands, and collar-like lobes on segments 1 and 3. 25–30 cm. long, 5–6 mm. wide. Rose coloured, yellow or brown. Branchiae red. Tentacles white. Fringed tubes stand erect in the sand. Common.

Polymnia. Head tentacles numerous, breaking easily. Collar-like lobes on peristomial segment. Seventeen thoracic segments with chaetae. Distinct ventral glands. Capillary chaetae with fine points from segment 4 onwards, uncini from segment 5 in a double row, but not back to back. Tube incrusted with sand.

P. nebulosa. F. 1927, p. 257; M. 1922, p. 129, pl. 114. Body round, soft, fragile, about 100 segments. 5–15 cm. long, 3–8 mm. wide. Greyish orange, rose, brown dotted with white. Branchiae red, often marked with white. Tentacles rose or white. Tubes fixed on small stones, shells, etc.

Nicolea. Head tentacles numerous, filiform. Thoracic region with fifteen to twenty-two chaetigerous segments. No lateral lobes on first few segments. Distinct ventral glands. Dorsal capillary chaetae commence on fourth segment, uncini on fifth. Tube soft, membranous, incrusted with sand.

1. Colour uniform. Branchiae short, without main axis to the numerous branches. Fifteen chaetigerous thoracic segments. **N. zostericola**

   Spotted with white on the back. Branchiae branching from a main trunk. Seventeen chaetigerous thoracic segments. **N. venustula**

N. venustula. F. 1927, p. 260; M. 1922, p. 150, pl. 112. Fifty to seventy segments, of which nineteen are thoracic, seventeen bearing chaetae and the first two branchiae only. Tube thin, transparent, agglutinated with sand and algae. 3–6 cm. long. Brick red, dotted with white on the back, tentacles violet, branchiae red.

N. zostericola (Pl. VII, fig. 31). F. 1927, p. 261; M. 1922, p. 150, as *N. venustula* var. Body small, delicate, forty to fifty segments. Eighteen thoracic segments, fifteen bearing chaetae. 15–20 mm. long, but larger in northern seas. Colour pale rose or brown, transparent. Tube of agglutinated sand fixed on algae or hydroids.

Males and females assist in making a cocoon of fine threads for the eggs.

Polycirrus. Uncini in one row. Thoracic and abdominal uncini of the same type.

P. caliendrum. F. 1927, p. 281. Thin, fragile worm, fragments easily. 70–90 segments. Tentacles very numerous. Prostomium enlarged to form the tentacle-bearing lobe. Ventral glands on thorax. Anus with a circlet of little lobes. 10 cm. long, 4 mm. wide. Orange to yellow, phosphorescent. Tube mucous. Among algae, hydroids, etc.

SABELLIDAE.

Body elongated, cylindrical or flattened, divided into two distinct regions. (1) Thoracic, with four to twelve segments. Bundles of capillary chaetae dorsally, flat lobes with uncini ventrally. No thoracic membrane. (2) Abdominal. Very numerous segments with inversion of the arrangement of the chaetae, the uncini being dorsal and the capillary chaetae ventral. A long ciliated faecal groove dividing into two the glandular areas of the ventral face of the abdomen, and passing to the dorsal line of the thorax. The collar usually well developed on the peristomial segment. Branchiae forming a terminal tentacular crown surrounding the mouth. They consist of two semicircular or spiral lobes bearing numerous feathered ciliated filaments. They may also bear eyes and dorsal appendages, and may be joined at the base of the filaments by a membrane. No operculum. Two fleshy palps. Chaetae always simple, the capillary type winged or spatulate, uncini avicular (bird's-head shaped) and often accompanied by small chaetae shaped like lanceolate leaves on a stalk. Pygidium with terminal anus, with or without papillae, and often with eyes. Tube cylindrical, mucous, membranous or horny, covered with fine mud, rarely with other material, never calcareous.

1. Ventral thoracic chaetae avicular uncini. 2
   Ventral thoracic chaetae uncini with long manubrium. (Stalk). 7
2. Thoracic ridges with one row of uncini only. Branchial filaments with numerous paired dorsal appendages. **Dasychone**
   Thoracic ridges with two rows of chaetae, one row of uncini and one row shaped like pick-axes with two joints. 3
3. Collar well developed. Large worms. 4
   No collar. Small worms. Pygidium with eyes. **Amphiglena**

4. Subterminal eyes on branchial filaments. **Branchiomma**
   No subterminal eyes on branchial filaments. **5**
5. Branchial lobes rolled in a spiral. **Bispira**
   Branchial lobes not spirally rolled **6**
6. Tube of fine mud, flexible. Eight thoracic segments. Worm brown. **Sabella**
   Tube horny, transparent, but agglutinated with sand or mud. Worm orange or red. **Potamilla**
7. Very small worms. Three pairs of free branchial filaments. **Fabricia**
   Large worms, 4 cm. or more. Numerous branchial filaments, joined by membrane at base. **Myxicola**

Sabella. Branchial crown of two semi-circular lobes, each bearing numerous feathered filaments without appendages or subterminal eyes. Collar two- or four-lobed. Tube membranous, flexible, incrusted with fine mud. Filter feeder.

S. pavonina. F. 1927, p. 298; M. 1922, p. 223, pl. 114, as *S. penicillus.* A large tube worm, from 10–25 cm. long and consisting of from 100 to 600 segments, of which six to twelve are thoracic. Each branchial lobe bears from eight to forty-five greenish filaments, thin, flexible and feathered, usually barred at intervals with brown or reddish brown, and about 3 cm. long. The tube often projects vertically from mud or sand and is attached to rocks or shells.

Bispira. The filaments of the spiral branchial lobes bear small eyes arranged in pairs at intervals along the filament. Collar with two or four lobes. The tube is membranous, flexible and incrusted with fine mud.

B. volutacornis. F. 1927, p. 307; M. 1923, p. 254, pl. 115. Shorter and stouter than *S. pavonina.* It consists of about 100 segments, of which eight are thoracic, and measures from 5–15 cm. in length. The branchiae are white or faintly coloured and are not barred. The worms are usually gregarious and occur attached to rocks in muddy sand.

Potamilla. Collar two to four lobed, branchial lobes equal, semi-circular, filaments united at base. Eyes usually present on filaments.

P. reniformis. F. 1927, p. 309. Fragile, elongated. About 200 chaetigerous segments, of which about ten are thoracic. Seven to eight grouped eyes in a single file on the dorsal side of some branchial filaments. 10 cm. long, 2 mm. wide. Orange or brick red. In rock crevices and old shells.

Dasychone. Body short, flattened ventrally, consisting of two regions tapering to a short tail posteriorly. Collar thin, split dorsally. Branchiae webbed at base, the filaments bearing dorsal appendages and often pairs of eye-spots.

D. bombyx. F. 1927, p. 319; M. 1923, p. 273, pl. 112, as *D. argus.* Fifty to seventy chaetigerous segments, of which eight are thoracic. Eyes violet. Tube membranous, incrusted with fine mud, shells and sand, brownish. Worm 3–5 cm. long, 4–5 mm. broad, variable in colour, green or brown with variegated branchiae. On all coasts from shore to shallow water.

Amphiglena. Small worms, often occurring in enormous numbers attached to rocks. The branchial lobes are equal and are not spirally rolled; they are without eyes or dorsal appendages and each ends in a long tapering process. Eyes are often present, both on the anterior end (two pairs on thoracic 1) and on the pygidium (four to six pairs).

A. mediterranea. F. 1927, p. 324; M. 1923, p. 270, pls. 113A and 115. Body short, thirty to forty segments, of which nine or ten are thoracic. Hermaphrodite. Branchial filaments ten. Tube mucous, transparent. 5–15 mm. White or yellow, with greenish branchiae. The worm often leaves its tube and wanders off over the irregular mass of inhabited and empty tubes.

Myxicola. Collar small. Branchial filaments with numerous fine barbules. Ventral thoracic uncini stalked, dorsal abdominal uncini avicular. Tube mucous, thick, gelatinous, transparent.

M. infundibulum. F. 1927, p. 342; M. 1923, p. 319, pl. 116. About 150 bi-annulate segments, of which nine are thoracic. 20 cm. long, 15 mm. wide. Branchiae yellow or orange, tinged with violet. In sand or mud.

M. aesthetica. F. 1927, p. 344. About fifty faintly bi-annulate segments, of which four are thoracic. Pygidium with many eyes. 40 mm. long, 3 mm. wide. Whitish yellow, branchiae pale green or brown. Among algae, hydroids and old shells. Often gregarious.

Fabricia. Small worms, without collar or branchial eyes. Thoracic uncini with long manubrium, abdominal uncini short, denticulate. Tube of mucus, impregnated with sand.

F. sabella. F. 1927, p. 325. Branchial filaments with barbules. Two long palps. Two eyes on first segment and two on pygidium. Ten to twelve chaetigerous segments, of which eight are thoracic. 3 mm. long, 0·25 mm. wide. Yellow or pink. Among algae and ascidians in sand or mud. Worm wanders free from tube.

Branchiomma. Collar two to four lobed. Branchial lobes equal, not spiral. Filter feeder.

B. vesiculosum. F. 1927, p. 315. About 200 segments, usually nine thoracic. First branchial filament longer and thicker than the others. 15 cm. long, 10 mm. wide. Colour yellow, brown or red. Tube horny, heavily encrusted with gravel, broken shells, etc. In sandy gravel, among rocks and shells.

SERPULIDAE.

Body divided into two distinct regions. (1) Thoracic, with few segments, usually three to seven, and with a thoracic membrane. (2) Abdominal, with numerous segments with the chaetae reversed in order (i.e. uncini dorsal and capillary bundles ventral). The first segment usually bears a collar, entire ventrally but open dorsally and continuous with the thoracic membrane. Eyes may be present. The branchiae form a terminal crown of two semi-circular or spiral lobes bearing two rows of ciliated barbules on the numerous filaments, and sometimes eyes. The first filament of one side is usually devoid of barbules and becomes flattened to form a calcareous, horny or membranous operculum, and the corresponding filament of the other side may form a smaller or false operculum. Palps reduced or absent. No faecal groove. Tube calcareous, opaque, often ornamented, usually bent, or spirally coiled with resulting asymmetry of the worm's body. Gregarious.

1. Tube rolled in a spiral, small. Worm with fewer than five chaetigerous thoracic segments. **Spirorbis**
   Tube not spiral. Worm with at least five chaetigerous thoracic segments. **2**
2. Tube cylindrical in section, not keeled. **3**
   Tube keeled often prolonged into a point over the aperture. Operculum spiny or calcareous. **Pomatoceros**
3. Tubes massed together, very fine, white. Opercular stem with barbules. **Filograna**
   Tubes often solitary, coloured. Operculum denticulate or smooth, stem devoid of barbules. **Serpula**

Serpula. Body symmetrical. Tube with trumpet-shaped aperture and rings of growth. Branchial filaments without eyes, united at the base by a palmar membrane. Collar three-lobed, thoracic membrane oblique. No palps.

S. vermicularis (Pl. VII, fig. 32). F. 1927, p. 351; M. 1923, p. 353, pl. 116. About 200 chaetigerous segments. Thorax of seven chaetigerous segments. Operculum with smooth stem and funnel-shaped, toothed head. Branchial filaments red, about thirty each side. The tube is rarely pure white. It is fixed at the base and often elevated near the aperture, usually irregular. 5–7 cm. Worm red or yellow.

Pomatoceros. Seven thoracic segments. Opercular peduncle smooth, without barbules but with wings just below the conical calcareous operculum. Two small palps. Three-lobed collar, and thoracic membrane well developed.

P. triqueter. F. 1927, p. 370; M. 1923, p. 362, pl. 117, as *Pomatocerus.* About 80–100 chaetigerous segments. Tube variable in shape but usually spiral at first. 15–25 mm. Worm red and brown. Fixed to rocks and stones. Common. Living specimens may be used for artificial fertilisations, abdomen red in female, whitish in male.

Filograna. Operculum retaining barbules on its stem. Two palps. Eyes, collar and thoracic membrane present. Sexes separate. Gregarious. Tubes fine, calcareous cylinders, forming dense, irregular, twisted masses.

F. implexa. F. 1927, p. 376; M. 1923, p. 339, pl. 116. Twenty-five to thirty-five segments, of which six to nine are chaetigerous thoracic. Two branchial lobes, each with four filaments. Operculum transparent, thin, chitinous, usually oblique. Grey, transparent. Worms reproducing by budding as well as sexually. 3–5 mm. On stones, in masses often about 15 cm. square.

Spirorbis. Body asymmetrical, small. Fewer than five chaetigerous thoracic segments (usually three). Collar, thoracic membrane, and two short palps. A long achaetous region between thorax and abdomen. Opercular peduncle without barbules, calcareous. Branchiae few, four or five a side. Tube rolled in a spiral, dextrally* or sinistrally, calcareous. Hermaphrodite. Incubation of eggs in tube or in operculum.

* The animal lies on its *dorsal* side, and in a dextral form it is coiled in a clockwise direction, with the right side on the concave side of the spiral. Looked at from above it will *appear* anti-clockwise, with its aperture on the left. A sinistral form will *appear* clockwise in this view, with its aperture facing right.

1. Tube dextral.* Eggs incubated in operculum. **S. pagenstecheri**
   Tube sinistral. Eggs incubated in tube. **2**
2. Operculum round or oval. **S. borealis**
   Operculum with a bifid process. **S. corallinae**

S. pagenstecheri. F. 1927, p. 394. About fifteen chaetigerous segments. Operculum hollowed to form an incubatory pouch, the calcareous lid of which is rejected after use and regenerated. Eggs red or orange. Worm red or orange posteriorly, colourless anteriorly. Tube dextral, calcareous, with longitudinal ridges, aperture irregular. Worm 2 mm. long, tube 2 mm. diameter. On stones, shells or algae. The most widespread and generally abundant species.

S. borealis. F. 1927, p. 399; M. 1923, p. 409. Twenty-one to thirty-five chaetigerous segments. Branchial lobes with four to five filaments each. Operculum with an oblique, calcareous, excentric plate, round or oval. Eggs brownish red, incubated in the tube. Branchiae dull yellow, tinged with green. Worm orange-red posteriorly, brown anteriorly. Tube sinistral, calcareous, often rough, with two to four nautiloid spirals, round a deep umbilicus. Worm 3 mm. long. Tube 2 mm. in diameter. Very common on *Fucus*, *Laminaria*, etc.

S. corallinae. de Silva and Knight Jones, *J. Mar. biol. Ass. UK.*, 1962, 601. On *Corallina officinalis.* Tube sinistral, white, small, 2·5 mm across outer whorl. Operculum with a characteristic bifid process or "talon". Abdomen of 16–25 segments. Body length 3·5 mm. About 20 embryos incubated in the tube. Worm orange.

ABERRANT WORMS

SIPUNCULOIDEA, ECHIUROIDEA, PRIAPULOIDEA

(Plate VII, fig. 33)

AUTHORITIES

CUENOT, L. 1922. *Faune de France.* 4. *Sipunculiens. Priapuliens.* Paris.

FISCHER, W. 1925. "Echiuridae, Sipunculidae, Priapulidae." In *Die Tierwelt der Nord und Ostsee*, Teil VI, pp. 1–55.

SOUTHERN, R. 1913. "Gephyrea of the Coasts of Ireland." *Sci. Invest. Fish. Br. Ire.* 1912, no. 3.

STEPHEN, A. C. 1934. "The Echiuridae, Sipunculidae and Priapulidae of Scottish and Adjacent Waters." *Proc. R. phys. Soc. Edinb.* XXII, pt. 4, 159–85.

Three orders of worms are of doubtful affinities, and it is not even certain that the orders are related to one another. All are detritus feeders and burrow in sand or mud. They are important economically, since they are eaten by many food fishes such as cod and haddock.

The three orders, the Sipunculoids, the Echiuroids and the Priapuloids, have few genera and species. There are few common characters. Some possess an introvert surrounding the mouth, others a proboscis dorsal to the mouth. In some the anus is anterior and dorsal, in others posterior and terminal. They are usually thick-skinned, with strongly banded dermal musculature, but without segmentation or appendages. Bristles may be present or not.

1. Anterior region with an introvert surrounding the mouth. 2
   Anterior region with a proboscis dorsal to the mouth. **Echiuroidea**
2. Introvert frilled. Anus antero-dorsal. **Sipunculoidea**
   Introvert without tentacles. Anus terminal. **Priapuloidea**

SIPUNCULOIDEA

Thick-skinned. Mouth in the centre of the introvert, which is often frilled, ciliated and used for food catching, and in the absence of special respiratory organs may be employed also for that purpose. Sexes separate. The thick body wall bears thirty or more longitudinal muscle bands and a large number of transverse bundles, so that the effect externally in many species is to divide the surface into innumerable rectangular areas. In others the longitudinal bundles are not separate but are fused into a continuous muscle sheath.

Golfingia (*Phascolosoma*) *elongatum*. Fischer, 1925, p. 16; C. 1922, p. 8. The longitudinal muscles in the body wall form a continuous sheath and the introvert is composed of about twelve short bifurcate tentacles, with four retractor muscles. Skin devoid of adhesive papillae. To 9 cm. In sand, often with Ascidians.

G. minutum. Stephen, 1934, p. 167; C. 1922, p. 9. Occurs in crevices, amongst worm tubes, etc., and measures only 6–15 mm

ECHIUROIDEA

Sausage-shaped. Anus terminal. Many possess hooks or bristles, often two in number, sometimes above the nephridial pores and sometimes perianal. Worms often brightly coloured, golden being common.

Thalassema neptuni (Pl. VII, fig. 33). C. 1922, p. 22. Proboscis pointed at the end, not bifurcate, grooved, edges slightly frilled. Two ventral hook-like bristles situated anteriorly, no perianal chaetae. Body covered with little papillae and often clothed with mucus. Highly contractile, anterior end bluish or yellowish, middle grey or rose, posterior opaque white. 2–7 cm. In clefts, cavities in stones. South and west coasts.

PRIAPULOIDEA

Cylindrical worms with anterior mouth and posterior anus, a swollen bulbous body, and spiny mouth region. The muscles are grouped in rings, giving the animals a transversely striated appearance. The anterior region can be retracted (introvert), but though it is papillose it bears neither tentacles nor proboscis.

Priapulus. The introvert bears longitudinal rows of spines and is about one-third of the body length, the remainder of the body is transversely ringed. The posterior end bears a large number of papillae on a ventral caudal appendage in *P. caudatus* and on two appendages in *P. bicaudatus*. Carnivorous.

P. caudatus. C. 1922, p. 25. Flesh coloured with metallic sheen. Burrowing in dirty mud. 6 cm. long. East and west coasts (Menai Straits). Local.

V. WORMS

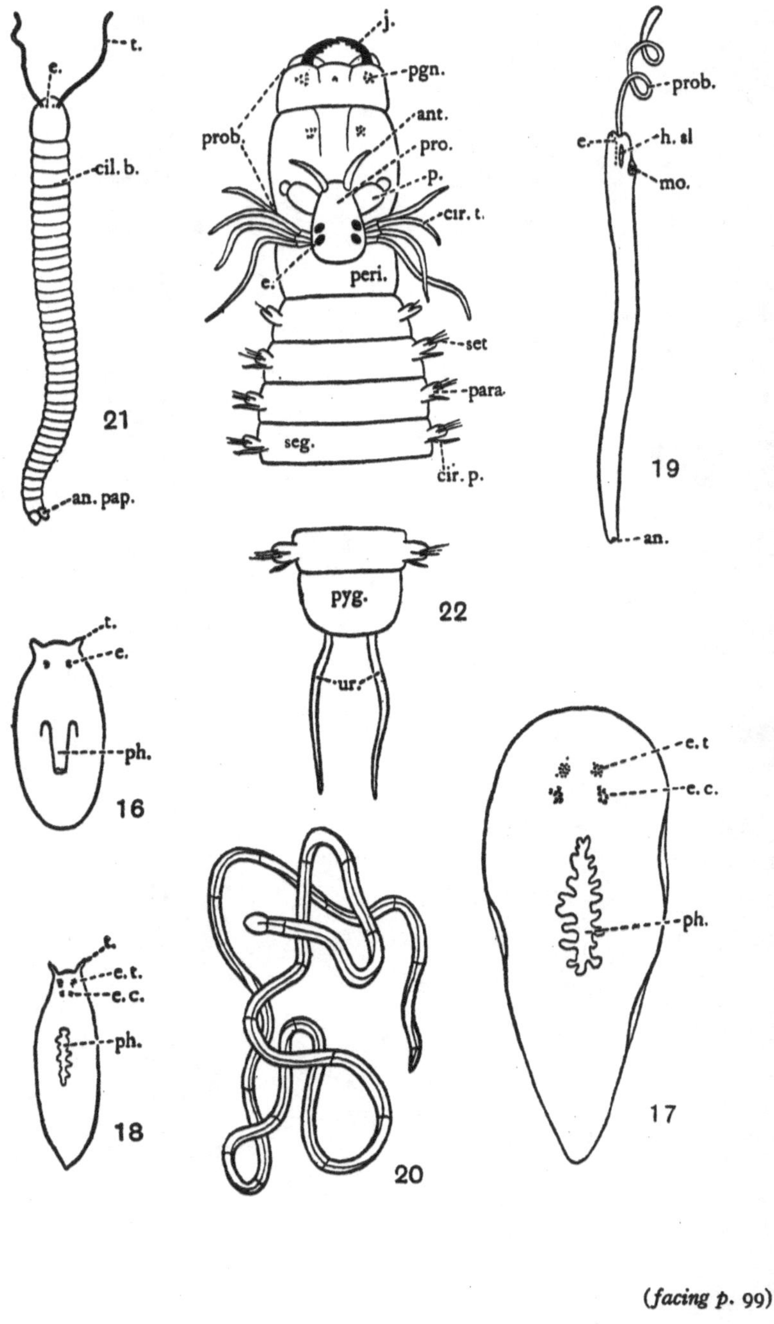

(*facing p.* 99)

EXPLANATION OF PLATE V

Worms (i), Figs. 16–22

Fig. 16. ***Procerodes ulvae***. Tricladida. Dorsal view. The pharynx is seen through the body. × 16.

Fig. 17. ***Leptoplana tremellaris***. Dorsal view. The shape of the pharynx is shown. × 4.

Fig. 18. ***Oligocladus sanguinolentus***. Dorsal view. × 2½.

Fig. 19. ***Lineus gesserensis***. Lateral view. To show the structure of a typical Nemertine. × 3.

Fig. 20. ***Tubulanus annulatus***. The species is eyeless, with a spatulate head. The colour varies from chocolate brown to red, with cream stripes and bars. × ½.

Fig. 21. ***Protodrilus flavocapitatus***. Dorsal view. × 7.

Fig. 22. ***Nereis fucata***. Anterior and posterior ends of a young specimen with proboscis protruded. The tentacular cirri are shorter than in the adult. Details of the parapodium are not shown. × 12.

an.	anus	p.	palp
an.pap.	anal papilla	para.	parapodium
ant.	antenna	peri.	peristomium
cil.b.	ciliated band	pgn.	paragnath
cir.p.	parapodial cirrus	ph.	pharynx
cir.t.	tentacular cirrus	pro.	prostomium
e.	eye	prob.	proboscis
e.c.	cerebral group of eyes	pyg.	pygidium
e.t.	tentacular group of eyes	seg.	segment
h.sl.	head slit	set.	chaeta
j.	jaw	t.	tentacle
mo.	mouth	ur.	urite

EXPLANATION OF PLATE VI

Worms (ii), Figs. 23–27

Fig. 23. Transverse sections to show chief types of parapodium. *a*, *Phyllodoce laminosa*. Uniramose type with dorsal and ventral foliaceous cirri. × 8. *b*, *Eunice harassi*. Uniramose with pectinate gill on dorsal cirrus. × 12. *c*, *Perinereis cultrifera*. Biramose. × 12.

Fig. 24. *Lagisca extenuata* (Aphroditidae). Dorsal view. Anterior region. On the right side the parapodia are seen through the elytra, on the left the elytra have been removed. Details of the parapodia are not shown. Note that the peristomium has a pair of parapodia. On the body segments elytra and dorsal cirri are borne alternately. × 18.

Fig. 25. *Phyllodoce maculata* (Phyllodocidae). Dorsal view. Anterior end. Tentacular cirri and dorsal cirri shown on left side only. × 10.

Fig. 26. *Syllis krohni* (Syllidae). Dorsal view. Anterior end of young specimen. The left antenna has been cut off to show the palp. × 20.

Fig. 27. *Glycera* sp. (Glyceridae). Anterior end, lateral view with proboscis extended. × 6.

acic.	aciculum
ant.	antenna
br.	branchia
cir.d.	dorsal cirrus
cir.d.fol.	dorsal (foliaceous) cirrus
cir.d.l.	long dorsal cirrus, alternating with
cir.d.s.	short dorsal cirrus
cir.t.	tentacular cirrus
cir.v.	ventral cirrus
el.	elytron
el.ph.	elytrophore
int.	intestine
j.	jaw
l.	lobe
m.d.l.	dorsal longitudinal muscle
m.obl.	oblique muscle
m.v.l.	ventral longitudinal muscle
ne.	nerve cord
neuro.	neuropodium
noto.	notopodium
p.	palp
pap.	papilla
para.	parapodium
para.b.	biramose parapodium (third segment)
para.u.	uniramose parapodium (first and second segments)
peri.	peristomium
pro.	prostomium
prob.	proboscis
seg.	segment
set.cap.	capillary chaeta
set.comp.	compound chaeta

VI. WORMS

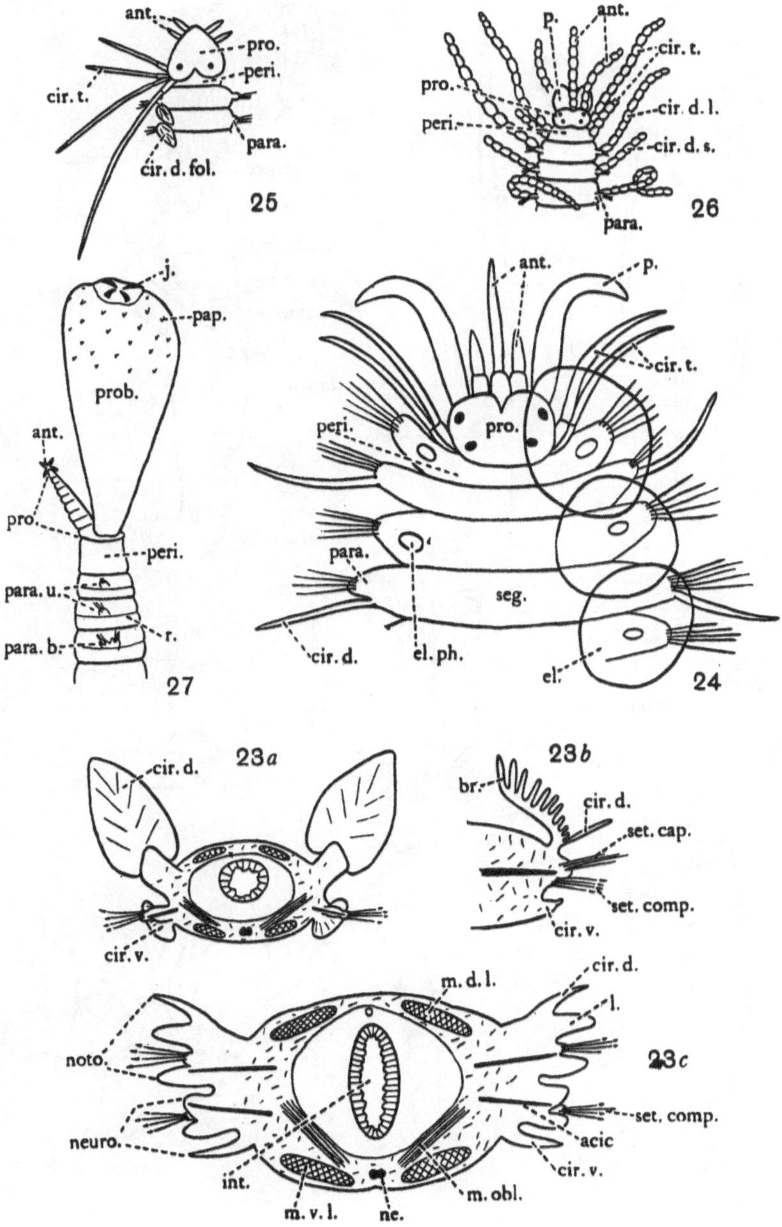

(*facing p.* 100)

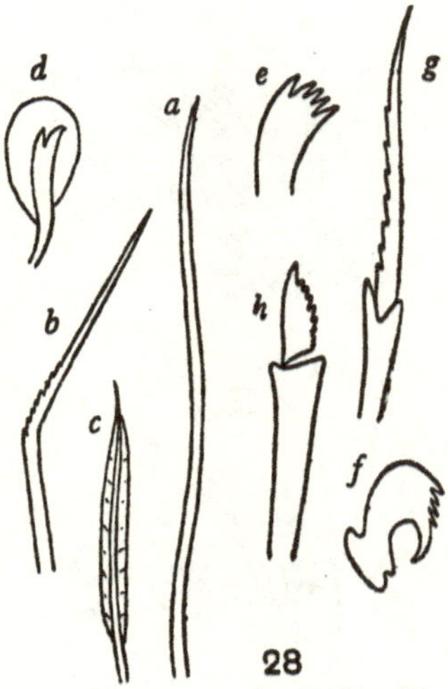

28

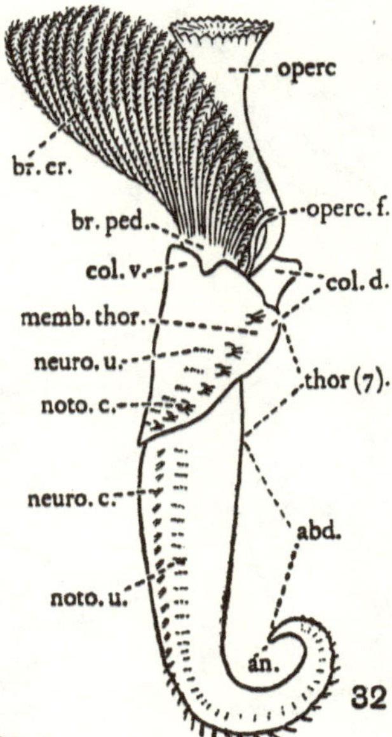

32

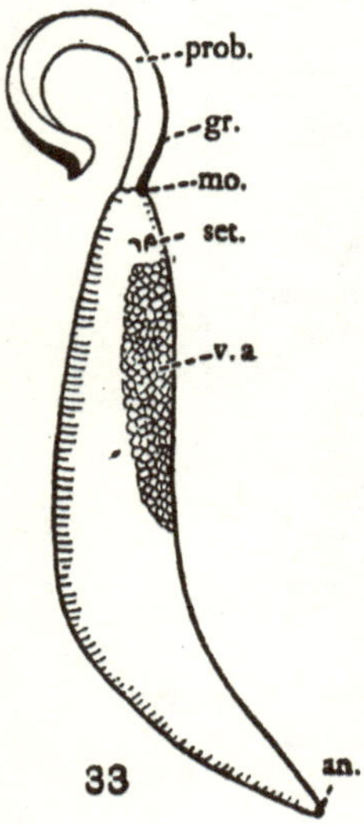

33

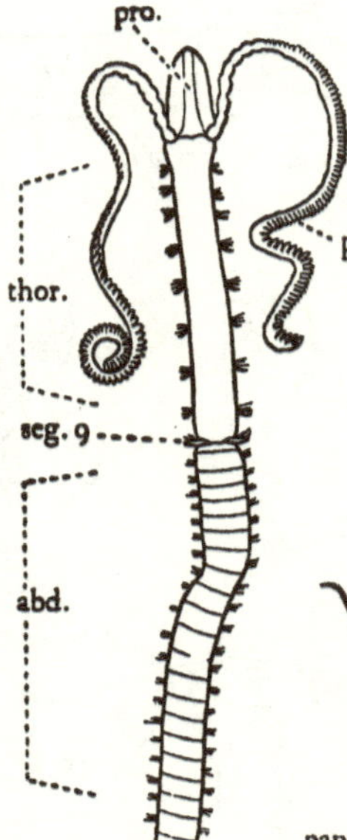

30

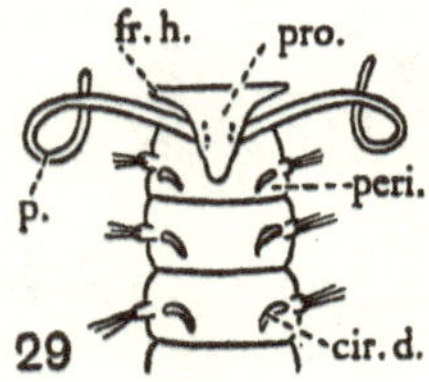

29

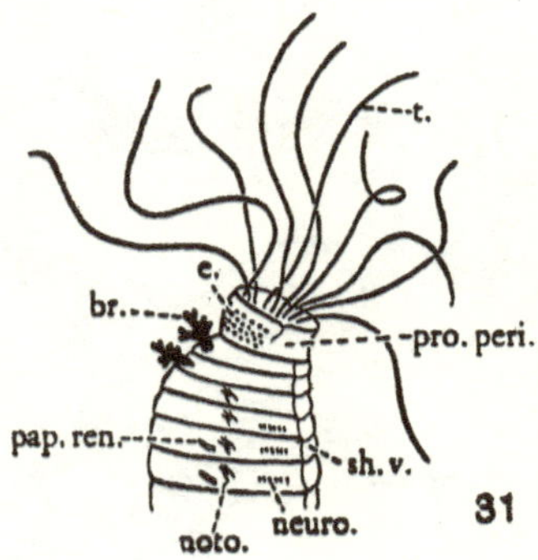

31

(*facing p.* 101)

EXPLANATION OF PLATE VII

Worms (iii), Figs. 28–33

Fig. 28. Chief types of chaeta. *a–f*, simple; *g* and *h*, compound. *a*, capillary (*Nephthys*); *b*, geniculate (*Nephthys*); *c*, winged (*Amphitrite*); *d*, hooded (*Polydora*); *e*, palea (*Sabellaria*); *f*, uncinus (*Amphitrite*); *g*, *Nereis*; *h*, *Syllis*.

Fig. 29. *Spiophanes bombyx* (Spionidae). Anterior end, dorsal view. The transverse bands of cilia have not been figured. × 7.

Fig. 30. *Magelona papillicornis*. Dorsal view of anterior portion. The body consists of eight thoracic segments marked by parapodia but not segmented externally, followed by a single specialised segment bearing enlarged dorsal cirri, leading to an elongated abdomen of many distinct segments. In the figure only one row of the papillae of the palp is shown. × 3.

Fig. 31. *Nicolea zostericola* (Terebellidae). Anterior end. Lateral view of a specimen extracted from its tube. × 12.

Fig. 32. *Serpula vermicularis*. Left lateral view of a specimen extracted from its tube. × 3.

Fig. 33. *Thalassema neptuni*. Half lateral view. The ventral groove is shown in black, the muscular rings are indicated on the left of the figure, and some of the ventral areas on the right. × 2.

abd.	abdomen
an.	anus
br.	branchia
br.cr.	branchial crown
br.ped.	branchial peduncle
cir.d.	dorsal cirrus
col.d.	dorsal lobe of collar (paired)
col.v.	ventral lobe of collar (median)
e.	eye
fr.h.	frontal horn
gr.	groove
memb.thor.	thoracic membrane
mo.	mouth
neuro.	neuropodium
neuro.c.	capillary chaetae of abdominal neuropodium
neuro.u.	uncini of thoracic neuropodium
noto.	notopodium
noto.c.	capillary chaetae of thoracic notopodium
noto.u.	uncini of abdominal notopodium
operc.	operculum
operc.f.	false operculum
p.	palp
pap-ren.	renal papilla
peri.	peristomium
pro.	prostomium
pro.peri.	prostomium fused with first body segment
prob.	proboscis
seg.9	ninth body segment
set.	chaeta
sh.v.	ventral shield
t.	tentacle
thor.	thorax
thor.(7).	thorax of seven segments
v.a.	ventral area

ARTHROPODA

AUTHORITIES

General

BATE, C. SPENCE and WESTWOOD, J. O. 1863–8. *A History of the British Sessile-eyed Crustacea*, vol. I, 1863; vol. II, 1868. London: Van Voorst.

BELL, T. 1853. *A History of the British Stalk-eyed Crustacea.* London: Van Voorst.

Cirripedia

DARWIN, C. 1851–4. *A Monograph of the subclass Cirripedia.* Ray Soc. Publ. vol. I, 1851; vol. II, 1854.

SMITH, G. 1906. "Rhizocephala." *Fauna. u. Flora Neapel*, XXIX.

Tanaidacea. Isopoda

HEWITT, C. G. 1907. "Ligia." *L.M.B.C. Mem.* XIV. Liverpool Univ. Press.

SARS, G. O. 1899. *An Account of the Crustacea of Norway.* Vol. II. *Isopoda.*

Mysidacea

TATTERSALL, W. M. and O. S. 1951. *The British Mysidacea.* London: Ray Society Publ. No. 136.

ZIMMER, C. 1927. "Schizopoden." In *Nord. Plankt.* XII, Lieferung VI, 1–178.

Amphipoda

CHEVREUX, E. and FAGE, L. 1925. "Amphipodes." *Faune de France*, IX, pp. 1–148.

Decapoda

BULL, H. O. 1937. "Notes on the British species of the Genus *Galathea* (Fab.)." *Rep. Dove Mar. Lab.* 1936.

KEMP, S. 1910. "The Decapoda Natantia of the coasts of Ireland." *Sci. Invest. Fish. Br. Ire.* 1908.

PALMER, R. 1927. "A Revision of the Genus *Portunus.*" *J. Mar. biol. Ass. U.K.* XIV, no. 4.

SELBIE, C. M. 1914. "The Decapoda Reptantia of the coasts of Ireland. Part I. Palinura, Astacura, and Anomura (except Paguridea)." *Sci. Invest. Fish. Br. Ire.* 1914, I.

—— 1921. "Part II. Paguridea." *Ibid.* 1921, I.

Pycnogonida

BOUVIER, E. L. 1923. "Pycnogonides." *Faune de France*, VII, 1–70.

HOEK, P. P. C. 1881. "Nouvelles études sur les Pycnogonides." *Arch. Zool. exp. gen. Paris*, IX, 445–542.

This phylum, the largest in the animal kingdom, contains creatures of all sizes, from minute pelagic species to giant crabs. They inhabit every zone. Some live on land (Arachnida and Myriapoda), some can fly (Insecta), others inhabit fresh or salt water (Crustacea). All possess a compound *head* and metamerically segmented body covered by a chitinous exoskeleton, moulted at intervals. The dorsal part of the exoskeleton of a segment is called the *tergum*, the ventral part the *sternum*, while laterally the overhanging flap of the tergum is called the *pleuron*. Typically each segment of head and body bears a pair of jointed appendages or legs, variously modified to act as sense organs (*antennules* and *antennae*), mouth parts (*mandibles*, *maxillae*, *maxillipedes*), walking or swimming limbs (*pereiopods*), swimmerets (*pleopods*), and tail appendages (*uropods*). The mouth is antero-ventral, the anus posterior.

The marine Arthropods belong to the Crustacea except a few littoral insects like *Petrobius*, and the eight-legged, spider-like Pycnogonids (p. 140).

Crustacea (Pls. VIII–XI) include barnacles, shrimps, lobsters, crabs, etc. At least five segments are fused to form a *head*, posterior to which lie the *thorax* (pereion) bearing the genital pores, and the *abdomen* (pleon). A backward fold of the head may form a free *dorsal shield* or *bivalved shell* (most Entomostraca, Leptostraca, larval forms of Malacostraca), or by fusion with the thorax a cephalothoracic shell or *carapace* (Decapoda). The abdominal segments are in most orders distinct. The chitinous exoskeleton is often calcified. The head may bear eyes, a median simple eye or group of eyes and a pair of compound eyes. The latter are of a type peculiar to the Arthropoda. The cephalic appendages are antennules, antennae, mandibles, maxillulae, and maxillae. The thoracic appendages vary greatly in number and structure. When turned towards the mouth to assist with the food, they are called maxillipedes. The posterior thoracic appendages may be used for swimming, walking or other purposes. The abdominal appendages or pleopods are usually the simplest, and consist of a stem or *protopodite* with mesial and lateral branches (*endopodite* and *exopodite*). Such an appendage is biramose; when the exopodite aborts, the appendage is described as uniramose. The last pair of abdominal appendages is often turned backwards alongside the *telson* (or terminal region of the abdomen). They are then called uropods, and uropods and telson together form the *caudal fan* of many Malacostraca.

Crustacea

1. Small Crustacea, usually less than 2 cm. long, with forked tail if free, degenerate if fixed. The egg hatches into a nauplius larva. Number of body segments variable, but not nineteen. **Entomostraca.** 2

   Larger Crustacea, usually more than 2 cm. long, without forked tails (*except in Nebalia*). Rarely parasitic, rarely with a nauplius stage in the life-history. The number of body segments and appendages is typically nineteen, five fused for the head, eight fused or free for the thorax, and six fused or free for the abdomen. **Malacostraca.** 3

2. A bivalved shell enclosing the whole body. Antennules and antennae used for locomotion and the trunk appendages reduced to two pairs. Pelagic. **Ostracoda** (not described here)

   Head skeleton fused with anterior thoracic skeleton, the remaining thoracic and abdominal segments being free. Antennules and antennae well developed, five or six pairs of trunk limbs, tail forked. Simple eyes. Some are ectoparasitic on fish, and many are pelagic. **Copepoda** (not described here)

   Attached forms fixed by the head and surrounded by a fold of the body wall called the mantle, often secreting a calcareous shell in pieces. Six pairs of biramose cirriform thoracic appendages. The parasitic forms are so degenerate that no characters of the group appear in the adult. **Cirripedia**, p. 105

3. Abdomen of seven segments. Thorax with distinct segments covered by a carapace and with foliaceous appendages. Tail forked. **Leptostraca**, p. 108

   Abdomen of six or fewer segments. Thoracic segments not all distinct and their appendages usually foot-like. **Eumalacostraca.** 4

4. Carapace present. 5

   No carapace. Eyes sessile. 10

5. Carapace coalescing dorsally with all thoracic segments. Eyes pedunculate. 6

   Carapace coalescing dorsally with anterior segments only, though it may cover all of them. Eyes sessile or pedunculate. 7

6. Thoracic appendages biramose, none modified as maxillipeds. Pelagic. **Euphausiacea** (not described here)

   Thoracic appendages uniramose. Three pairs of maxillipeds and five pairs of walking legs. **Decapoda**, p. 123

7. First thoracic segment fused with the head. Abdomen more slender than thorax. 8

   Segments bearing pedunculate eyes and antennules free from carapace. Abdomen larger and broader than thorax. **Stomatópoda** (not littoral)

8. Body compressed laterally. 9
Body depressed dorso-ventrally. Thoracic segments three to eight free. **Tanaidacea**, p. 109
9. Uropods slender, pleopods reduced or absent. A pseudorostrum of paired plates which meet in front of head. Eyes sessile. **Cumacea** (not littoral)
Uropods lamellar, forming with the telson, a tail fan. Rostrum median, shrimp-like. Eyes pedunculate. **Mysidacea**, p. 121
10. Body usually laterally compressed. Three pairs of pleopods and three pairs of uropods. Branchiae on the thoracic appendages. **Amphipoda**, p. 116
Body depressed. Pleopods all alike, usually bearing branchiae. **Isopoda**, p. 109

CIRRIPEDIA

(Pl. VIII, fig. 34)

Barnacles and their allies. Fixed or parasitic Crustacea, whose larval stage is free and pelagic. Body of few segments, enclosed in a fold of integument known as the mantle, which may be strengthened with calcareous plates. Fixation by the head, through cement organs in the antennules. Mandibles and two pairs of maxillae are present, also six pairs of thoracic legs, which in the fixed condition of the animal act as food catchers. Thus Huxley described the barnacle as a little animal that stands on its head and kicks its food into its mouth with its legs. The abdomen is reduced and bears no appendages. Usually hermaphrodite.

The fixed forms belong to the suborder Thoracica and include the stalked barnacles and acorn barnacles. The mantle usually secretes calcareous plates. In the acorn barnacles the calcareous plates of the mantle form a cup-shaped wall or *shell* whose plates are immovable, and an *operculum* or lid with movable scutes. The wall is composed typically of six pieces, median dorsal *carina*, median ventral *rostrum*, paired *laterals* and paired *carino-laterals*. There may also be paired *rostro-laterals*, or these may be fused with the rostrum. The mode of overlapping varies in different families. Each plate is thickened in the middle and has a thin extension on either side of this. If the extension overlaps the next plate, it is called a *radius*; the part overlapped is an *ala*. Thus in Balanidae the rostrum has two radii, the carina two alae, the lateral and carino-lateral an ala ventrally and a radius dorsally. The operculum is composed of two pairs of triangular plates, the *scuta* and *terga*, which are freely movable and form a lid for the cup-shaped shell. The body of the animal is attached to the scuta above and to the calcareous or membranous base.

The parasitic members are degenerate forms without appendages or alimentary canal in the adult, though the larval stage is the typical pelagic nauplius larva. They attack Decapod Crustacea (e.g. the shore crab, *Carcinus maenas*), and live partly within and partly outside the body. The portion within the host consists of nutritive roots, while outside lies a rounded chitinous sac containing the reproductive, nervous and muscular systems. These constitute the suborder Rhizocephala.

1. Fixed, non-parasitic forms with six pairs of cirriform thoracic appendages. **Thoracica.** 2
   Parasitic forms living in Decapod Crustacea. **Rhizocephala.** 5
2. Stalked barnacles, with laterally compressed mantle bearing calcareous plates. **Pedunculata** (not littoral)
   Sessile barnacles, with broad basal disk, wall of strong calcareous. scutes and movable operculum covering the body. **Operculata.** 3
3. Shell asymmetrical, of six pieces including operculum. **Verrucidae**, p. 106
   Shell symmetrical, usually of six pieces excluding the operculum 4
4. Side pieces of shell (laterals and carino-laterals) overlapping the end pieces (rostrum and carina). **Chthamalidae**, p. 107
   Rostrum overlapping laterals, carino-lateral overlapping carina. **Balanidae**, p. 107
5. Adults attached to abdomen of Decapod Crustacea. **Peltogastridae**, p. 108

OPERCULATA

VERRUCIDAE (Pl. VIII, fig. 34*c*).

Wall incomplete, consisting of carina and rostrum meeting in a zigzag suture on one side of an asymmetrical barnacle. Both scutal and tergal plates are present on the other side of the animal, but the right scutum and tergum are fixed to rostrum and carina, and only the left scutum and tergum are movable. Not two of the pieces are alike in size or shape.

Verruca stroemia. D. 1854, p. 518. Occurs between tide marks and down to deep water, on shells, stones, etc. The shell is ribbed longitudinally and is about 6 mm. in diameter and white in colour.

CHTHAMALIDAE (Pl. VIII, fig. 34*a*).

Depressed, symmetrical, of ovate shape and on a membranous base. Both carina and rostrum are winged and are overlapped by the laterals and carino-laterals respectively. The lateral also overlaps the carino-lateral. Aperture kite-shaped.

Chthamalus stellatus. D. 1854, p. 455. White or greyish, about 1 cm. in diameter, on rocks. Wall plates often fluted, overlapping areas (radii) slight so that the thickened parts of the plates are in contact. Cirri brown. West and south-west coasts. Upper shore on wave washed coasts. Breeds April to September.

BALANIDAE (Pl. VIII, fig. 34*b*).

Symmetrical. Shell wall of six plates, scuta and terga movable. Carina with two alae (i.e. overlapped by the carino-laterals), rostrum with two radii, lateral and carino-lateral with ala ventrally and radius dorsally. Base calcareous or membranous. Six pairs of cirri. In *Balanus* the base is usually calcareous and flat, or nearly so, the shell aperture lozenge-shaped.

1. Orifice oval, not toothed. Base perforated by pores. Cirri purple. On rocks in large numbers. **Balanus perforatus**
   Orifice toothed. **2**
2. Base calcareous, with crenate edges. Cirri yellow. On stones, seaweed, etc. Isolated or in small groups. **B. crenatus**
   Base membranous, with crenate edges. Cirri white. On rocks in large numbers. **B. balanoides**

Balanus balanoides. D. 1854, p. 267. The most abundant acorn barnacle, with the widest range, occurring all round the British coasts. Middle to lower shore. To 13 mm. diameter. Shell dirty white in colour and ridged longitudinally. Breeds in winter. One brood, c. 10,000 eggs.

B. perforatus. D. 1854, p. 231. Shell almost smooth, aperture not toothed, base perforated by pores. Colour often purplish, and the conical shell is often as tall as broad, though in crowded situations specimens have been found measuring 1·2 cm. in basal diameter and nearly 4 cm. high. The range is more restricted than that of the former species; it is common on south and west coasts, on rocks, piles, etc.

B. crenatus (Pl. VIII, fig. 34*d*). D. 1854, p. 261. Although this species is typically a shallow-water one, it is frequently found on stones, shells, seaweeds, Crustacea, etc., at low-water mark. Shell white, conical, with elliptical toothed orifice. Valves usually rugged, with longitudinal folds. About 12 mm. in diameter.

Other species occur locally. An Australasian Barnacle, *Elminius modestus*, D. 1854, p. 350, was introduced about 1940, and is spreading rapidly along the British coasts. It has four plates only in the shell. Breeds all the year round.

RHIZOCEPHALA

Members of the Peltogastridae infest crabs, hermit crabs and other Decapod Crustacea. They occupy the part of the host where the eggs are normally attached, i.e. the ventral side of the abdomen. The presence of the parasite has a marked effect on the sexual condition and moulting habits of the host.

Peltogaster paguri. G. Smith, 1906, p. 109. Infects the hermit crab, *Eupagurus bernhardus*. The sac is curved and sausage-shaped, usually yellow or bright red in colour, with an anterior mantle aperture. Often two parasites are present. Height about 1 cm.

Sacculina carcini. G. Smith, 1906, p. 111. The sac is transversely elongated and occurs on the abdomen of *Carcinus maenas*, the shore crab. It may be 18 mm. wide and 12 mm. long, and is usually yellow in colour. The mantle aperture is posterior.

LEPTOSTRACA

The Leptostraca are intermediate in structure between the small Crustacea and the Malacostraca. The body is composed of twenty segments, of which nineteen bear appendages; five cephalic, eight thoracic, and seven abdominal segments, with five, eight and six pairs of appendages respectively. The last abdominal segment has no appendages. The telson is bifurcate. The carapace is not attached to the body in the thoracic region; there is an adductor muscle between the two halves, and the underlying thoracic segments are distinct. The thoracic limbs are all alike and foliaceous. The eyes are pedunculate, the sexes separate, with sexual dimorphism.

Nebalia bipes (Pl. IX, fig. 38). Shrimp-like. From 8–12 mm. long, of a pale orange or greenish yellow colour. Hides under stones or swims in the shaded parts of shore pools. Eyes red.

Gregarious. Antennae of male as long as the body, shorter in female.

TANAIDACEA

This order is an aberrant offshoot from the Isopoda. Body slender, depressed. Head coalesced with the first two thoracic segments, the cephalothorax so formed overhanging lateral branchial cavities. Eyes on short, immovable stalks, or absent. First pair of thoracic limbs are maxillipedes, second pair chelate, three to eight are pereiopods. Biramose pleopods used for swimming. Uropods slender, terminal. Many burrow in sand or mud.

Tanaidae.

Body slender. Antennae uniramose. Third thoracic appendage like the others and used for walking.

Tanais cavolini. B. and W. 1868, p. 125, as *T. vittatus*. Inhabits crevices in limestone or wood piles, where it makes a tube of mud. Claws strong and polished. Abdomen covered with fine setae in bundles. Reaches 4 mm. in length and is gregarious.

Apseudidae.

More elongated, with biramose antennae and long biramose styliform uropods. Digging forms.

Apseudes talpa, the mole-like *Apseudes*. B. and W. 1868, p. 148. Second thoracic appendage strong and chelate, third flattened for digging. Back keeled, and there is a short, sharp keeled rostrum between the eyes. 8–12 mm. long. Yellowish white.

ISOPODA

(Pl. VIII, figs. 35–37)

The familiar example of this order is the common woodlouse. Body depressed. First thoracic segment fused with the head but the second to eighth segments are free. No distinct carapace. In many Isopods the coxa of the pereiopod is expanded to form a *coxal plate*, which may fuse with the lateral portion of the segment (pleuron) and may even extend so that it is visible on the dorsal side. In Isopods possessing coxal plates the coxopodite of the leg is immovable and the basipodite is the first flexible segment. Telson usually not well defined. Eyes sessile. Antennules usually small, uniramose. Antennae large. One pair of maxillipedes, seven pairs of pereiopods, uniramose, all alike. Pleopods biramose, usually with branchiae. A brood pouch present. A few are parasitic. Terrestrial, fresh water and marine.

KEY TO SUBORDERS

1. Free-living, occasionally epizoic. 2
   Parasitic, living in Crustacea. **Epicaridea**
2. Coxopodites of thoracic legs expanded into coxal plates. Abdominal segments not all fused. 3
   No coxal plates. All abdominal segments are fused. Pleopods branchial. Uropods terminal. **Asellota**
3. Uropods lateral. 4
   Uropods styliform, terminal. **Oniscoidea**
4. Uropods forming, with terminal segments of abdomen, a caudal fan. **Flabellifera**
   Uropods forming an operculum, like folding doors, over the branchial pleopods. **Valvifera**

ASELLOTA

Pleopods covered by a thin cover plate formed by the *first* pair. Uropods usually styliform and biramose. The Janiridae have broadly ovate bodies, and dorsal eyes. The Munnidae resemble spiders, with long legs and lateral eyes.

JANIRIDAE.

Antennules fairly well developed, antennae very long, often as long as the whole body.

1. Uropods elongated, projecting beyond the abdomen, whose border is serrated. **Janira maculosa**
   Uropods very small, only just visible beyond the abdomen, whose border is often incised. **Jaera. 2**
2. Lateral borders of thoracic segments smooth. **J. marina**
   Lateral borders of thorax and abdomen fringed with setose hairs. **J. nordmanni**

Janira maculosa. B. and W. 1868, p. 338. The broad flat body is ash coloured, usually with dark brown dots. Antennae longer than the body. Females measure 7 mm. in length and males 10 mm. Breadth 3–4 mm. The species resembles the fresh-water louse, *Asellus aquaticus*. Between tide marks to shallow water.

Jaera. Body broader, antennae shorter, uropods smaller than in *Janira*. Abdomen notched mesially on its posterior border, and the minute uropods appear in the notch.

J. marina (Pl. VIII, fig. 35). B. and W. 1868, p. 317, as *J. albifrons*. Antennae at least one-half body length. Body dull ash coloured, often with a white margin to the head. Female 4 mm., male 3 mm. Tolerates reduced salinities. Lower shore.

J. nordmanni. B. and W. 1868, p. 320. Antennae barely one-third body length. 4 mm. Upper shore.

MUNNIDAE.

The grotesque members of this family resemble spiders owing to the extreme length of their slender legs. Body short and narrow, antennae half as long again as the length of the body, legs nearly twice as long. Cephalothorax broad and relatively large, eyes projecting sideways on short rigid stalks. The second thoracic appendage (first pereiopod) strong, short and prehensile, thoracic appendages three to eight very long and slender, with two claws. Uropods small and do not project beyond the pear-shaped abdomen.

Munna kröyeri. B. and W. 1868, p. 326. Occurs between tide marks and in shallow water. It is a yellowish or brown, very active creature, running fast amongst weed and corallines. Its slender legs are spiny. 3 mm.

ONISCOIDEA

Woodlice. Abdominal segments usually free. The antennules are minute, the antennae stout with a peduncle of five segments. Semiterrestrial.

Ligia oceanica (Pl. VIII, fig. 36). B. and W. 1868, p. 444. A large Isopod measuring 28 mm. in the male and 20 mm. in the female. Inhabits the zone above high-water mark, where it may be splashed by the waves, but is not often submerged. Occurs under stones and in crevices. A scavenger and vegetable feeder. Antennae stout and equal two-thirds the length of the body. Uropods long, slender, and biramose.

VALVIFERA

The abdominal segments are more or less coalesced, though the anterior ones are usually distinguishable. Exopodites of the uropods vestigial or absent.

IDOTEIDAE.

Body ovate or oblong, tending to become linear, elongated. Thoracic segments subequal and distinct, but in the abdomen at least three segments are fused. Antennules short, antennae long, often very long. Pereiopods nearly alike, but the first three pairs (2, 3, 4 thoracic) may be subchelate, and the last two pairs (7 and 8) multiarticulate.

Idotea. Nocturnal, gregarious scavengers.*

1. Telson with posterior border straight or concave. **2**
   Telson with posterior border produced mesially. **3**
2. Body small, narrow. Antennae more than half body length. **I. linearis**
   Body large, ovate. Antennae less than one-third body length. **I. emarginata**
3. Telson tridentate posteriorly. **I. baltica**
   Telson with rounded lateral borders. **4**
4. Antennae stout and short. Telson with short blunt apex. **I. pelagica**
   Telson with distinct median process. **5**
5. Dorsal surface pitted and granular, reddish brown. **I. granulosa**
   Body slender, smooth, green or brown. **I. viridis**
   Body broad. Antennae slender, about half body length. **I. neglecta**

I. pelagica. B. and W. 1868, p. 379. Posterior border of abdomen rounded, but with a faint mesial keel. Female 9 mm., male 13 mm., breadth 3–4 mm. Brown. Lower shore.

I. baltica (Pl. VIII, fig. 37). B. and W. 1868, p. 379, as *I. tricuspidata*. Antenna one-half body length. Dirty greenish grey, variable, sometimes striped longitudinally with a pale line, or blotched with yellow, red or greenish brown. 35 mm.

I. emarginata. B. and W. 1868, p. 386. A large handsome species, 18 mm. in female, 30 mm. in male. Deep chocolate brown, sometimes spotted. Common amongst weed.

* E. NAYLOR: 1955. British species of *Idotea*. *J. mar. biol. Ass. U.K.* **34**, 467–493.

I. granulosa. Sars, 1899, p. 82. Body narrow, abdomen ending in a long plate with median projection. Antennae one-quarter body length. 11–15 mm. long. On brown weeds.

I. linearis. B. and W. 1868, p. 388. Body long and narrow. Legs strong and hooked. A common species, reaching 38 mm. in length, coloured red or brown according to the colour of the weed on which it lives. On drifting weed.

I. neglecta. Sars, 1899, p. 84. A common gregarious species. Abdominal plate rounded, ending bluntly. Dark brown or black, occasionally light specimens. Female 16 mm., male 25 mm. Amongst algae.

I. viridis. Sars, 1899, p. 83. Body very slender, almost linear. Antennae slender, one-third body length. Green. Often on *Zostera.* Female 10 mm., male 12 mm. In brackish water or in pools.

FLABELLIFERA

Abdominal segments free or only partly coalesced. The mouth parts are often suctorial and many of the members are ectoparasitic on fish. The uropods are generally biramose, lamellar and forming a caudal fan.

1. All pereiopods alike and used for walking. **2**
   Three pairs of pereiopods modified for prehension. Eyes small, not faceted. **Cirolanidae**, p. 114
2. Abdomen of six distinct segments. **3**
   Abdominal segments more or less fused. Body broadly ovate. **Sphaeromidae**, p. 115
3. Head large, with distinct mandibles. Abdomen narrower than thorax. Only five pairs of pereiopods. **Gnathiidae**, p. 113
   Head small. Abdomen as wide as thorax. In wood. **Limnoriidae**, p. 114

GNATHIIDAE.

Ectoparasitic on fish during their larval stage, but probably free-living when adult. Strong sexual dimorphism. Head large and flattened. The thorax has only five free segments owing to fusion of the others. The mandibles of the male are very large and hook-like and project forwards beyond the head; in the female they are vestigial. The five pairs of pereiopods alike, slender.

Gnathia maxillaris. B. and W. 1868, p. 187, as *Anceus.* This grotesque creature is found at low-tide mark, in shallow or deep water, under stones or in rock crevices. It reaches only 6 mm. in length in the male and a little longer in the female. Variable in colour, being green, blue, or grey, often spotted. The male with his great mandibles, and the female with her weak legs and thorax swollen with eggs are very much unlike one another. In both sexes the second thoracic appendages are short, strong and hairy, but are not used for walking.

CIROLANIDAE.

Isopods resembling woodlice at first sight, but the first three pairs of pereiopods are stout and hooked for grasping the host, since the members are usually ectoparasitic on living fish, or scavengers on dead fish. Body ovate or elongated, rounded anteriorly and arched or vaulted (i.e. not flattened). Antennules very small, antennae long. The six abdominal segments are distinct, the last being the longest.

Eurydice pulchra. B. and W. 1868, p. 310. This pretty species occurs in pools, where it attacks and feeds on other animals. It is about 7 mm. long, semi-transparent, of a pale grey colour, with dark, star-shaped markings. It swims actively. The body is about twice as long as broad, ovate, convex, with large, rounded terminal abdominal segment. Antennae more than half the body length.

Cirolana cranchi. B. and W. 1868, p. 296. A whitish form, about 10–15 mm. long, whose habits are unknown. Sars claims that it is not parasitic. Body about three times as long as broad, oval in shape, with antennae barely half the body length. All the appendages are setose and bear a few spines. The last abdominal segment is triangular.

LIMNORIIDAE.

The gribbles are troublesome Isopods burrowing in woodwork exposed to the action of the sea.

Limnoria lignorum. The gribble. B. and W. 1868, p. 351. A small brown Isopod about 5 mm. long and less than 2 mm. broad. Body oblong and depressed. The cephalothorax, which bears two small eyes, is narrower than the thorax. Abdomen six-jointed, with the last segment very large and broad. Antennules and antennae very short, pereiopods all alike, short and slender. The uropods are small, fixed laterally, and have very short spiny exopodites. The distribution is world-wide.

SPHAEROMIDAE.

Many members of this family can roll themselves into a ball like certain woodlice, which they resemble. They can also run and swim. Body short and broad, highly convex. Cephalothorax broad, with large lateral eyes. Antennules and antennae strong but usually not very long. The pleopods form five pairs of delicate, membranous, setose abdominal appendages. Last abdominal segment, which bears the uropods, large, very broad, and either rounded or notched posteriorly. Uropods well developed, with large exopodites and endopodites.

1. Body contractile into a ball. Last abdominal segment with rounded posterior margin. **Sphaeroma.** 2
   Body straight. Abdomen notched posteriorly. A pair of backwardly directed processes on the seventh thoracic segment of the male, whose uropods are long, like two tails. **Naesa (bidentata)**
2. Exopodite of uropod serrated laterally. Abdomen smooth. **S. serratum**
   Exopodite of uropod not serrated. Abdomen rough. **S. rugicauda**

S. serratum. B. and W. 1868, p. 405. A gregarious form found under stones intertidally and often in estuaries. 13 mm. long and about 8 mm. broad. Colour grey or white, marked with red. West and south-east coasts, but its range has extended northwards of recent years.

S. rugicauda. B. and W. 1868, p. 408. An active swimming form, grey, with streaks of black, and often with a light stripe down the back. It is typically estuarine. 10 mm. long, 6 mm. broad. On all coasts.

Naesa bidentata. B. and W. 1868, p. 431. There is a marked sexual dimorphism in this species. The female is so different from the male that it was given another name, and appears as *Dynamene rubra*, B. and W. 1868, p. 49. This female resembles a *Sphaeroma* but differs in having a notched abdomen. 5 mm. long, reddish or grey, with black or red specks. It occurs on rocky coasts in the south and west. The male is relatively larger than the female and reaches 8 mm. The seventh thoracic segment is enlarged and produced backwards into a strong bifid process which overhangs the notched abdomen. The uropods have their exopodites greatly elongated and curved towards one another, so that they look like two tails. The colour is grey with red or blue streaks. The male is common between tide-marks on rocky shores, but chiefly on the west and south coasts. Feeds on *Fucus*.

EPICARIDEA

Ecto- or endo-parasites in the adult stage, showing varying degrees of degradation of the typical Isopod structure. The host is usually a Crustacean, often a Decapod, or a Cirripede. Hermaphroditism is common, and it is possible that in many cases there is marked Protandry, the animal passing through a male phase, which is often Isopod-like, followed by a female phase which is asymmetrical and degenerate. The life-history of many species is imperfectly known. The mouth is suctorial.

Bopyrus squillarum. B. and W. 1868, p. 218. Forms a pale green swelling in one of the branchial chambers of the prawn, *Leander serratus*. The female stage is 11 mm. long, broadly oval in shape but asymmetrical, with reduced or very small appendages. The male stage is only 2 mm. in length, and is attached to the pleopods of the female. Male with eyes, and fairly well-developed appendages.

Hemioniscus balani. B. and W. 1868, p. 267, as *Cryptothiria balani*. More degenerate still. Lives in *Balanus balanoides*. The female stage is reduced to a sac filled with ova, and is devoid of antennae, legs or branchial appendages. The male stage, which precedes the female (i.e. the animal is a protandrous hermaphrodite), has a segmented body and seven pairs of pereiopods. Local.

AMPHIPODA

(Plate X)

Body laterally compressed, eyes sessile. No carapace. Head fused with the first thoracic segment, but segments 2–8 are free and so are the six abdominal segments, except in aberrant families. The thoracic appendages have coxal plates attached to the pleura of their segments (see under Isopoda). The appendages are:

Antennules (usually biramose) with multiarticulate flagellum.
Antennae (usually uniramose) with multiarticulate flagellum.
Mandibles, maxillulae and maxillae.
One pair of maxillipedes.

Two pairs of broad subchelate gnathopods.

Five pairs of pereiopods or walking legs.

Three pairs of biramose pleopods.

Three pairs of biramose uropods.

Many are of very small size. Sand shrimps, sand hoppers, skeleton shrimps, etc.

1. Amphipods with full complement of appendages. 2
Amphipods with a number of appendages reduced. Body extremely attenuated, moving like a leech. **Caprellidae**, p. 119
2. Telson bifid, or split posteriorly. Antennule with accessory flagellum. Uropods all biramose. 5
Telson entire. 3
3. Hopping forms with small curved abdomen, uniramose posterior uropod and minute telson, short antennules and long antennae. **Talitridae**, p. 118
Swimming or walking Amphipods. 4
4. Antennules and antennae subequal, uropods all biramose. Eyes red. **Amphithoidae**, p. 118
Antennae very long. Last uropods uniramose. Eyes small, black. **Corophiidae**, p. 119
5. Pereiopods 5–7 dilated and used for digging. Antennule shorter than antenna. **Haustoriidae**, p. 120
Pereiopods 5–7 not dilated for digging. Antennule longer than antenna. **Gammaridae**, p. 117

GAMMARIDAE.

Sand shrimps. Antennule with accessory flagellum well developed. Gnathopods 1 and 2 subchelate, robust. The four anterior coxae belonging to thoracic segments 2–5 well developed, 6, 7 and 8 small.

Gammarus. Telson usually split right up to the base so that it looks double. Body not keeled, smooth, except for small spines dorsally on posterior regions. Eyes linear or reniform, black.

Green or buff. Endopodite of third uropod less than half the length of the exopodite. **G. marinus**
Spotted with red. Rami of third uropod sub-equal. **G. locusta**

G. (*Marinogammarus*) *marinus*. C. and F. 1925, p. 250; B. and W. 1863, p. 370. Ten spines on fourth abdominal segment; twelve on fifth; five on sixth. Antenna half as long as the antennule, which is half the length of the body. Gregarious, amongst weed, often estuarine. 13 mm. Common. Intertidal.

G. locusta (Pl. X, fig. 42). C. and F. 1925, p. 257; B. and W. 1863, p. 378. Eyes reniform, black. Antenna nearly as long as antennule, flagellum with tufts of setae. Bluish green with red spots. Female 18 mm., male 20 mm. In pools at low water.*

TALITRIDAE.

Generally resembling Gammaridae. Sand and shore hoppers. Coxal plates large, those of the sixth thoracic segment bilobed. Antennule shorter than antenna, without accessory flagellum. Sexual dimorphism in thoracic 3 (second gnathopod). Uropods getting progressively smaller. Abdomen without spines but the small telson may have a few.

Both gnathopods subchelate, second strong in male.	**Orchestia**
First gnathopod not subchelate, second weak.	**Talitrus**

Talitrus. Antennule shorter than the peduncle of the antenna. Pleopods often reduced. Telson simple, reduced.

T. saltator. Sand hopper. C. and F. 1925, p. 271; B. and W. 1863, p. 16, as *T. locusta*. Antenna of male two-thirds length of body. Pleopods weak, first uropods twice as long as the third. Last three segments of abdomen very short and folded under body; their sudden extension assists in the springing habit. Exoskeleton polished. Eyes large, irregularly round, black. Never go into water voluntarily but live in sand amongst decaying weed at highest tide mark. Feed on carrion, even cannibalistic. On all sandy shores. Greenish buff or fawn with line of black down back. Female 18 mm., male 25 mm., more slender. Very common.

Orchestia. Resembling *Talitrus* but both pairs of gnathopods subchelate. Telson single and well developed, entire with spiny margin.

O. gammarella (Pl. X, fig. 43). C. and F. 1925, p. 274; B. and W. 1863, p. 27, as *O. littorea*. The shore hopper. Eyes black, irregularly round, rather large. Antennules as in *Talitrus*. Antennae barely half body length. Greenish or reddish, spotted brown. Occurs on more rocky ground than *Talitrus*, amongst weed and decaying refuse, not submerged. Male 18 mm., female 15 mm.

AMPHITHOIDAE.

Antennule with short third joint and long flagellum. Antenna with four joints in peduncle, the last two subequal, the whole a

* For estuarine and brackish water species, see *J. Mar. Biol. Ass. U.K.* xxv, no. 3 and xxvii, no. 1.

little shorter than the antennule. Resembles *Gammarus* but has heavier appendages and claws.

Amphithoe rubricata. C. and F. 1925, p. 334; B. and W. 1863, p. 418. Hook-like spines on third uropods. Body compressed. Eyes small, oval, red. Variable; bright red or reddish brown, if amongst red algae; clear green, spotted with brown, amongst green algae. Makes a nest of weed fastened by fibres. 14 mm. long. Littoral to shallow water.

COROPHIIDAE.

Body not compressed. All the segments distinct and well developed. Eyes lateral. Antennules and antennae usually long, with or without accessory flagellum. Second gnathopod different in the two sexes, its terminal joint robust and elongated in the male.

Corophium. Head subquadrangular, with very small eyes. No accessory flagellum on antennule, antenna longer and more robust. Appendages very hairy. Many species.

C. volutator. C. and F. 1925, p. 364; B. and W. 1863, p. 493, as *C. longicorne.* Male. 6–8 mm. Eyes small and black. Antennule equals one-half body length. First joint is both longer and wider than the second and third together, flagellum of twelve to fourteen joints. Antenna equals body length. Pereiopods 3 and 4 long, equal, 5 and 6 shorter, with expanded ischiopodites, 7 very long, extending backwards beyond the telson, with expanded ischiopodite. Female. Antennule more than one-third body length. Antenna equals one-half body length. White or grey to pink or brown. Forming galleries in muddy sand, whence it can be obtained in large numbers by sieving. Attacks and feeds on worms and molluscs. For other species, see Crawford, *J. Mar. biol. Ass. U.K.* XXI, No. 2, 1937.

CAPRELLIDAE.

The "skeleton shrimps" are very slender-bodied Amphipods which attach themselves to Hydroids, algae and Bryozoa by their hind legs and extend the anterior portion of their bodies in an attitude resembling that of the praying mantis. They walk with a leech-like movement. They are usually coloured exactly the same as the animal or plant on which they rest, and are therefore difficult to detect unless on the move. The head and first two segments of the thorax are fused together. The remaining six thoracic segments are cylindrical and very long, but the abdomen is greatly reduced. The head bears a pair of small laterally placed eyes and the usual appendages are well developed. Antennules and antennae uniramose. There is one pair of maxillipedes. The second and third thoracic appendages are subchelate gnathopods and are always very unequal in size. The appendages of the fourth and fifth thoracic segments are absent; their place is taken by two pairs of branchiae. 6, 7, 8 bear long pereiopods, by means of which the Caprellid attaches itself. The pleopods are vestigial in the male and usually absent in the female.

Caprella. Two pairs of incubatory lamellae on thoracic segments 3 and 4. The male has two pairs of vestigial pleopods, the female none. Specimens of the two littoral species can usually be taken by collecting red weeds, *Tubularia* and *Crisia*, and putting them in a bowl of sea water. The Caprellids after a time drop from the weed to the bottom of the bowl.

C. acanthifera. C. and F. 1925, p. 446; B. and W. 1868, p. 65. Body ornamented with tubercles on the dorsal side. Antennae only slightly hairy and less than half the body length. The inflation of the head gives it the name skull-headed skeleton shrimp. Greenish yellow, may be marked with black or orange. Amongst algae. Variable. 8–10 mm. long.

C. linearis (Pl. X, fig. 44). C. and F. 1925, p. 456; B. and W. 1868, p. 52. Body smooth or only feebly tuberculate. Antennae hairy and only half as long as the antennules, the latter being half the body length. Colour wine red. 10–20 mm. long, smaller in the female. The female carries the young on her back after incubation in the pouch formed by the four plates on thoracic segments 4 and 5.

HAUSTORIIDAE.

Body usually compressed. Pereiopods 5–7 often enlarged for digging purposes. Uropods biramose, the third pair different from the first two. Telson flat, more or less incised.

Haustorius. Body robust, with large coxal plates. Pereiopods 5–7 with the meropodite and carpopodite greatly dilated for digging. Third uropods with the exopodite two-jointed. Telson very broad and incised.

H. arenarius. C. and F. 1925, p. 95. Male unknown. Female

9 mm. long, pale yellow. Eyes small and weak. All the legs are hairy and broad. On sandy shores.

Urothoe brevicornis. C. and F. 1925, p. 100. Eyes large, reniform, black. Telson narrow, deeply incised. Female 8 mm. long.

MYSIDACEA

(Pl. IX, figs. 39 and 40)

Shrimp-like Crustacea inhabiting rocky pools, sandy bays and the open sea. They have large black eyes on stalks, and their bodies are either transparent, or partly transparent and partly coloured by chromatophores. The anterior part of the body bears a loose, soft carapace which barely covers the thoracic segments, has a median rostrum anteriorly and is usually excavated posteriorly. It can be raised from the thorax, showing some free thoracic segments below. A Mysid can be readily recognised by its appendages. The antennae have a flat scale as exopodite. All the thoracic appendages are biramose or split (hence the old name Schizopoda for the two distinct groups Mysidacea and Euphausiacea). The pleopods of the male are biramose swimmerets or they may be reduced. In the female they are small and often uniramose. The uropods form, with the telson, a broad tail-fan, and the endopodite of the uropod bears a statocyst. The littoral forms belong to the family Mysidae.

MYSIDAE.

1. Exopodite of uropod divided into two by a transverse suture. Proximal segment with spines, distal with bristles on lateral border. **Siriella**
   Exopodite of uropod undivided, with bristles but no spines. Telson incised posteriorly. 2
2. Antenna scale with a spine. 3
   Antenna scale without a spine. Body bright red. **Hemimysis**
3. Antenna spine subterminal. Body green or brown. Tarsus of from 4–7 segments. **Praunus**
   Antenna spine lateral, body almost or quite transparent. **Schistomysis**

Siriella. Exopodite of the uropod with two joints, the first with spines, the second with feathered bristles on its lateral border. Three pairs of brood lamellae in the female. In the male the pleopods are well developed and some have curved or spiral gill processes at the base of the endopodite. Telson long and tongue-shaped, with spiny border, not excavated posteriorly.

S. armata. Zimmer, 1927, p. 69. Tattersall, 1951, p. 145. Body slender, 21 mm. long, with eyes projecting laterally on long stalks. Rostrum long. Antenna scale five times as long as broad. In pools.

Hemimysis. Telson broadly triangular, with a deep incision posteriorly. The first pair of pleopods are vestigial in both sexes. Tarsus of foot of four, five or more segments.

H. lamornae. The midge shrimp. Z. 1927, p. 143. T. 1951, p. 331. Body short and strongly built. Telson only half the length of the uropod, strongly incised, its distal half beset with ten to twelve spines laterally, the incision with smaller spines. Colour bright red or orange. 10 mm. long. In pools.

Praunus. Eyes well developed on rather long thick stalks. Antenna scale very long, four to nine times as long as broad, its mesial border set with bristles, its lateral smooth except for a subterminal spine. In the male the first, second and fifth pleopods are vestigial, the third and fourth broad and powerful.

P. flexuosus (Pl. IX, figs. 39 and 40). Chameleon shrimp. Bell, 1853, p. 336, as *Mysis chamaeleon*; T. 1951, p. 378. This Mysid, as its popular name implies, is very variable in colour. It is a brown, green or brownish red form, reaching 25 mm. in length. Body slender and bent. Endopodite of the uropod shorter than the telson, the latter with twenty to twenty-four spines on each side. In pools and in shallow water. Often estuarine. Gregarious. A darker and smaller form, regarded by Zimmer as the same species, but by Keeble and Gamble as a separate species, to whom they gave the name *P. neglectus*, is also at times very common. It differs from *P. flexuosus* in having the endopodite of the uropod longer than the telson, the latter having a curved apical pair of spines in addition to eighteen or more lateral spines.

Schistomysis. Ghost shrimps. Delicate transparent Mysids, with or without chromatophores. The telson is triangular and incised posteriorly. The antenna scale has a spine situated below the apex.

S. spiritus. T. 1951, p. 361. A clear, transparent species, almost without pigment, 18 mm. in length. Eyes on long stalks. Tarsus of seven to nine segments. Telson constricted near its base, with twenty to thirty lateral spines. Endopodite of the uropods bent inwards at the tip, its border with long spines like a comb.

S. ornata. T. 1951, p. 369. Body large, robust. 19 mm. Eyes short, broad. Carapace short. Endopodites of the uropods straight. Chromatophores branched, red, brown and yellow. Estuarine.

DECAPODA

(Pl. IX, fig. 41 and Pl. XI)

Usually large Crustacea with five pairs of pereiopods. They include the swimming forms like the shrimps and walking forms like the crabs.

The exoskeleton is usually calcified, and all the thoracic segments, or all but the last one or two, are united to form a cephalothorax covered by a *carapace*. Compound stalked eyes are present. Antennules and antennae are well developed; the mouth parts consist of mandibles and two pairs of maxillae. The first three pairs of thoracic appendages are turned towards the mouth as maxillipedes, the last five pairs are the pereiopods. The five pairs of pleopods are usually swimming appendages of simple biramose type, and the last pair of abdominal appendages, or uropods, form with the telson a tail fan. In some families the abdomen and its appendages are reduced.

NATANTIA (Pl. IX, fig. 41)

Swimming Decapods. The body is usually laterally compressed, the abdomen well developed. The *rostrum* is compressed and often serrated. Both antennules and antennae have a flat lobe or scale springing from the base. On the antennule it is called the *stylocerite*, on the antenna it is the *scale*. The pereiopods are long and slender, the pleopods strong. The first abdominal segment is not markedly smaller than the rest. British species of Natant Decapods belong to the subdivision Caridea. In these the first and second pereiopods are usually chelate and the pleura of the second abdominal segment overlaps those in front of and behind it, so that the abdomen is often humped or gibbous. Shrimps and prawns.

REPTANTIA (Pl. XI, figs. 45–48)

Walking Decapods. The body is usually depressed. The abdomen is either well developed or reduced, with small first segment, a broad tail fan or none. The first pair of pereiopods are usually much larger than the others and strongly chelate, the basipodite and the ischiopodite are fused. They are called the *chelipeds*. The rostrum is depressed or absent. The group is a large one and contains the lobsters, squat lobsters, hermit crabs and crabs.

NATANTIA

1. Carpus of second pereiopod divided into a number of segments. 2
   Carpus of second pereiopod not divided. 4
2. First pereiopod slender, not, or only microscopically chelate. **Pandalidae**, p. 125
   First pereiopod strongly chelate. 3
3. Rostrum short, simple, partially or wholly covering the eyes. **Alphaeidae**, p. 126
   Rostrum well developed, serrate. Abdomen humped. **Hippolytidae**, p. 125
4. First pereiopod chelate, rostrum well developed serrate, antennules triramose. **Palaemonidae**, p. 126
   First pereiopod subchelate, rostrum short simple, antennules biramose. **Crangonidae**, p. 127

REPTANTIA

1. Abdomen armoured and capable of extension, with broad tail fan. 2
   Abdomen soft, or if armoured usually bent under the thorax. 3
2. Rostrum small or absent. Exopodite of uropod not segmented. No chelipeds. **Palinura.** 4
   Rostrum well developed. Exopodite of uropod segmented. Chelipeds present. **Astacura.** 5
3. Uropods usually present. **Anomura.** 6
   Uropods rarely present. No tail fan. True crabs. **Brachyura.** 9
4. No chelae. Antennae with long flagellum. Rock lobsters. **Palinuridae**, p. 127
5. First three pairs of pereiopods chelate. Lobsters. **Nephropsidae**, p. 128
6. Tail fan well developed. Body symmetrical. 7
   Tail fan reduced. Body asymmetrical. Abdomen soft. Hermit crabs. **Paguridae**, p. 131
7. Abdomen extended. Abdominal pleura small. Burrowing forms. **Callianassidae**, p. 130
   Abdomen flexed, but not degenerate. 8
8. Abdomen capable of extension. Fifth pereiopod slender and folded in the branchial chamber. Carapace oval. Squat lobsters. **Galatheidae**, p. 128
   Abdomen pressed under sternum but with a tail fan. Carapace circular, crab-like. **Porcellanidae**, p. 130

9. Front of carapace projecting as a rostrum, single or bifid. 10
Front entire, lobed or toothed, but without a median rostrum. 11
10. Antennae long, at least half as long as the carapace. Chelipeds small, compressed. **Pirimelidae**, p. 133
Antennae short, usually shorter than the bifid rostrum. Carapace triangular. **Maiidae**, p. 135
11. Front entire. Carapace partly membranous, almost circular. **Pinnotheridae**, p. 135
Front lobed or toothed. 12
12. Front toothed. Fifth pereiopod usually flattened for swimming. **Portunidae**, p. 132
Front with flat lobes. 13
13. Front with three to five lobes. Antero-lateral margin similarly lobed. **Cancridae**, p. 134
Front with two lobes. Antero-lateral margin toothed. **Xanthidae**, p. 134

NATANTIA

PANDALIDAE.

The rostrum is long and spiny. First pair of pereiopods not chelate, second pair chelate, both pairs slender. The carpus of the second pereiopod is many jointed.

Pandalus montagui. Aesop prawn. Bell, 1853, p. 297, as *P. annulicornis*; Kemp, 1910, p. 86. This species resembles the prawn and is eaten as such. The rostrum has ten to twelve teeth dorsally and five or six ventrally; it is curved upwards and is as long as the carapace. Antennae longer than the body. Second pereiopods unequal, with minute chelae and about twenty joints in the carpus. Semi-transparent with red stripes and patches. Antennae ringed with red. 5–7 cm. long, but usually rather small. On all coasts, in rock pools, but also in deeper water. Carnivorous.

HIPPOLYTIDAE.

Humped prawns with large rostrum and prominent eyes. The first pereiopods are chelate, short and robust, the second pair chelate and more slender, with the carpus divided into two or more joints. In *Hippolyte* the carpopodite of the second pereiopod has three joints, in *Spirontocaris* six or seven joints.

1. Rostrum shorter than carapace. Eye-stalks broad and short. **Hippolyte varians**
Rostrum longer than carapace. Eye-stalks long and slender. **H. prideauxiana**

H. varians. Bell, 1853, p. 286. This species shows a remarkable colour change between light and darkness. During the day specimens exhibit an opaque green, red or chocolate brown colour, but at night they become blue and transparent. The rostrum is str aght, with a prominent dorsal tooth at the base and another near the apex. Ventrally there are two teeth. The keel of the rostrum extends down the carapace. Between tide marks to deep water. 18–20 mm.

H. prideauxiana. Kemp, 1910, p. 101. On the west coast a larger species of *Hippolyte* occurs; it attains a length of 40 mm. Rostrum without dorsal tooth at its base. Eyes rather small. Antennal scale very long, though not quite as long as the rostrum. The colour is uniform bright green, often with a longitudinal white stripe down the middle of the carapace and abdomen. Occasionally brown and red specimens are found.

Spirontocaris cranchi (Pl. IX, fig. 41). Bell, 1853, p. 288, as *Hippolyte*; Kemp, 1910, p. 106. Rostrum short, bidentate at the tip, with five teeth dorsally and none ventrally. The keel of the carapace is not toothed. Carpopodite of second pereiopod with six joints. Semi-transparent. Green, purplish brown. 18 mm. West and south-west coasts. Common.

ALPHAEIDAE.

The first pereiopods are stout, often with asymmetrical chelae. The second pereiopods have minute chelae and are very slender, the carpus with two or more segments.

Athanas nitescens. Bell, 1853, p. 281; Kemp, 1910, p. 122. A small lobster-like form, about 20 mm. long, red in colour, with a white stripe down the middle of the back. Antennule with three flagella, one shorter and stouter than the rest. The carpopodite of the second pereiopod has six joints. Local, chiefly on west coast.

PALAEMONIDAE.

Prawns. The rostrum is large, compressed and serrated. The antennules have two or three flagella. First pereiopod small, chelate, second usually larger, also chelate.

Palaemon. Rostrum curved upwards and bearing serrations both dorsally and ventrally. Antennule triramose, two flagella being very long, one flagellum short and curved. Abdomen large, with well-developed pleopods and triangular telson.

1. Without red chromatophores. Dorsal spines of rostrum confined to posterior two-thirds, apex bifid. **P. serratus**
With red chromatophores. Dorsal spines of rostrum all along its length. **P. squilla**

P. serratus. Bell, 1853, p. 302; Kemp, 1910, p. 128. Rostrum long and curved upwards, strongly serrated, with six to eight dorsal teeth and four or five ventrally, all the dorsal teeth in the posterior two-thirds, apex often bifid. Antennules as long as the body, antennae 1½ times as long. Transparent grey, spotted with purple. Turns red when boiled. To 10 cm. long. West and south coasts. In pools at low tide. Common.

P. squilla. Bell, 1853, p. 305; Kemp, 1910, p. 129. A smaller species, about half the size of *P. serratus.* Rostrum slightly curved, spiny all along its length, with seven to ten dorsal and three ventral teeth. Antennae about as long as the body. 5 cm. Littoral to shallow water. A rock pool prawn in spring and summer.

CRANGONIDAE.

Shrimps. Rostrum short and flat, the first pereiopod strong and subchelate, the second and third slender and reduced, fourth and fifth slender but longer.

Crangon vulgaris. Common brown shrimp. Bell, 1853, p. 256; Kemp, 1910, p. 137. The whole aspect is clumsy. Antennal scale very large and used for burrowing. Pleopods well developed, with the endopodite half as long as the exopodite. Grey or black, mottled with red. To 7 cm. Littoral, on sand. Often estuarine. Moves off shore in autumn. Turns brown when boiled.

REPTANTIA

PALINURA

PALINURIDAE.

The rock lobsters are distinguished from the true lobsters by the absence of chelae and by the spiny carapace. They inhabit rock clefts, and though they usually live in deeper water they may occasionally be found in pools at extreme low-tide level.

Palinurus vulgaris. Crawfish. Bell, 1853, p. 213. Rostrum short with triangular lateral spines. Carapace spiny. Antennules with long peduncles and short flagella. Antennae longer than the body and spiny, but without a scale. Colour reddish or purplish brown with white spots. 30–45 cm. long. South and west coasts only. Used for food but not as good as the true lobster.

NEPHROPSIDAE.

True lobsters and crayfish. Antennule with a short peduncle and two fairly long flagella. Antenna with very long flagellum and a scale as exopodite. The first, second and third pereiopods are chelate, the first pair being long and very strong (chelipeds).

Homarus vulgaris. The lobster. Bell, 1853, p. 242. Chelipeds large, flattened and smooth except for a few serrations on the edge of the propodite, the carpopodite spiny. Right and left claws usually unequal in size. Carapace almost smooth. Colour blue, brownish blue or reddish, spotted with white. Length 25–50 cm. On rocky shores, in deep pools or under rocks, to shallow water. On all coasts.

ANOMURA

GALATHEIDAE.

The squat lobsters have symmetrical, depressed bodies with oval cephalothorax. Abdomen well developed, usually flexed, but capable of full extension, with uropods and a telson forming a tail fan. Only the first pereiopods chelate, the fifth pair slender and folded within the branchial chamber. Antennules resemble those of *Palinurus*. Antennae without a scale.

Galathea. Rostrum large and triangular, with the median spine the longest, and four spines on each side. Each abdominal tergum has at least one transverse furrow, but the telson has a median longitudinal groove, scales, bristles and spines. Chelipeds very large, equal to twice the length of the body.

1. Basal joint of antennule with two sharp spines. Each abdominal tergum with one transverse furrow. **G. intermedia**
   Basal joint of antennule with three sharp spines. 2
2. Propodite of cheliped spiny all over dorsal surface. Colour of body red and blue. **G. strigosa**
   Propodite of cheliped spiny on side only. 3

3. Meropodite of third maxilliped much longer than ischiopodite. Chelipeds covered with scaly tubercles. Colour greenish brown. **G. squamifera**
   Meropodite of third maxilliped about the same length as ischiopodite. 4
4. Chelipeds very hairy. Abdominal segments with one transverse groove. Rostrum concave, not very hairy. Colour greenish red with blue markings on side of body and on labrum. **G. nexa**
   Chelipeds not very hairy. Abdominal segments with three transverse grooves. Rostrum flat, thickly covered with setae. With red, yellow or orange but no blue colour. **G. dispersa**

G. squamifera (Pl. XI, fig. 45). Bell, 1853, p. 197; Selbie, 1914, p. 69. The most scaly species, the carapace being crossed by transverse ridges covered with scaly tubercles. Rostrum short, with nine spines, of which the three central ones are enlarged and subequal. Chelipeds covered with scaly tubercles and spiny along their edges. Colour greenish brown. Joints dark coloured and the spines dark green, tipped with red. The species attains a length of 7·5 cm. but is usually about 5 cm. long. Common off all coasts and migrates inshore in spring.

G. strigosa. Bell, 1853, p. 200. The largest and most spiny species. Red with blue transverse lines and spots, handsome but very ferocious. Chelipeds very spiny, with large flat claws. Second, third and fourth pereiopods also spiny. 8 cm. long or more. Under stones. Littoral to deep water.

G. nexa. Bell, 1853, p. 204; Bull, 1937, p. 42. Larger and more heavily built than *G. dispersa.* Carapace fringed with fine setae. Rostrum tipped with red. General appearance smooth and glossy. Chelipeds vary with sex and age but differ from those of *G. dispersa*, being thickly covered with long plumose setae, which give them the furry appearance peculiar to the species. Colour greenish red with a brilliant blue labrum and a blue stripe on the pleural fold. The blue colour is very constant. From the tip of the rostrum to the end of the abdomen 30–40 mm. in male, less in female. Total length including chelipeds 50–80 mm. Very local in distribution. Usually in shallow water but may be taken on the shore.

G. dispersa. Bull, 1937, p. 46. Lateral margin of carapace irregularly spiny. General appearance dull and rough. Chelipeds rather long, with irregular spines, tubercles, scales and fine setae, but not woolly or furry. Red, yellow, orange, but without greenish red or blue colouring. Body 35 mm. including abdomen, 72 mm. including chelipeds. Shallow water. Common.

G. intermedia. Selbie, 1914, p. 66. Basal joint of antennule with two spinous processes (all others have three). Chelipeds of male very large, twice the length of the body with abdomen folded, unequal. Both fingers end in a broad curved tooth, and above this is a smaller and sharper tooth in line with the outer margin. Smallest British species. 12–20 mm. long. From littoral zone to shallow water. On all coasts.

PORCELLANIDAE.

Porcelain crabs. Small circular crabs with a terminal swimming fan on the flexed abdomen, long slender antennae and reduced fifth pereiopods. They resemble the Galatheidae in most characters.

Porcellana. Antennules small, antennae long whip-like, rostrum reduced, chelae large and broad, abdomen large, nearly as long as the rather rounded carapace. Pleopods in male on second abdominal segment only, in female four pairs.

P. platycheles (Pl. XI, fig. 46). Bell, 1853, p. 190; Selbie, 1914, p. 87. Hairy porcelain crab. Carapace broadly pear-shaped, grooved and polished, but often hairy in young individuals. Three blunt teeth between the eyes. Antennae longer than the carapace, slender. The most marked feature of the species is the large flattened chelae, which are fringed with dirty greyish setae and have a denticulate knob on the carpopodite. Reddish brown to dirty grey, usually coated with mud. Carapace 12 mm. Chelipeds 30 mm. Under stones in muddy situations. Local.

P. longicornis. Bell, 1853, p. 193; Selbie, 1914, p. 87. Minute porcelain crab. Carapace smooth, almost circular, not grooved. Antennae nearly twice as long as the carapace, very slender. Chelipeds unequal, grooved and keeled. Colour pale, often with a reddish tinge, and the body is free from mud. Carapace 4 mm. long. Under stones. Generally distributed and very common.

CALLIANASSIDAE.

Rare burrowing forms which occur in sandy situations on the south and east coasts. Carapace short and laterally compressed. Abdomen extended, large and broadest about the middle. Rostrum small or absent. Tail fan well developed. Antennules and antennae of moderate size, third maxillipeds foot-like, chelipeds large, equal or unequal, pereiopods 2–5 small, equal. The exopodite of the uropod has no suture. Clumsily built forms.

1. Chelipeds equal. **Upogebia**
   Chelipeds unequal. **Callianassa**

Callianassa. Rostrum and eyes small. Propodite of third pereiopod much flattened and expanded for burrowing.

C. subterranea. Bell, 1853, p. 219 (page wrongly numbered 217). Carapace smooth, less than half the length of the abdomen. The larger cheliped may be three times as long and four times as broad as the smaller one, and may belong to either side. Second pereiopod chelate. Bright red. 5 cm. long. Rare, south and west coasts. Littoral to deep water. In sand.

Upogebia. Rostrum of moderate size, pointed. Carapace very narrow anteriorly. Abdomen thick, narrowing towards the tail fan. Chelipeds imperfectly chelate.

U. stellata. Bell, 1853, p. 223, as *Gebia.* Antennae as long as the body. Carapace well calcified. Body hairy. 3–5 cm. Burrowing in sand. North and east Scottish coasts, English Channel. Not common. Another species, *U. deltaura,* occurs on the south and east coasts.

PAGURIDAE.

Hermit crabs. Asymmetrical, with soft, swollen, unsegmented abdomen adapted for holding on to the Molluscan shell in which the animal lives. Carapace partly membranous, shorter than the abdomen and covering only a portion of the thorax dorsally. The abdomen is membranous, tapering and coiled, with vestigial terga. As in the Callianassidae, the third maxilliped is foot-like. Chelipeds large and unequal, the second and third pereiopods with nail-like dactylopodites, the fourth and fifth pairs reduced. The abdominal appendages are reduced, and the pair of uropods act as prehensile organs for holding on to the Molluscan shell. Most of the hermit crabs inhabit deep water, but young specimens of two species of *Eupagurus* are to be found on the shore. *Eupagurus* is a right-handed crab with four pleopods in the female, three in the male, on the left side only.

E. bernhardus. Soldier crab. Bell, 1853, p. 171, as *Pagurus*;

Selbie, 1921, p. 15. This pugnacious species is red and yellow in colour and attains a length of 12 cm. Chelae tuberculate or granulated but not hairy. Second and third pereiopods spiny. Adults often associated with the Hydroid *Hydractinia*, or the sponge *Suberites domunculus* (*Ficulina ficus*), and with the Annelid *Nereis fucata*. Living in whelk shells. Young specimens found on shore in winkle and dog whelk shells. On all coasts.

E. prideauxi. Bell, 1853, p. 175, as *Pagurus*; Selbie, 1921, p. 34. Chelipeds finely granulated but not tuberculate, and the second and third pereiopods are not spiny. Adults usually associated with the Anemones *Adamsia palliata* or *Calliactis parasitica*, or with the Hydroid *Hydractinia echinata*, also with *Nereis fucata*. Light reddish brown. 6 cm. Young specimens on all coasts.

BRACHYURA

PORTUNIDAE.

Swimming or fiddler crabs. In nearly all, the fifth pereiopods are flattened for swimming so that they resemble oars. Carapace depressed, with spiny antero-lateral margin, usually widest in the region of the last spine. Eye-stalks short and narrow. The two chief genera are *Carcinus*, in which the dactylopodite of the last pereiopod is pointed, and *Portunus*, in which the last dactylopodite is flattened.

Portunus. The broad carapace has three, eight or ten teeth between the eyes, and five spines on each antero-lateral border. Carpopodite of the cheliped usually with a strong spine. Dactylopodite of the last walking leg ovate or lanceolate. Abdomen of male five-jointed, of female seven-jointed.

1. Front with eight to ten teeth. Joints bright blue. Carapace covered with fine setae. **P. puber**
   Front with three teeth. **2**
2. Carapace scaly and hairy. Dactylopodite of fifth pereiopod smooth, tipped with violet. **P. depurator**
   Carapace convex, marbled with coloured lines. **P. marmoreus**

Portunus puber. Velvet crab. Bell, 1853, p. 90; Palmer, 1927, p. 882. This handsome and pugnacious species is common on the south and west coasts, rarer on the east coast. It is the largest swimming crab, measuring 5–7 cm. in length, 10 cm. in breadth, and in colour is bright reddish brown, with bright blue joints, and soft downy hairs on the body and legs. Chelipeds often unequal, with spiny carpus, and tuberculate teeth on the chelae. Fifth pereiopods are ridged longitudinally. Under stones at low water and in shallow water, on rocky shores.

P. marmoreus. Marbled swimming crab. Bell, 1853, p. 105; Palmer, 1927, p. 888. Carapace not hairy but somewhat inflated, marked with coloured lines which give it a marbled appearance. The front is regularly three-lobed, with obtuse teeth. Chelipeds rather short, angular, with a strong spine on the carpopodite, and a keeled chela with tuberculate fingers. Buff, brown and brownish red. Length 3 cm. Breadth 4 cm. Usually a local species, occasionally littoral but more often in shallow water on sand, commonest on the south and west coasts.

P. depurator (Pl. XI, fig. 47). Cleanser or wrinkled swimming crab. Bell, 1853, p. 101; Palmer, 1927, p. 893. Carapace irregular, with granules, scales and setae. Chelipeds sculptured, the chelae keeled and tuberculate. Dactylopodite of fifth pereiopod broadly oval, smooth, and tipped with violet. Pale reddish brown. 4 cm. long, about 5 cm. broad. Common locally, often in shallow water, but sometimes taken inshore.

Carcinus. Carapace rather flattened, broader than long, the three-toothed front projecting and forming a continuous curve with the antero-lateral margin, which has five teeth. Chelipeds slightly unequal, with a strong spine on the carpus. Fifth pereiopod somewhat compressed but not flattened as in *Portunus*, and with a narrow lanceolate dactylopodite. Abdomen of male five-jointed, of female seven-jointed.

C. maenas. The shore crab. Bell, 1853, p. 76. This dingy blackish green crab is the commonest of the shore crabs. The carapace is finely tuberculate, with distinct regions and straight postero-lateral margin. Length 4 cm. Breadth 5 cm. Range from shore to shallow water.

PIRIMELIDAE.

Front projecting as a shelf or blunt rostrum.

Pirimela denticulata. Bell, 1853, p. 72. Carapace nearly as long as broad, convex, smooth, with strongly marked elevations. The projecting front has three teeth, the orbital margin two teeth and the antero-lateral margin five sharp teeth. Chelipeds small and compressed, the chelae keeled. The other pereiopods compressed and hairy, with pointed dactylopodites. A small pretty species, green, mottled with brown, purple and red. 12 mm. long and a little more in breadth. On clean coarse gravel.

CANCRIDAE.

Carapace strong, front with from three to five lobes, orbit with a strong tooth mesially and laterally. Antero-lateral margin toothed or lobed, ending in a sinuous margin which is elevated over the smooth postero-lateral margin. Posterior margin short, almost straight.

Cancer pagurus (Pl. XI, fig. 48*a*). The edible crab. Bell, 1853, p. 59. Carapace broadly elliptical, convex. The three anterior and the ten antero-lateral lobes rounded or quadrate, not pointed. Shell granulated, pinkish or reddish brown. Chelipeds smooth, with tuberculate teeth on the black-tipped chelae. In holes in rocks, under stones at low tide to shallow water. Shore forms 10–12 cm. long and about 15 cm. wide. Shallow water forms may weigh up to 12 lb.

XANTHIDAE.

Small, dark crabs somewhat resembling edible crabs. Front with a median fissure, projecting like a shelf, antero-lateral margin toothed. Both antennules and antennae short, chelipeds very large and heavy, with curved and pointed chelae. Second to fifth pereiopods short, compressed and clumsy.

1. Chelae equal. Front not toothed. **Xantho**
   Chelae unequal. Front finely denticulate. **Pilumnus**

Xantho. Antennae very short, chelae heavy. Abdomen of male five jointed and of female seven jointed. Carapace not hairy, but deeply furrowed. Front straight, with median groove.

1. Dactylopodites of chelipeds smooth, not grooved. Chelae black. **Xantho incisus**
   Dactylopodites of chelipeds longitudinally grooved. Chelae brown. **X. hydrophilus**

X. incisus. Bell, 1853, p. 51, as *X. florida*. Front straight, smooth. Chelipeds 10 cm. long, chelae usually black, not grooved.

Carapace of male 4 cm. long; 6·5 mm. broad, reddish brown. Common between tide marks on rocky shores of west coast, but local.

X. hydrophilus (Pl. XI, fig. 48*b*). Bell, 1853, p. 54, as *X. rivulosa*. A rarer species. Chelae brown with grooved dactylopodite. Legs hairy. Front beaded. Yellow and red. Only 2·5–5 cm. long. Under stones at low water to shallow water.

Pilumnus. Antennule with a club-shaped tip. Chelipeds unequal, stout, rounded. Front projecting, toothed. Abdomen of both sexes seven-jointed.

P. hirtellus. Bell, 1853, p. 68. Carapace rather hairy, especially the antero-lateral border, which is downy. Front broad, divided mesially by a deep fissure, with numerous minute denticulations on each side. Second to fifth pereiopods setose with clubbed hairs. Brownish red with yellowish spots. Chelae light brown. Carapace 15–18 mm. long, 25 mm. broad. Under stones at low water, to deep water. South and west coasts.

PINNOTHERIDAE.

Pea crabs. Minute semi-parasitic crabs not exceeding 12 mm. in length, inhabiting the mantle cavity of bivalve Molluscs and the tests of Tunicates. Carapace nearly circular, without a rostrum, soft and largely membranous. Eyes widely spaced and very small, antennae very short. Chelipeds equal and small, the walking legs compressed with strong curved dactylopodites. Abdomen seven-jointed. Males smaller than females.

Pinnotheres pisum. Bell, 1853, p. 121. Inhabits the shells of mussels, cockles, oysters, etc. Male smaller than the female, with a breadth of only 6 mm. Colour pale yellow or grey. Female 8 mm. in breadth, carapace brownish with yellow markings. Abdomen very broad and yellow in colour. Mainly on south and west coasts.

MAIIDAE.

Spider crabs. Carapace triangular or pear-shaped, longer than broad, narrowing anteriorly to a rostrum which is usually bifid. Legs often slender and spider-like, in many species provided with serrated and hooked hairs for holding algae, sponges, Coelenterates, etc., with which the crab clothes itself protectively. Eyes sometimes retractile, sometimes not.

1. Eyes retractile. Dactylopodite of walking legs pointed. **2**
   Eyes not retractile. Dactylopodite of walking legs 2 and 3 pointed, 4 and 5 with flat curved nails. Legs very slender. Abdomen six-jointed. **Macropodia**
2. Abdomen seven-jointed. Carapace tuberculate, depressed. Rostral horns converging. Legs shorter than in *Inachus*. **Hyas**
   Abdomen six-jointed. Carapace inflated, tuberculate. Legs slender, long. **Inachus**

Hyas. Legs relatively shorter and stouter than in many other spider crabs, though still longer than the chelipeds. Carapace covered with tubercles. Rostrum short and bifid, the horns converging.

H. araneus (Pl. XI, fig. 48*c*). Bell, 1853, p. 31. Colour greyish or dull red, according to the surroundings, usually with seaweeds and zoophytes on the forked setae of the carapace and legs, so that the body is often completely masked by them. Second pereiopod 1½ times body length. Carapace to 7 cm. long. Small specimens in deep pools, larger in shallow water. South, west and east coasts.

Inachus. Small spider crabs with very long slender legs and inflated tuberculate spiny carapace. Rostrum short and bifid, the horns slightly diverging. Eyes retractile. Abdomen six-jointed, keeled.

I. dorynchus. Bell, 1853, p. 16. Carapace with three spines on the gastric region, two on each branchial region and three small tubercles on the cardiac region. Chelipeds heavy and rounded; chelae short. Second pereiopods three times as long as the body. 16 mm. long, often covered with sponges, seaweeds, etc. Littoral to shallow water.

Macropodia. The most slender-legged of the spider crabs. Carapace triangular, prolonged into an attenuated bifid rostrum with parallel horns. Chelipeds shorter than the very slender walking legs, rather heavy in the male. Pereiopods 2–5 very long, the first walking leg (2) at least three times as long as the body, the others gradually diminishing in size. All the legs hairy.

M. rostrata. Bell, 1853, p. 2, as *Stenorhynchus phalangium*. Carapace spiny, with one spine on the gastric region, one on each hepatic region (these three forming an equilateral triangle), two on

VIII. CRUSTACEA

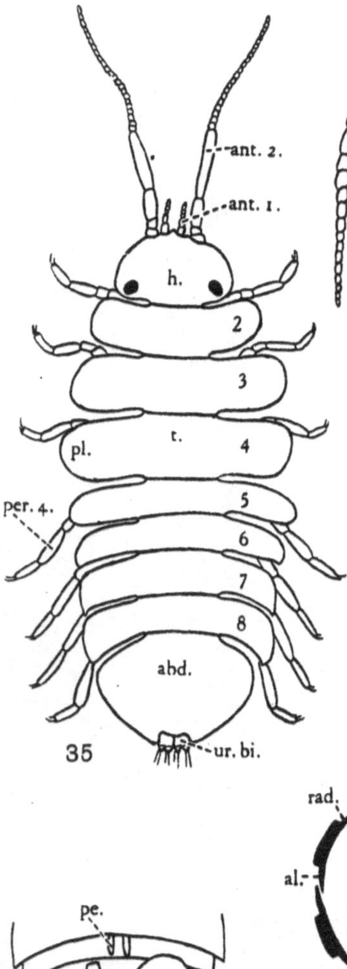

35

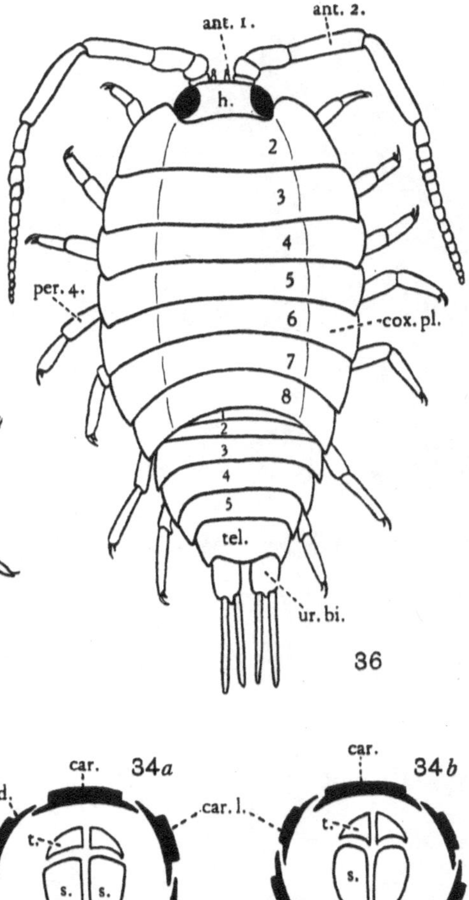

36

34a

34b

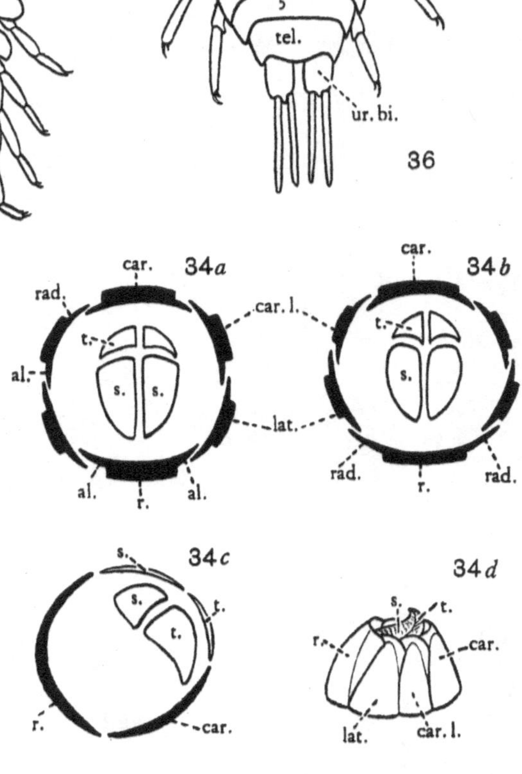

34c

34d

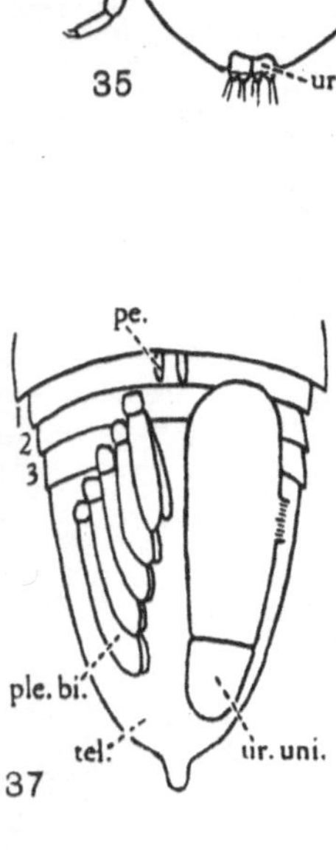

37

(*facing p.* 137)

each branchial region and one large one on the cardiac region. Usually covered with algae or sponges on legs and body. Reddish. 16 mm. long. Under stones at low water down to deep water. Common on all coasts.

EXPLANATION OF PLATE VIII

Crustacea (i), Figs. 34–37

Fig. 34. *a–d*, Cirripedia. *a*, *b* and *c* are schematic transverse sections of the shell of three families of acorn barnacle to show the arrangement of the plates. The operculum is represented in surface view.

a. Chthamalidae. Both carina and rostrum are inset (i.e. have two alae), the lateral is wholly outside (i.e. has two radii), but the carinolateral is partly within and partly without (i.e. has one ala for the lateral and one radius for the carina). The scuta of the operculum are larger than the terga, and the aperture is kite-shaped.

b. Balanidae. The carina is inset (i.e. has two alae), the rostrum wholly external (i.e. has two radii). The lateral and carinolateral have each a ventral ala and a dorsal radius. The scuta of the operculum are larger than the terga, and the aperture is diamond-shaped.

c. Verrucidae. The shell is composed of two wall plates and two pieces from the right side of the operculum. These are carina, rostrum, right tergum and right scutum. The plates do not overlap, and the small tergum and large scutum are fixed. The left scutum and tergum form a movable one-sided operculum, the tergum being larger than the scutum.

d. Lateral view of *Balanus crenatus*. × 2½.

Fig. 35. *Jaera marina*. Asellota. The head is fused with the first thoracic segment. Note the unsegmented abdomen and minute terminal uropods. There are no coxal plates. × 14.

Fig. 36. *Ligia oceanica*. Oniscoidea. Large coxal plates, segmented abdomen, styliform uropods. The head is fused with the first thoracic segment, the telson with the sixth abdominal segment. After G. Hewitt. × 2.

Fig. 37. *Idotea baltica*. Valvifera. Ventral view of abdomen to show partial segmentation, five biramose pleopods and the left lid-shaped uropod. The right uropod has been removed to expose the pleopods. The telson is fused with three abdominal segments. × 5.

abd.	abdomen (pleon)	per. 4	fourth pereiopod
al.	ala	pl.	pleuron
ant. 1	antennule	ple. bi.	biramose pleopod
ant. 2	antenna	r.	rostrum
car.	carina	rad.	radius
car. l.	carinolateral plate	s.	scutum
cox. pl.	coxal plate	t.	tergum
h.	head (cephalon) one or more thoracic segments may be fused with the head	tel.	telson. One or more abdominal segments may be fused with the telson
lat.	lateral plate	ur. bi.	biramose uropod
pe.	penis	ur. uni.	uniramose uropod

EXPLANATION OF PLATE IX

Crustacea (ii), Figs. 38–41

Fig. 38. *Nebalia bipes.* Female. Lateral view to illustrate Arthropod structure. The thorax is represented as seen through the transparent carapace. The details of the foliaceous thoracic limbs are not shown. Note the seven abdominal segments, telson and caudal furca. ×8.

Fig. 39. *Praunus flexuosus.* The biramose pereiopod. ×12.

Fig. 40. *Praunus flexuosus.* Caudal fan. ×12.

Fig. 41. *Spirontocaris cranchi*, a member of the Decapoda Natantia. ×3.

abd.	abdomen (pleon)	mp.3	third maxilliped
ant.1	antennule	ot.	otocyst
ant.2	antenna	p.	propodite
b.	basipodite	per.1	first pereiopod
c.	carpopodite	per.2	second pereiopod
c.f.	caudal furca	pl.	pleuron
cox.	coxopodite	ple.bi.	biramose pleopod
cp.	carapace	ple.uni.	uniramose pleopod
d.	dactylopodite	pr.	protopodite
e.c.	compound eye	r.	rostrum
endo.	endopodite	sc.	scale of antenna (squame)
exo.	exopodite	sty.	stylocerite of antennule
h.	head (cephalon). One or more thoracic segments may be fused with the head	t.	tergum
		tel.	telson. One or more abdominal segments may be fused with the telson
i.	ischiopodite		
m.	meropodite	th.	thorax (pereion)
max.	maxilla	th.ap.	thoracic appendages
md.	mandible	ur.bi.	biramose uropod

EXPLANATION OF PLATE X

Crustacea (iii), Figs. 42–44

Fig. 42. *Gammarus locusta*, to illustrate Amphipod structure in a swimming type. The head is fused with the first thoracic segment, the remaining body segments being free. ×10.

Fig. 43. *Orchestia gammarella*, a hopping form. Segmentation as in Fig. 42. ×5.

Fig. 44. *Caprella linearis.* Female to show incubatory lamellae. The head is fused with two body segments. ×7.

IX. CRUSTACEA

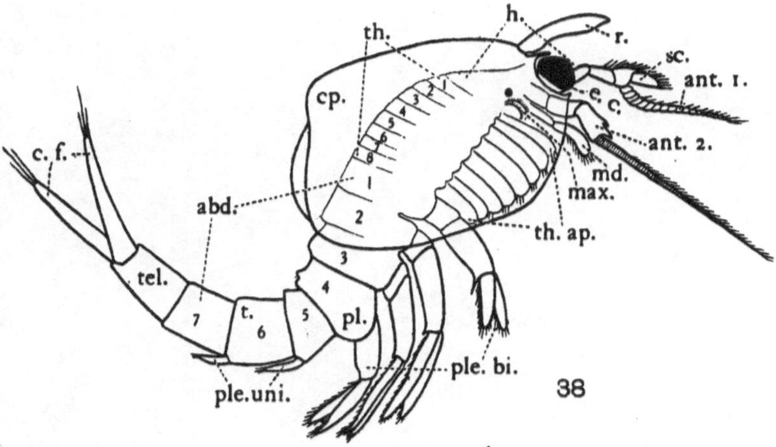

38

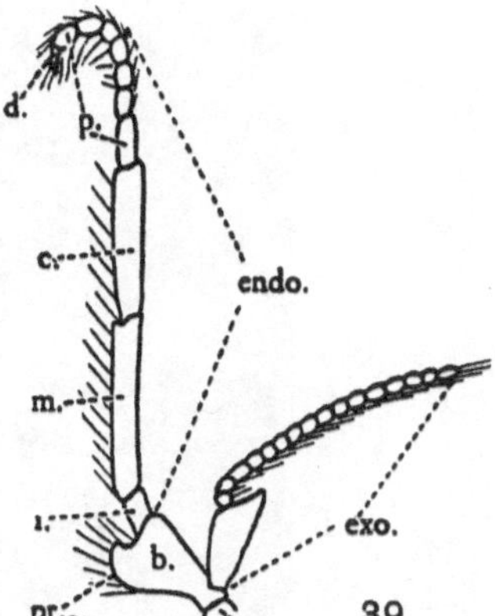

39

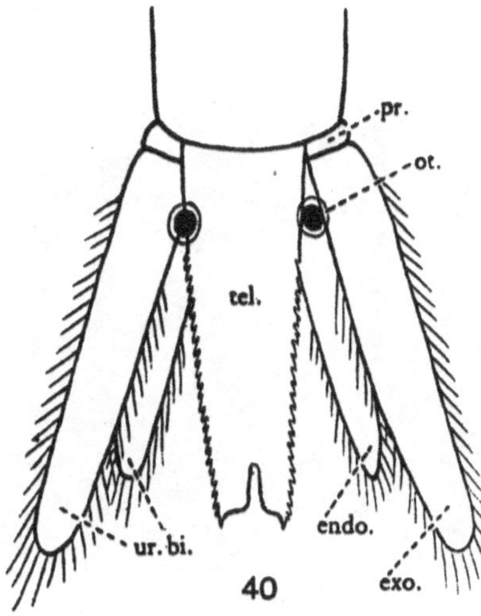

40

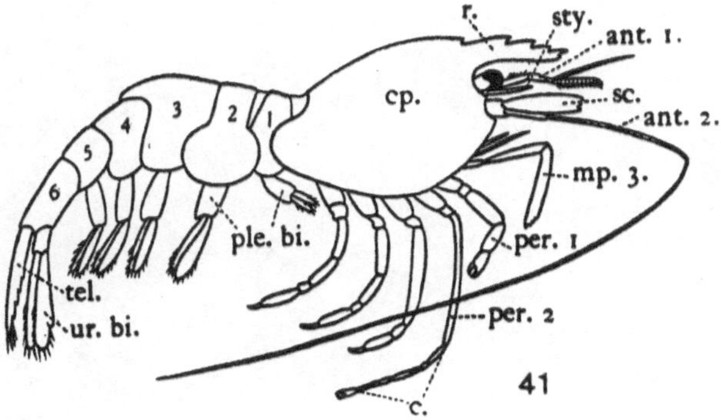

41

(*facing p.* 138)

X. CRUSTACEA

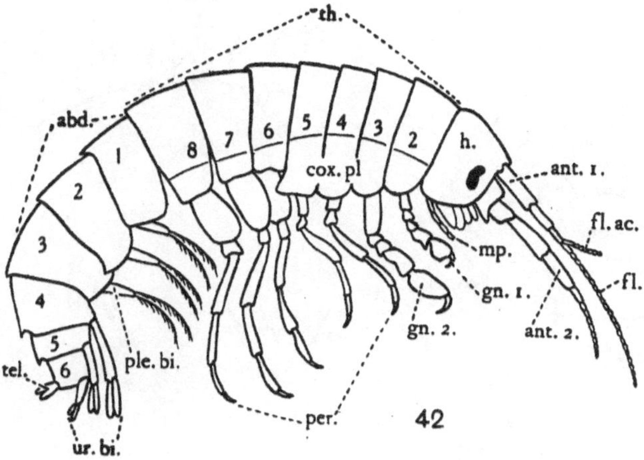

42

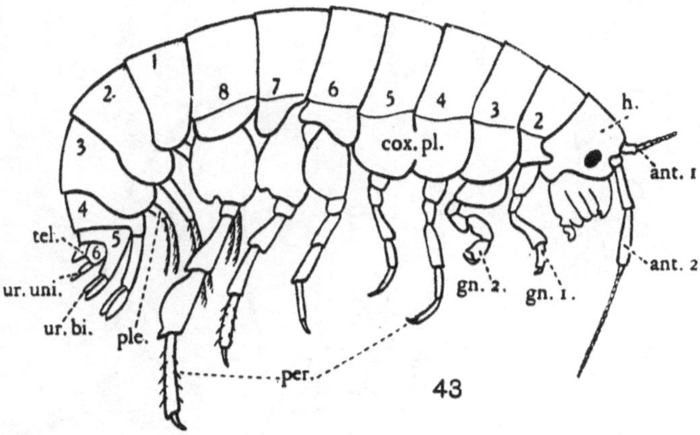

43

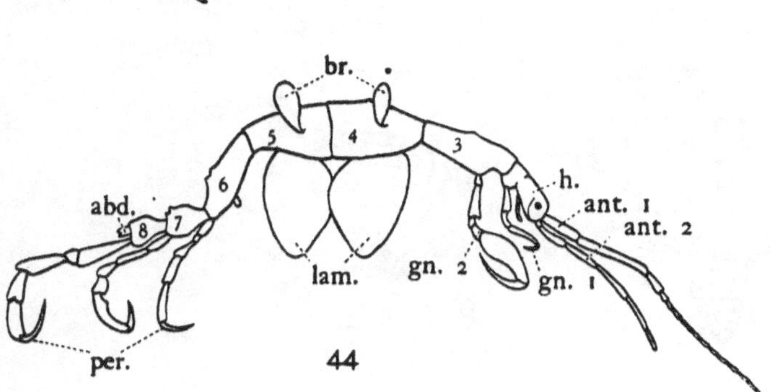

44

XI. CRUSTACEA

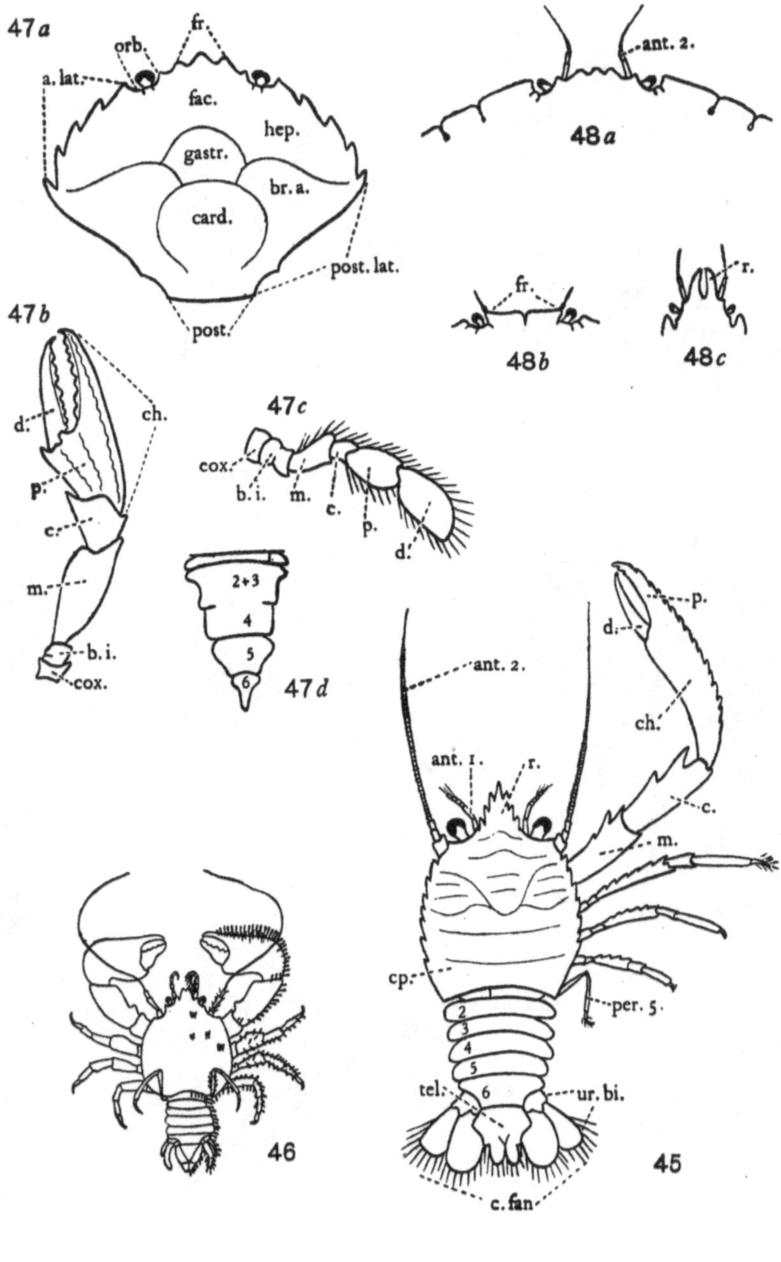

(*facing p. 139*)

abd. abdomen
ant.1 antennule
ant.2 antenna
br. branchia
cox.pl. coxal plate
fl. flagellum
fl.ac. accessory flagellum
gn.1 first gnathopod (thoracic appendage 2)
gn.2 second gnathopod (thoracic appendage 3)
h. head (cephalon). One or more thoracic segments may be fused with the head
lam. incubatory lamella
mp. maxilliped
per. pereiopod
ple. pleopod
ple.bi. biramose pleopod
tel. telson. One or more abdominal segments may be fused with the telson
th. thorax (pereion)
ur.bi. biramose uropod
ur.uni. uniramose uropod

EXPLANATION OF PLATE XI

Crustacea (iv), Figs. 45–48

Fig. 45. ***Galathea squamifera.*** Anomura. Dorsal view. × 1½.

Fig. 46. ***Porcellana platycheles.*** Anomura. The abdomen is normally flexed, but has been extended to show the caudal fan. The spatulate setae are shown on the right side. × 1.

Fig. 47. ***Portunus depurator.*** Brachyura. *a*, The carapace to show borders and areas. *b*, A Cheliped. *c*, The fifth pereiopod. *d*, Abdomen of the male. × 1.

Fig. 48. Three types of front and orbit. *a*, ***Cancer pagurus***, tri-lobed. × ½. *b*, ***Xantho hydrophilus***, flat. × ½. *c*, ***Hyas araneus***, with bifid rostrum. × ½.

a.lat. antero-lateral border of carapace
ant.1 antennule
ant.2 antenna
b.i. fused basi- and ischiopodites
br.a. branchial area of carapace
c. carpopodite
c.fan. caudal fan
card. cardiac area of carapace
ch. chela
cox. coxopodite
cp. carapace
d. dactylopodite
fac. facial area of carapace
fr. front border of carapace
gastr. gastric area of carapace
hep. hepatic area of carapace
m. meropodite
orb. orbit
p. propodite
per.5 fifth pereiopod
post. posterior border of carapace
post.lat. postero-lateral border of carapace
r. rostrum
tel. telson
ur.bi. biramose uropod

PYCNOGONIDA

(Plate XII)

Spider-like eight-legged Arthropods with very slender bodies, divided into a *rostrum*, *cephalon*, *trunk* and *abdomen*. The *rostrum*, which is often relatively large, bears the mouth at its extremity, and a joint separates it from the cephalon. The *cephalon* bears the eyes, and, typically, three pairs of appendages. There are four eyes on a median *ocular tubercle*. The cephalic appendages are *chelicerae*, *palps* and *ovigerous legs*. The *trunk* bears four pairs of walking legs in British species. The segments in the trunk may be distinct or fused, and the first segment may be joined to the cephalon. Laterally the segments bear prolongations for the attachment of the legs, which are eight-jointed. The *abdomen* is greatly reduced and bears neither joints nor appendages. At its apex lies the anus. The male carries the eggs on the ovigerous legs. The joints of the walking legs are first, second, third coxae, femur, first and second tibiae, tarsus and propodite. The propodite may bear specialised spines and claws.

Pycnogonids are all marine and carnivorous. Most species feed on Coelenterates, and their larvae are parasitic in the coelenteron of Hydroids.

1. Chelicerae and palps present. Ovigerous legs in both sexes. **2**
   Palps absent. Ovigerous legs in male only. **3**
2. Body and legs very slender. Chelicerae chelate. Palps five-jointed. **Nymphonidae**
   Body and legs strongly built. Chelicerae not chelate in the adult. Palps eight- or nine-jointed. **Ammotheidae**
3. Chelicerae present. Abdomen ovate. **Phoxichilidiidae**
   Neither chelicera nor palps. Abdomen truncate. **4**
4. Legs long and slender. Ovigerous legs seven-jointed. **Phoxichilidae**
   Legs strongly built. Ovigerous legs nine-jointed. **Pycnogonidae**

The Pycnogonidae and Ammotheidae are heavily built, with short, stout legs, the other families are lightly built, with long, slender legs, especially in Nymphonidae and Phoxichilidae.

PYCNOGONIDAE.

Cephalon short and wide. Trunk short, with three segments. Abdomen truncate, horizontal. Neither chelicerae nor palps.

Ovigerous legs in male only, very short, nine-jointed, clawed but not spiny. One claw on walking legs. Egg mass simple, cake-shaped.

Pycnogonum littorale (Pl. XII, fig. 52). Bouvier, 1923, p. 61. Rostrum conical. A dorsal tubercle on the cephalon and on each trunk segment. Large lateral processes on trunk. Legs thick, strongly clawed, scarcely as long as the body, knobbed. White or dirty yellow, the surface covered with punctate scale-like markings. About 15 mm. long. Under stones, on old shells, etc.

NYMPHONIDAE.

Cephalon elongated, with distinct neck. Ocular tubercle set far back. Chelicerae and palps present. Ovigerous legs in both sexes, ten-jointed, with leaf-like spines on the last four joints and terminal claws. Legs and body very slender. Eggs in compact balls.

Nymphon. Legs very slender, smooth or slightly setose. Palps five-jointed, 1½ times the length of the rostrum, but shorter than the chelicerae. Abdomen reduced and usually held upright.

1. Second joint of palp longer than third joint, fourth and fifth sub-equal. **N. gracile**
   Second and third joints of palp sub-equal, fourth twice as long as the fifth. **N. rubrum**

N. gracile (Pl. XII, fig. 49). Bouvier, 1923, p. 30. Rostrum cylindrical, twice as long as broad. Clear rose colour to yellow, gut darker red or brown, showing through the test, body sometimes marked with transverse bands of a darker colour. Length of body 8 mm., walking legs about 24 mm. On algae, calcareous sponges, etc. Littoral to shallow water. Common. Eggs in spring and summer.

N. rubrum. Bouvier, 1923, p. 30. Body segments much elongated. Length of body 4–5 mm., legs 16–20 mm. Colour deep red, with darker bands. Among algae. Eggs in spring and summer.

PHOXICHILIDIIDAE.

Cephalon small. Ocular tubercle anterior, rostrum inserted obliquely. Palps absent or vestigial. Ovigerous legs without special spines or terminal claws, usually absent in female. Legs long and slender. Larvae live parasitically in Coelenterates.

1. Ovigerous legs with six joints. Chelicerae slender and weak. **Anoplodactylus**

Ovigerous legs with five joints. Chelicerae well developed. **Phoxichilidium**

Anoplodactylus. Chelicerae weak with a rather long simple scape and unarmed crossed fingers. Legs long and narrow, with dilated femur in the female and cement glands with pores in the male, terminated by a long slender claw. Several species.

A. petiolatus. Bouvier, 1923, p. 40. Elongated cephalon separated from trunk by a long neck. Legs three times as long as body, which is 2–3 mm. long. White or greenish. Common on Hydroids, also amongst Bryozoa, algae, old shells, etc. Larvae in manubrium of *Obelia* medusae.

Phoxichilidium. Cephalon truncate anteriorly, broader than long, with ocular tubercle arising from its middle. Chelicerae fairly strong, with slightly crossed fingers. Legs long, slender, naked. Femur dilated in the female.

P. femoratum (Pl. XII, fig. 51). Bouvier, 1923, p. 43. Red, brown, or reddish brown. Femur greatly enlarged. Body 2–3 mm. Clawed legs three times as long. On Gymnoblastic hydroids. Larvae form galls on *Syncoryne eximia*.

PHOXICHILIDAE.

Body narrow, legs very long. Cephalon short, broader than long, with posteriorly placed ocular tubercle. Neither chelicerae nor palps. Legs with principal and auxiliary claws, spiny. Femur dilated in female, with numerous cement pores in male. Abdomen small, truncate, upturned.

Endeis spinosa. Bouvier, 1923, p. 45, as *Chilophoxus*. Rostrum long, a little shorter than trunk, turned downwards. Cephalon short. Legs long and slender, armed with spines, three to four times as long as the trunk, which may reach 5 mm. in length. Green, brown or rose coloured. Amongst algae and Hydroids. Littoral to deep water. Larvae in *Obelia* hydranths.

AMMOTHEIDAE.

Stoutly built. Chelicerae much shorter than rostrum. Chelae occur in larvae but are reduced in adult. Palps present. Rostrum oblique. Walking legs with auxiliary claws.

XII. PYCNOGONIDA

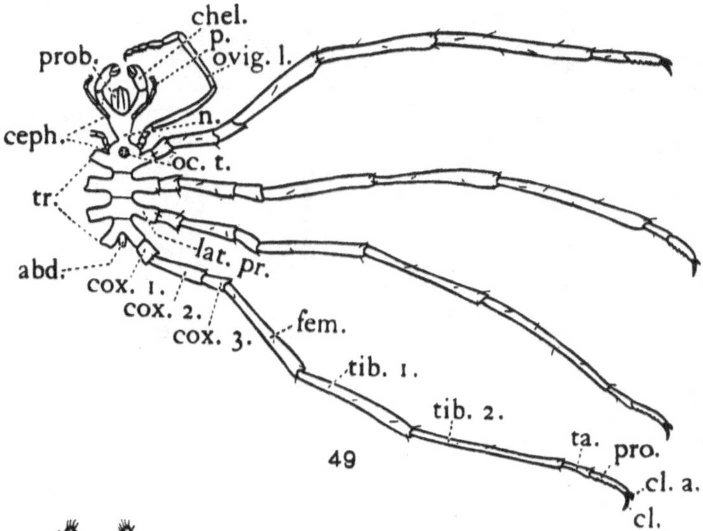

49

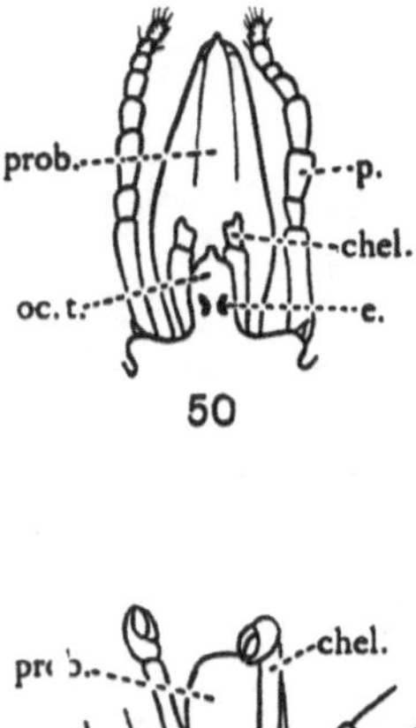

50

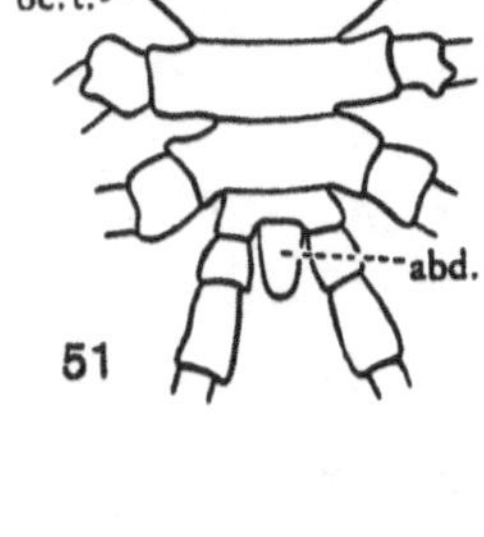

51

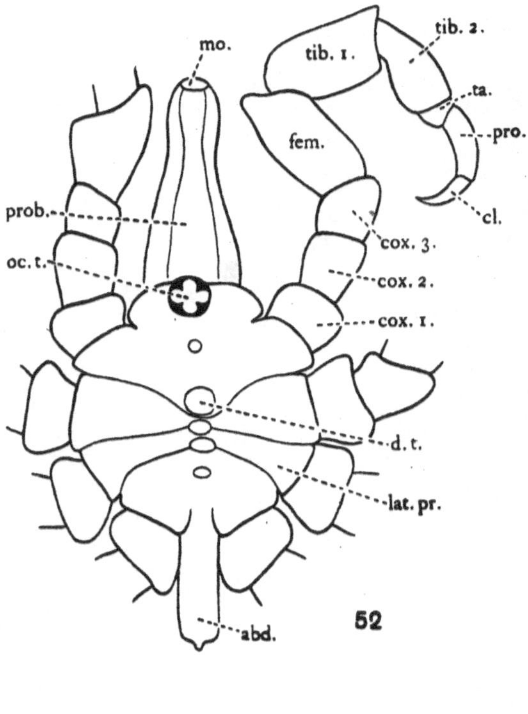

52

Ammothea. Palp with eight or nine joints. Ovigerous legs of nine joints, the last four small and armed with a few leaf-like spines. Walking legs thick, spiny.

A. echinata (Pl. XII, fig. 50). Bouvier, 1923, p. 55. Rostrum pear-shaped with acute apex, nearly twice as long as cephalon. Palp with eight joints. Spines on summit of tubercles. Body 2 mm. long, legs 6 mm. Littoral, amongst Campanularian Hydroids.

For other British species, see M. V. Lebour, 1945. "Notes on the Pycnogonida of Plymouth." *J. Mar. Biol. Ass. U.K.* XXVI, no. 2.

EXPLANATION OF PLATE XII

Pycnogonida. Figs. 49–52

Fig. 49. *Nymphon gracile.* A Pycnogonid with chelicerae, palps and clawed ovigerous legs. The appendages of one side only are shown. Dorsal view. ×6.

Fig. 50. *Ammothea echinata.* Proboscis and cephalon from the dorsal side. Palps and non-chelate chelicerae present. Four eyes fused to form two. ×25.

Fig. 51. *Phoxichilidium femoratum.* Dorsal view with legs cut short. No palps, but unarmed chelate chelicerae present. ×20.

Fig. 52. *Pycnogonum littorale.* Dorsal view. All the legs except one cut short. Neither palps nor chelicerae. ×7½.

abd.	abdomen	mo.	mouth
ceph.	cephalon	n.	neck
chel.	chelicera	oc.t.	ocular tubercle
cl.	claw	ovig.l.	ovigerous leg
cl.a.	auxiliary claw	p.	palp
cox. 1, 2, 3.	first, second and third coxa	pro.	propodite
		prob.	proboscis
d.t.	dorsal tubercle	ta.	tarsus
e.	eye	tib.1.	first tibia
fem.	femur	tib.2.	second tibia
lat.pr.	lateral process	tr.	trunk

MOLLUSCA

(Plates XIII–XVII)

AUTHORITIES

ALDER, J. and HANCOCK, A. 1845–55. *A Monograph of the British Nudibranchiate Mollusca.* Ray Soc. Publ.

ELIOT, SIR C. 1910. Supplement to Alder and Hancock. Ray Soc. Publ.

ELLIS, R. A. 1926. *British Snails.* Oxford.

FORBES, E. and HANLEY, S. 1853. *British Mollusca.* London.

JEFFREYS, J. G. 1862–9. *British Conchology.* London.

MATTHEWS, G. 1953. "A Key for use in the identification of British Chitons." *Proc. malacol. Soc. Lond.* XXIX, 6, 241–8.

WHITE, K. M. 1938. "The nomenclature of British Nudibranch Mollusca by Alder and Hancock and by Eliot correlated with that of British Marine Mollusca by Winckworth." *J. Conch.* XXI, no. 1, 14–19.

WINCKWORTH, R. 1922. Note on the British species of *Anomia. P. Malac. Soc.* XV, pt. 1, 32–34.

—— 1932. "The British Marine Mollusca." *J. Conch.* XIX, no. 7, 211–52.

The Molluscs form a large but well-defined phylum of soft-bodied, bilaterally symmetrical, unsegmented animals, whose skeleton, when present, takes the form of a morphologically external shell. In many the whole of the body is retractile within the shell, in some the shell is reduced so that it covers only a portion of the body. The form of the Mollusc varies considerably in the different classes, its very softness admitting of great plasticity, but three structures are characteristic of the group. These are the *mantle*, the *foot* and the *ctenidia* or "gills". The *mantle*, or pallium, is a dorsal projection of the body wall which protects the abdominal viscera and often the ctenidia, and which secretes the shell. The *foot* is a median ventral organ of locomotion, capable of considerable expansion, and in correlation with its shape used as a plough (Lamellibranchs), a flat creeping sole (Amphineura, Gastropods), or a swimming organ (Cephalopods). The *ctenidia* are vascular plate-like or much-branched outgrowths of the body, situated under the mantle (i.e. in the mantle cavity), and usually bearing filaments covered with ciliated epithelium for causing a strong current of water to pass over the surface.

The shell adapts itself to the shape of the animal, and consists of a horny substance allied to chitin and known as conchiolin, impregnated with calcium carbonate and other mineral salts. There are usually three layers, an external horny covering or *periostracum*, a middle *prismatic layer* composed of calcite, and an inner smooth *pearly layer*. Most Gastropods have one shell, Lamellibranchs a shell of two valves, and Amphineura of eight valves. In Nudibranchs the shell is absent.

A great many Molluscs are marine, only a few live in fresh water or on land. They are usually free-living, though a number are commensal with other organisms and some are fixed temporarily or permanently. They are adapted to every variety of feeding habit, ciliary feeding, herbivorous and carnivorous habits being common.

The phylum is divided into five classes: Lamellibranchiata, Amphineura, Gastropoda, Scaphopoda, Cephalopoda. The first, second and third have littoral representatives.

LAMELLIBRANCHIATA (BIVALVIA)

(Plates XIII and XIV)

Bivalve Molluscs. The symmetry is perfectly bilateral in most; there are a pair of shell valves, a pair of mantle folds and a pair of ctenidia. The foot is adapted for burrowing or ploughing through sand and mud. The right and left valves of the shell are closed by muscular action anteriorly and posteriorly, the *anterior* and *posterior adductor muscles* running transversely so that they unite the two mantle lobes, and through them the two shell valves secreted by the mantle. These and other muscles leave impressions on the shell which are important for identification. There is no distinct head in Lamellibranchs. The anterior end of the body bears the mouth and a pair of flat *labial palps*, but neither tentacles nor sense organs. Eyes and statocysts may occur but they are not situated in this region. Eyes may be present on the mantle rim and more rarely at the base of the gills. Statocysts lie within the foot.

The *mantle* usually covers the whole body and consists of a pair of thin lobes, attached dorsally to the body but often free ventrally. Posteriorly and ventrally right and left mantle lobes may be joined in one, two, three, or even four *sutures*, giving rise to apertures for the entrance and exit of the respiratory and food current and for the foot respectively. Of these the postero-dorsal is always the first to be formed; it is exhalant in action and may be drawn out to form a tube called the anal or *exhalant siphon*. Ventral to it, but still posterior (except in Erycinidae), the mantle edges may join to form a second tube, the branchial or *inhalant siphon*. In a few families a third and ventral suture occurs, so that the mantle is almost closed; this suture provides an aperture for the foot, called the *pedal aperture*, while in the razor shells and their allies a fourth aperture may occur which in some is both exhalant and inhalant. The inhalant and exhalant siphons may be of considerable length and may be free from one another, or partially or completely fused in different families. In a few families the gonads may extend into one mantle lobe (Anomiidae), or both mantle lobes (Mytilidae), but usually the mantle is very thin, though the edges are thickened. This edge provides the shell-secreting epithelium and the muscles are attached to the shell in a primitively orbicular line, the *pallial line*. When siphons are present on the mantle, their presence is usually indicated on the pallial line by the formation of an indentation called the *pallial sinus*, formed by the siphonal muscles. The mantle may bear tentacles, glands and eyes.

The chief *muscles* are the transverse *anterior* and *posterior adductors* of the shell, the *pallial muscles*, the *paired protractors* and two pairs of *retractors* of the foot, all of which leave their impressions on the shell. The impressions of the adductors and the pallial muscles are typical of certain families, and their shape and arrangement are used in classification. The anterior adductor lies antero-dorsal to the mouth, the posterior adductor antero-ventral to the anus. In some Lamellibranchs these two are equal and almost symmetrical, in others the anterior adductor is reduced in size and strength, with consequent increase in the posterior adductor muscle, and in several families the anterior adductor has disappeared altogether. An attempt to classify these as Dimyarian and Monomyarian Lamellibranchs, however, is not satisfactory, as other anatomical features do not agree with this arrangement.

In general, however, a bivalve with a posterior adductor muscle only, or with extremely unequal adductor muscles, is of sedentary habits.

The *shell* of Lamellibranchs is formed of two *valves*, each valve being secreted by the corresponding mantle lobe, starting from an initial point, the *umbo*. The valves are right and left in position, fastened together mid-dorsally by a *hinge*, but free ventrally, where the valves can either gape or be closed on the body of the Mollusc. In a few cases, the valves do not fit perfectly, so that a gap is left when the adductor muscle contracts, as in *Pholas*. Some Lamellibranchs have equal valves, in others one valve may be larger than the other, but this *inequivalve* condition is derived from the *equivalve* type. The shell may be *equilateral* or *inequilateral*; in the latter case the part anterior to the umbo is either smaller or larger than that posterior to the umbo. The shell measurements are length, breadth and depth, the length being measured along the antero-posterior axis, the breadth from hinge to gape (i.e. along the dorso-ventral axis), and the depth represents the greatest valve-to-valve measurement when the shell is lying on its side. The shell exhibits concentric lines of growth which encircle the umbo. The umbo is typically near the hinge, which may be weak, or very strong owing to the development of interlocking *hinge teeth* (*cardinals* below the umbo, *laterals* on either side). In most Molluscs an additional attachment along the dorsal side is given by a chitinous *elastic ligament* which may be visible externally or not. When the ligament relaxes, the shell opens. Some bivalves possess a strengthening element in the form of a calcareous projection extending inwards from the umbo and known as the *apophysis* (see Pholadidae, Myidae, etc.). In addition accessory shell plates may be developed (Pholadidae).

The *foot* is a median muscular expansion of the ventral body wall which may or may not contain portions of the viscera (gut, liver and gonads). In most bivalves it is flattened and is used for ploughing through the sand or mud; in fixed or burrowing forms it may become greatly reduced. Temporary fixation is brought about by means of a special gland in the foot known as the byssus gland. The *byssus* which this gland secretes may be thread-like and horny (*Mytilus*) or solid and calcified (*Anomia*).

The *ctenidia* or "gills" are paired outgrowths of the mantle between the foot and mantle flaps, extending in the more advanced forms from anterior to posterior ends of the body. A ctenidium consists of a vascular axis bearing an *inner* and *outer gill plate* with a row of parallel *filaments* on each plate. Each plate, and consequently each gill filament, except in primitive Lamellibranchs, is bent back or reflected on itself away from the ctenidial axis, to form apparently four gill surfaces shaped more or less like the letter W, but with a tall central stem where the axis lies at the junction of the gill plates. The ctenidium will therefore have two descending portions in the middle and two ascending portions forming the outer parts of the W. These are known as the *descending* and *ascending lamellae* or the direct and reflected lamellae. The descending lamellae are always distinct, but between the descending and ascending lamellae of a gill plate is an interlamellar space which in some Lamellibranchs forms an incubatory pouch for the eggs. Between the *filaments* of a lamella is the interfilamentar space. In order to strengthen the ctenidial apparatus, junctions unite the lamellae and the filaments across the interlamellar and interfilamentar spaces. These are known as the *interlamellar* and *interfilamentar junctions* respectively. In those Lamellibranchs in which the gill filaments are easily disarranged, the interfilamentar junctions are nothing more than patches of interlocking cilia, and the group is called *Filibranchia*. In others, known as *Eulamellibranchs*, the interfilamentar junctions are stronger tissue junctions and are usually vascular. In the *Pecten* family the interlamellar bridges are composed of connective tissue without blood vessels, but in most other Lamellibranchs these junctions are vascular. Finally the ascending lamella is rarely free at its ends. The ascending lamella of the outer gill plate may be attached for part of its length to the mantle, the ascending lamella of the inner gill plate may be attached to the base of the foot, or to its fellow of the opposite side. Any such attachment will shut off a portion of the mantle cavity as a *suprabranchial chamber*, which, since it is connected with the exhalant aperture, is itself exhalant, and drains the water out of the gill meshes, as well as the renal excretion from the kidney, and the genital products. It is probable that the "gills" in Lamellibranchs are more important for food sorting and collecting than for respiration, the respiratory function being carried on in the mantle.

1. Branchial filaments joined together by cilia. Usually with a byssus. Mantle open or with one suture, but no siphons. No pallial sinus and no true hinge teeth. **Filibranchia.*** 2
Branchial filaments with vascular connections. Mantle folds usually united by one or more sutures to form siphons. Byssus may or may not be present. Usually two adductor muscles. **Eulamellibranchia.** 4
2. Two equal adductor muscles. Pallial eyes present. Mantle open. Not attached by a byssus. **Arcidae**, p. 151
Anterior adductor muscle smaller than posterior. Attached by a thread-like byssus. Shell equivalve, inequilateral. **Mytilidae**, p. 152
One large posterior adductor muscle. Shell inequivalve. 3
3. Calcareous byssus present. Hole in right or lower valve. **Anomiidae**, p. 150
No byssus. Right valve entire. Processes or "ears" on either side of the hinge on each valve. **Pectinidae**, p. 153
4. Mantle open. Adductor muscles very unequal in size, or one only. 5
Mantle with one suture. 6
Mantle with two sutures or closed. 7
5. Shell inequivalved. One adductor muscle. No byssus. **Ostreidae**, p. 154
Shell equivalved, elongated. Small anterior and large posterior adductor muscles. Byssus present. **Pinnidae**, p. 154
6. Mantle fringed with twenty-five long and forty short processes. Small Molluscs associated with *Upogebia*. **Leptonidae**, p. 156
Mantle fringed with short processes only. Associated with Echinoderms. **Montacutidae**, p. 156
7. Mantle with two pallial sutures. 8
Mantle closed, with aperture for foot. 15
8. Siphons at opposite ends of animal. **Erycinidae**, p. 155
Siphons sessile or short. No pallial sinus. 9
Siphons long. Pallial sinus present. 11
9. Siphons sessile. 10
Siphons short. Shell with radiating ribs and crenate margin. Foot very large. **Cardiidae**, p. 157
10. Shell squarish with strong growth lines. Anterior adductor impression elliptical. **Diplodontidae**, p. 155
Shell roundish with numerous fine concentric ridges; anterior adductor impression remarkably elongated. **Lucinidae**, p. 155
11. Siphons joined. 12
Siphons separate. 13
12. Siphons fused throughout their length. Pallial sinus deep. Apophysis present in both valves. **Lutrariidae**, p. 166
Siphons free distally. Pallial sinus shallow. Hinge with three strong teeth. **Veneridae**, p. 158
13. Shell with a keel and a flat portion posterioriy. **Garidae**, p. 163

* See note on p. 153.

Shell nearly equilateral. Mantle edge thickened, but not double. External gill lamella directed dorsalwards. 14
Shell inequilateral. Mantle margin double. External gill lamella directed ventralwards. **Donacidae**, p. 161

14. Byssus present. Ligament of shell external. **Tellinidae**, p. 161
No byssus. A slight external ligament but a large internal one. **Scrobiculariidae**, p. 162

15. Apophysis present. 16
No apophysis. 17

16. Apophysis in left valve only. Striations on entire shell. **Myidae**, p. 167
Apophysis in both valves. Shell with striations confined to, or most marked on, anterior region. **Pholadidae**, p. 168

17. Shell reduced, not covering the vermiform animal. Wood borers. **Teredinidae**, p. 169
Shell not reduced, covering the animal. 18

18. Shell equivalve; burrowers in sand. 19
Shell inequivalve. Burrowers in limestone or free. 20

19. Shell elongated, open at both ends. **Solenidae**, p. 164
Shell shaped like an isosceles triangle. **Mactridae**, p. 165

20. Shell inequilateral, gaping, coarsely wrinkled. **Hiatellidae**, p. 167
Shell with left valve convex, right flat, smooth and pearly. **Pandoridae**, p. 170

FILIBRANCHIA

Gills of parallel reflected filaments, ventrally directed, and joined together by cilia. They easily become ragged by separation of the filaments. Byssus usually present.

ANOMIIDAE.

Saddle oysters. Inequivalved. Shell thin, rounded. One large posterior adductor muscle. Animal fixed by a calcareous byssus which passes through a hole in the flattened right valve.

1. Upper valve (left) with three muscle scars. Gill W-shaped with a short membranous flap, or fifth lamella, on the outer lamella. **Anomia (ephippium)**
Two muscle scars on the left valve. 2
One muscle scar on left valve, obviously formed by the union of two scars. **Monia squama**

2. Two scars separated on shell. Gill W-shaped but without the fifth lamella. **Monia patelliformis**
Two scars adjacent. Gill Λ-shaped, no ascending filaments. **Heteranomia squamula**

Anomia ephippium (Pl. XIII, figs. 55 and 56*a*). F. and H. 1853, II, 325; J. 1863, II, 30; W. 1922, p. 32. Shell white, glossy inside.

Umbo near margin. Two byssus muscle scars, one adductor scar. Gill with ascending lamella and a short flap, or fifth lamella. From low-water mark, very common, attached to shells, seaweeds, etc., and usually about 30 mm. across, but on some shores, such as the south Irish coast, it grows as large as 70 mm.

Monia. Shell generally roundish, with twenty to thirty ribs radiating from the umbo. Crenate margin. Yellowish white, sometimes with reddish brown. Two muscle scars.

M. patelliformis (*Anomia*) (Pl. XIII, fig. 56*c*). F. and H. 1853, II, 334; J. 1863, II, 34; W. 1922, p. 32. Shell flattened, often spotted or streaked. 35 × 33 mm. On hard ground and on shells.

M. squama (*Anomia striata*) (Pl. XIII, fig. 56*d*). F. and H. 1853, II, 336; J. 1863, II, 35; W. 1922, p. 33. Shell more convex, ribs fine and often raised into scales and prickles; colour, when present, streaky. The two muscle scars are united and green in colour. About 26 mm. Habitat as above.

Heteranomia squamula (*Anomia aculeata*) (Pl. XIII, fig. 56*b*). F. and H. 1853, II, 332; J. 1863, II, 32, as var. of *A. ephippium*; W. 1922, p. 33. Smaller. Striations regular and prickly. Two muscle scars distinct but contiguous. On all coasts.

ARCIDAE.

Shell oblong, rhomboidal or round; equivalve; inequilateral, ligament external. Numerous small teeth on the straight hinge. Muscle scars distinct.

1. Shell round. **Glycymeris (glycymeris)**
   Shell oblong or rhomboidal. **Arca (tetragona)**

Glycymeris glycymeris (*Pectunculus*). F. and H. 1853, II, 245; J. 1863, II, 166. Shell suborbicular, convex, nearly equal-sided. Margin smooth or slightly crenate. Umbones separated. Teeth oblique, six to twelve, small. Shell solid, opaque, with dark markings on a lighter ground. 50 mm. long. No byssus.

Arca tetragona. F. and H. 1853, II, 234; J. 1863, II, 180. Noah's ark shell. Shell boat-shaped, very tumid. Yellow or reddish brown colour. Forty to fifty hinge teeth. Umbones widely separated. Anterior side of shell short, upper corner almost a right angle. Lower edge obliquely rounded. 38 × 20 mm. Brown, cup-shaped deciduous byssus.

MYTILIDAE.

Mussels. Shell oblong, oval or rhomboidal; equivalve, inequilateral. Long narrow internal ligament. Mantle open anteriorly, joined posteriorly. Byssus present.

1. Shell rhomboidal with a row of striae radiating from the umbones. Mantle folded anteriorly. Small. **Musculus.** 2
   Shell oblong or oval, without striae. Mantle not folded. 3
2. Shell yellow or green, mottled. Hinge plate finely denticulated. Fifteen to eighteen striae anteriorly, twenty to twenty-five posteriorly. Muscle scars faint. **Musculus marmoratus**
   Shell yellowish brown, not mottled. Hinge plate coarsely denticulated. Ten to twelve striae set wide apart anteriorly, thirty to forty close set posteriorly. Muscle scars distinct. **M. discors**
3. Mantle fringed. Umbones at anterior extremity. **Mytilus edulis**
   Mantle not fringed. Umbones near anterior extremity. **Modiolus.** 4
4. Hinge plate denticulated. Umbones nearly terminal. Simple external shell projections present. **M. phaseolinus**
   No denticulations. Shell projections plain. **M. modiolus**
   No denticulations. Shell projections barbed. **M. barbatus**

Mytilus edulis (Pl. XIV, figs. 58 and 59). Edible mussel. F. and H. 1853, II, 170; J. 1863, II, 104. Shell triangular, rather thin and glossy, with concentric growth lines; dark blue with brownish yellow epidermis. Inside blue and white; muscle scars distinct, pallial line distinct with serrated edges. Mantle edge three lobed. Up to 100 mm. long. Gregarious. Abundant from high-water mark to a few fathoms, on rocky, stony or muddy bottoms.

Musculus. Shell rhomboidal, sculptured by striae which radiate from the umbones anteriorly and posteriorly, leaving the middle of the valves smooth. Umbones incurved.

M. marmoratus. F. and H. 1853, II, 198 (*Crenella marmorata*); J. 1863, II, 122 (*Modiolaria marmorata*). Shell oval, swollen, somewhat glossy. Colour yellow or green, with purple or reddish brown blotches and streaks. The muscle scars are difficult to make out. 20 mm. long, 10 mm. broad. Embedded in Tunicates or attached to old shells on lower shore. Common.

M. discors. F. and H. 1853, II, 195, as *Crenella*; J. 1863, II, 126 (*Modiolaria*). Shell oval, much broader than above, rather compressed and thin, not glossy, bright green. Muscle scars distinct. 12 mm. long, 9 mm. broad. Gregarious, at the roots of seaweeds between tide marks.

Modiolus. Differs from *Mytilus* in the position of the umbones, which are not terminal. The mantle has an entire margin. Shell with long filaments except in the adult *M. modiolus*. Body reddish brown.

M. phaseolinus. F. and H. 1853, II, 186 (*Modiola phaseolina*); J. 1863, II, 118 (*Mytilus*). Stiff beard-like processes on the shell. Hinge plate with fine denticulations. Rather common on rocky and hard ground. 16 mm. long, 10 mm. broad.

M. modiolus. Horse mussel. F. and H. 1853, II, 182 (*Modiola*); J. 1863, II, 111 (*Mytilus*). Long thorn-like filaments with plain edges in the young form. No denticulations on the hinge plate. 125 mm. long, 62 mm. broad. Commoner in the north.

M. barbatus (*Modiola*). F. and H. 1853, II, 190; J. 1863, II, 114, as *Mytilus*. Shell filaments with serrated or barbed ends. No denticulations on hinge plate. 46 mm. long, 26 mm. broad. Common.

PECTINIDAE.*

Scallops. Shell inequivalve. Umbones small, with triangular processes or "ears". Ligament strong. Shell often ribbed. Mantle edge thick and bearing ocelli. Active Molluscs, which swim by flapping the valves.

1. Both valves more or less convex, scaly. **Chlamys. 2**
   Upper valve nearly flat and concave towards the umbo. Lower convex. **Pecten (maximus)**
2. Ears nearly equal. Shell circular and equilateral. About twenty ribs. **C. opercularis**
   Ears unequal. More than twenty ribs. **3**
3. Shell oval, equilateral. Twenty to thirty ribs. Muscle scars distinct. **C. varia**
   Shell nearly equilateral. About seventy narrow ribs which are alternately large and small. Attached by lower valve. **C. distorta**

Pecten. Right valve convex, left valve flat.

P. maximus. Edible scallop. F. and H. 1853, II, 296; J. 1863, II, 73. Shell nearly circular, yellow or reddish brown. Ears equal. Thirty to thirty-five ocelli, greenish blue or purple, in two rows, on the edge of the mantle. 110 mm. long, 125 mm. broad. Common in shallow water, occasionally at low-water mark. Hermaphrodite.

* Some authors group the Pectinidae, Ostreidae, and Pinnidae as Pseudo-lamellibranchia.

Chlamys. Both valves of the shell convex.

C. opercularis (*Pecten*). Queen scallop. F. and H. 1853, II, 299; J. 1863, II, 59. Shell circular, rather thin, hinge straight. Colour very varied. Mantle cirri white, irregular, in two or three rows, outermost the longest. Thirty-five to forty ocelli with white centres surrounded by a black ring. 60 mm. long, 64 mm. broad. Common in shallow water; washed ashore occasionally. Hermaphrodite.

C. varia (*Pecten varius*). F. and H. 1853, II, 273; J. 1863, II, 53. Shell oval, nearly equilateral and broader ventrally than at hinge; colour varied. Ears unequal. Twenty-five to thirty smooth ribs. About thirty small black ocelli. 46 mm. long, 42 mm. broad. From low-water mark to shallow water. Temporary hermaphrodite.

C. distorta (*Pecten pusio*). F. and H. 1853, II, 278; J. 1863, II, 51. Shell irregular in shape, often greatly distorted, attached, colour variable. About seventy narrow, alternately long and short sharp ribs. A few large ocelli. 42 mm. long, 36 mm. wide. Common on rocky coasts to shallow water, with young specimens at low-water mark.

EULAMELLIBRANCHIA

Ostreidae.

Oysters. Shell irregular and solid, flat, inequivalve, without hinge teeth, composed of imbricating scales. Shell closes tightly, attached by left valve. No byssus.

Ostrea edulis. Edible oyster. F. and H. 1853, II, 307; J. 1863, II, 38. Young shell round but becomes very irregular as it matures; colour yellowish brown. Mantle has two rows of processes, the outer row irregularly double. 75 mm. long, 100 mm. broad. Common, usually gregarious in beds. From low-tide mark to shallow water.

Pinnidae.

Shell equivalve, an elongated triangle, more than twice as long as broad, thin and brittle, gaping widely, hinge toothless. Mantle open. Byssus present. Anterior adductor muscle small.

Pinna fragilis. F. and H. 1853, II, 255 (*P. pectinata*); J. 1863, II, 99 (*P. rudis*). The largest British Lamellibranch. Inequilateral. Mantle with two rows of cirri on the posterior margin and one on the anterior. Foot conical with a byssal groove; fine silky byssus. 270 mm. long, 127 mm. broad. Local, gregarious, from low-water mark to shallow water, buried in mud or fine gravel.

LUCINIDAE.

Shell roundish, inequilateral, ridged; ligament partly external. Inside white, with distinct muscle scars. Foot very long and narrow; mantle margin wavy; siphons sessile.

Lucinoma borealis. F. and H. 1853, II, 46 (*Lucina*); J. 1863, II, 242 (*Lucina*). Shell tumid and opaque, umbones nearly at right angles to dorsal margin, nearer the anterior end. Teeth: two cardinals, anterior double, and two plate-like laterals, in each valve. 35 mm. long, 32 mm. broad. In muddy sand and gravel from low-water mark.

DIPLODONTIDAE (UNGULINIDAE).

Shell squarish with strongly marked growth lines; ligament short, yellowish, external. Muscle scars large and well marked. Two cardinal teeth, one of which is bifid, in each valve. Siphons sessile, two pallial sutures.

Diplodonta rotundata. F. and H. 1853, II, 66; J. 1863, II, 254. Shell moderately glossy, white; umbones small, scarcely above the dorsal margin and not contiguous. Teeth: two cardinals and two plate-like laterals in each valve. Anterior adductor muscle scar elliptical, posterior oval. 23 mm. long, 25 mm. broad. Sandy mud.

ERYCINIDAE.

Shell thin, equivalve, inequilateral. Ligament internal, muscle scars large and distinct. Two pallial sutures with siphons at opposite ends of the animal. Viviparous. Fine byssus present.

1. Shell usually globular, white. Mantle folded anteriorly to form a bag-shaped inhalant tube. Exhalant tube visible. Foot protruding ventrally. **Kellia (suborbicularis)**
   Shell oval, white but usually tinged, often deeply, with red or brown. Inconspicuous exhalant tube. Anterior fold less complete. Foot long, protruding anteriorly. **Lasaea (rubra)**

Kellia suborbicularis. F. and H. 1853, II, 87; J. 1863, II, 225. Shell usually globular, rather thin and glossy; white; umbones incurved, nearly in centre of dorsal margin, but not touching. Left valve with a thick, erect cardinal tooth and both valves have two strong lateral teeth. Gills pale yellow and triangular. Foot tongue-shaped and bent, in the middle of the ventral margin. 10 mm. long,

12 mm. broad. Fine mud, in dead shells and under stones. Low water to deep water.

Lasaea rubra. F. and H. 1853, II, 94 (*Kellia*); J. 1863, II, 219. Shell oval with close-set concentric striae and distant lines of growth. White, tinged more or less densely with red or brown. Umbones broad, prominent and touching one another, towards the anterior end. Right valve has two triangular lateral teeth, left valve similar but with a minute cardinal and an erect tooth below the umbo. 3 mm. long, 2 mm. wide. Abundant in rock crevices, amongst weeds, etc., usually between tide marks.

LEPTONIDAE.

Shell roundish oval, equivalve, inequilateral. Mantle protruding, with long filaments and short cirri. One suture giving an exhalant siphon. Ligament internal.

Lepton squamosum. F. and H. 1853, II, 98; J. 1863, II, 194. Shell very flat and thin, semi-transparent and glossy, white; umbones small, a little nearer the anterior than the posterior end. Single cardinal and two lateral teeth in each valve. Mantle very large with a row of twenty-five long filaments and forty strong, close-set cirri, one of which, in front, is long and tentacle-like. Foot bluish white, transparent. 14 mm. long, 11 mm. broad. In sand and shelly gravel at the mouth of the burrows of *Upogebia*.

MONTACUTIDAE.

Shell oval or elongated, equivalve, inequilateral. Mantle protruding and bearing short cirri. One suture to form an exhalant siphon. Commensal with Echinoderms. Fine byssus present.

1. Shell triangular, oval, rather solid and opaque, very inequilateral. Margin of mantle with brilliant silvery indentations. With *Acrocnida brachiata*. **Mysella bidentata**
   Shell oval or triangular, thin and semi-transparent. Mantle with short cirri. 2
2. Mantle very large, edges reflected over shell. On *Leptosynapta*. **Entovalva (perrieri)**
   Mantle not reflected. **Montacuta.** 3
3. Shell oval, white, with faint ribs. With *Spatangus*. **M. substriata**
   Shell triangular, white, stained reddish brown, thin and glossy. With *Echinocardium*. **M. ferruginosa**

Entovalva (*Devonia*) *perrieri*. J. 1863, II, 216; 1869, V, 178 (*Montacuta donacina*). Foot very large and when expanded equals the rest of the animal in length. On the ventral side is a sucker which is used for attachment to *Leptosynapta inhaerens*, usually at the posterior end. 8 mm. long, 3 or 4 mm. broad.

Mysella bidentata (*Montacuta*). F. and H. 1853, II, 75; J. 1863, II, 208. Two strong lateral teeth in each valve, those of the right somewhat shorter. Margin of mantle protruding and fringed with brilliant close-set silvery indentations. Very short exhalant siphon; foot large, long, sprinkled with opaque white spots. About 3 mm. In sand, with *Acrocnida brachiata*.

Montacuta. Small bivalves. Mantle protruding with short cirri. Exhalant siphon short; foot large and tongue-shaped.

M. substriata. F. and H. 1853, II, 77; J. 1863, II, 205. Shell oblique, oval, rather convex, semi-transparent, with about twelve indistinct ribs, yellowish white. Umbones small, blunt, and nearly straight, projecting dorsally nearer anterior end. Two strong lateral teeth. Coarse byssal threads present. 3 mm. In sand with *Spatangus purpureus*.

M. ferruginosa. F. and H. 1853, II, 72; J. 1863, II, 210. Shell elongated, thin and glossy; white, but deeply stained with iron. Umbones oblique, blunt and about one-third of the distance from the anterior end. An erect narrow and oblique lateral tooth. 5 mm. long, 7 mm. broad. In sand in the anterior ambulacrum of *Echinocardium cordatum*.

CARDIIDAE.

Cockles. Shell coarse, solid and opaque, dull, equivalye, with radiating costae (ribs) and external ligament, margin crenate, hinge with two primary teeth. No pallial sinus, two pallial sutures. Gills much folded; foot cylindrical, geniculate, with byssus, used for creeping.

1. Shell triangular, compressed towards the margin. Twenty-four to twenty-eight ribs. Pallial scar slight. **Cardium edule**
   Shell round, not compressed towards the margin. Sculpture coarse, twenty-one or twenty-two ribs. Pallial scar broad and entire. **C. tuberculatum**

Cardium (*Cerastoderma*) *edule* (Pl. XIII, fig. 57). Edible cockle. F. and H. 1853, II, 15; J. 1863, II, 286. Shell triangular, convex, twenty-four to twenty-eight ribs, with crests, which are traversed by fine longitudinal ones. White or yellow, with some reddish brown. Inside of shell glossy with a slight pallial scar and very strong muscular scars. Mantle pale yellow, fringed. Siphons short, separated and conical, white or pale yellow, sometimes with pigment spots, inhalant the larger, with ten long and twenty to thirty short cirri, exhalant with a valve; sides of both with curly white filaments. 55 mm. long, 50 mm. broad. Common in sand and sandy mud.

Cardium (*Acanthocardia*) *tuberculatum*. F. and H. 1853, II, 11 (*C. rusticum*); J. 1863, II, 273. Shell roundish, very convex, globular and solid. Twenty-one or twenty-two ribs, the crests of which have tubercle-like processes. Yellowish brown with a tinge of red, inside rather glossy. Body reddish yellow, mantle thick and notched posteriorly, siphons fringed; foot very long, crimson. Low-water mark to shallow water. 70 mm. long, 65 mm. broad.

VENERIDAE.

Shell thick, equivalve, umbones incurved and turned towards the anterior side, ligament external. Teeth: three or four cardinals in each valve. Pallial scar well defined, with deep sinus. Siphons cylindrical. Foot thick, used for burrowing.

1. Shell very solid, opaque, with crenate inside margin. **Venus.** 3
   Shell variable, but with smooth inside margin. 2
2. Shell nearly orbicular. Siphons united throughout their whole length. **Dosinia.** 4
   Shell triangular, with wedge-shaped posterior portion. Siphons partly united or separate. **Venerupis (Paphia).** 5
3. Shell with a few strong concentric ribs. **Venus fasciata**
   Shell small, resembling a cockle, with forty to fifty fine ribs radiating from the umbones and crossed by twenty coarser concentric ones. **V. ovata**
   Shell with numerous regularly imbricated concentric ribs. **V. striatula**
4. Shell yellowish white, with rays and blotches of colour. **Dosinia exoleta**
   Shell white, without coloured markings. **D. lupinus**
5. Siphons separate, pigmented. Shell with coarse ribs bearing tubercles. **Paphia decussata**
   Siphons partly united. 6
6. Shell oblong, with flattened irregular, concentric ribs and reddish brown markings. **P. rhomboides**
   Shell oval, with fine concentric ribs. 7
7. Shell regularly oval. Free-living in sand or mud. **P. pullastra**
   Shell longer than broad. Inhabiting rock crevices. **P. saxatilis**

Dosinia. Shell with concentric flat laminae; inside margin smooth with large and well-developed scars. Siphons united throughout, apertures brown, inhalant with eight short cirri.

D. exoleta. F. and H. 1853, I, 428 (*Artemis*); J. 1863, II, 327 (*Venus*). Shell yellowish white with rays or blotches of red, purple or pink; inside polished and spotted, sometimes tinged purple. Foot very large and white with vertical bands of brilliant opaque white. 50 mm. long, 50 mm. broad, in sandy bays.

D. lupinus (*lincta*). F. and H. 1853, I, 431 (*Artemis*); J. 1863, II, 330 (*Venus lincta*). Shell white, tinged pale yellow; inside white. Siphons are extraordinarily extensile in that they can extend to three times the length of the shell. Foot large, wedge-shaped. 37 mm. long, 37 mm. broad. Sand and mud from low-water mark to shallow water.

Venus. Shell solid and opaque, inside margin crenate. Siphons short and almost united, apertures fringed.

V. ovata. F. and H. 1853, I, 419; J. 1863, II, 342. The oval *Venus*. Shell an oval triangle, rather solid; forty to fifty fine ribs radiate from the umbo and are crossed by about twenty coarser concentric ones. Yellowish, tinged pink or orange and blotched brown. Inside dull white with a lilac stain near the umbones. Siphons very short, equal, united except at their apices, bearing twenty alternately long and short cirri with a pigment spot at the base of each. 20 mm. long, 16 mm. broad. Common on sandy bottoms and amongst weed, from lower shore onwards.

V. fasciata. F. and H. 1853, I, 415; J. 1863, II, 334. The banded *Venus*. Shell a roundish triangle, somewhat compressed, very solid. The young have numerous concentric ridges which become fewer but stronger in the adult. Colour very varied, red, pink, yellow or brown, with rays, blotches or zigzag markings; inside dull white with purple towards hinge. Mantle edge muscular, fringed and bordered by a reddish line. Siphons short, separated at extremity, with red spots; inhalant aperture with twenty cirri, exhalant with fifteen. 20 mm. long, 23 mm. broad. Abundant in gravel and sand from shore to deep water.

V. striatula. F. and H. 1853, I, 408; J. 1863, II, 344 (*V. gallina*). Shell triangular, ribs concentric, irregular, crossed by fine longitudinal striae. Yellowish white with three reddish brown longitudinal rays, inside chalky white. Mantle edge with short filaments arranged in tufts and making it look sinuous. Siphons slender, almost united, equal in length, with fine red-brown cirri, those of the inhalant aperture being alternately long and short. Foot white and rather thick. 30 mm. long, 28 mm. broad. Abundant in sandy ground from low-water mark to deep water.

Venerupis (*Paphia*). Shell triangular, with wedge-shaped posterior portion. Inside margin smooth.

V. rhomboides (*Tapes virginea*). F. and H. 1853, I, 388; J. 1863, II, 352. Shell glossy, with flattened, irregular, concentric ribs. Yellowish white with reddish brown zigzag lines, inside glossy, frequently coloured yellow. Mantle edges denticulate anteriorly scalloped posteriorly with a few short, yellow filaments at both ends. Siphons united for about three-quarters of their length, pale yellow, tinged brown at division, and tipped with purple, equal; inhalant with fourteen pointed, alternately large and small cirri, exhalant curves up, with sixteen short white cirri. 48 mm. long, 40 mm. broad. In sand, etc., from low-water mark to deep water. Common. Filter feeder.

V. pullastra (*Tapes*). F. and H. 1853, I, 382; J. 1863, II, 355. Shell oval, not glossy as above, with numerous fine ribs which are much stronger posteriorly, intercepted by longitudinal ones; yellowish white, stained brown, yellow or purple. Umbones prominent. Inside white, finely tuberculated near hinge. Exhalant siphon a little longer and smaller than inhalant and curved up, both wrinkled and pigmented; single row of cirri on exhalant and double on inhalant aperture. 43 mm. long, 38 mm. broad. Common in muddy gravel or sand to shallow water.

V. saxatilis (*Tapes perforans*). F. and H. 1853, I, 383; J. 1863, II, 357, as *T. pullastra* var. Shell irregular in shape, but otherwise resembles *P. pullastra*. Inhabits rock crevices and holes of *Pholas* and *Hiatella*. Usually attached by a byssus.

V. decussata (*Tapes*). F. and H. 1853, I, 379; J. 1863, II, 359. Shell like *P. pullastra* but squarer and with coarser concentric and longitudinal ribs forming tubercles at their intersections. Colour more uniform. Siphons separate with speckles near apertures, inhalant with twelve long and twelve short, brownish cirri, exhalant with about twenty equal ones. 45 mm. long, 50 mm. broad. In muddy gravel and sand at low-water mark.

DONACIDAE.

Wedge shells. Shell equivalve, inequilateral, with anterior side larger than posterior. Hinge with conspicuous teeth; ligament internal. Mantle open for two-thirds of its length, margin double, with cirri and filaments on outer, and scalloped edges to inner portion. Two siphons separate, transparent, equal in length, with cirri.

Donax vittatus. F. and H. 1853, I, 332 (*anatinus*); J. 1863, II, 402. Shell oblong, compressed, opaque and glossy. Numerous fine longitudinal striae which radiate from the umbo but do not extend all over the shell. Brightly coloured, yellow, brown or purple, often with three rays of white. Inside highly polished, stained purple. Margin strongly notched. Body purplish red or yellow. Inhalant siphon with eight opaque white lines which terminate at the aperture in processes which close the tube. Foot very large. 30 mm. long, 16 mm. broad. In sand from low-water mark to shallow water.

TELLINIDAE.

Shell oval, triangular or long, slightly inequivalved and inequilateral; ligament prominent. Inside margin smooth; hinge teeth present. Pallial scar large and distinct. Mantle usually has tentacular fringe and bears two well-developed elongated and separated siphons. Two pallial sutures. (*Pharus*, p. 164, belongs here.)

1. Shell convex, somewhat triangular, opaque and glossy. Siphons nearly transparent, exhalant aperture plain, inhalant with four to six minute dentations. **Macoma balthica**
   Shell flattened. **Tellina. 2**
2. Shell roundish oval, strong, with coloured stripings. Siphons each with six tentacular cirri on apertures. **T. crassa**
   Shell oblong or suboval. Siphons with plain apertures. No radial stripings on shell. **3**
3. Shell oblong. Oblique striae on right valve. **T. fabula**
   Shell suboval, right valve slightly larger than left but with no striae. **T. tenuis**

Tellina tenuis. F. and H. 1853, I, 300; J. 1863, II, 379. Shell flattened, thin, very glossy; white or yellow to crimson. Inside glossy and coloured like the outside. Siphons white and nearly equal, the exhalant usually curved upwards. Foot large, broad and compressed. 25 mm. long, 16 mm. broad. Common in sand from low-water mark. Intertidal in Firth of Clyde. See p. 285.

T. crassa. F. and H. 1853, I, 288; J. 1863, II, 373. Shell somewhat glossy, solid, with numerous strong concentric ridges, white, tinged yellow, with red streaks. Mantle thick, with fimbriated edges; siphons funnel-shaped and long, with six tentacular cirri at apertures, exhalant three or four times longer than inhalant. Foot very large. 54 mm. long, 46 mm. broad. Common in sand from low-water mark.

T. fabula. F. and H. 1853, I, 302; J. 1863, II, 382. Shell white, with faint pink or yellow markings but no stripings. Pallial scar larger than in *T. tenuis*; muscle scars irregular. Mantle finely fringed; siphons white, nearly equal in length; gills large, pale brown. 20 mm. long, 13 mm. broad. Common in sand from low-water mark.

Macoma balthica. F. and H. 1853, I, 304 (*T. solidula*); J. 1863, II, 375 (*Tellina*). Shell opaque and glossy, with fine, close-set but irregular concentric striae. All colours. Inside glossy, large pallial scar extending on a line parallel with the margin. Mantle fringed with fine short white processes; siphons nearly transparent, exhalant curves up, plain, inhalant with four to six very small dentations. Foot white, very large. 26 mm. long, 22 mm. broad. Sandy gravel or mud. Common between tide marks, often estuarine.

SCROBICULARIIDAE (SEMELIDAE).

Shell oval or oblong, thin, white, gaping at the posterior ends, slightly striated. Umbones directed posteriorly; two small cardinal teeth on the right valve, and one on the left, laterals not always present. Siphons very long, separate, and each covered with a membranous sheath.

1. Shell large, to 50 mm. long. Yellow or greyish. **Scrobicularia plana**
   Shell small, to 20 mm. long, thin, white, usually glossy. **Abra.** 2
2. Shell almost equilateral; length and breadth nearly equal. **A. tenuis**
   Shell inequilateral. 3
3. Shell elongated, twice as long as broad. Siphonal aperture with cirri. **A. prismatica**
   Shell oval, slightly longer than broad. Siphonal aperture plain. **A. alba**

Scrobicularia plana (*S. piperata*). F. and H. 1853, I, 326; J. 1863, II, 444. Varied in shape, but length always greater than breadth,

from triangular to suborbicular. Shell opaque, fragile, greyish white, often stained. Inside polished white, pallial scar distinct, with large sinus, muscle scars rather deep; no lateral teeth. Mantle irregularly denticulated, bordered by a dark line, siphons yellow, not frilled. Foot large and white. 51 mm. long, 38 mm. broad. Common in sand and mud and likes an admixture of fresh water, from low-water mark to shallow water.

Abra tenuis. F. and H. 1853, I, 323 (*Syndosmya*); J. 1863, II, 442 (*Scrobicularia*). Shell almost equilateral, breadth and length nearly equal; thin, opaque, not glossy but sometimes iridescent; greyish white. Mantle edge fringed, siphons cylindrical and narrow, the exhalant capable of great extension, the other usually shorter and curved. 8 mm. long, 10 mm. broad. Estuaries and brackish water, in mud, from low-water mark.

A. prismatica. F. and H. 1853, I, 321 (*Syndosmya*); J. 1863, II, 435 (*Scrobicularia*). Shell elongated, inequilateral, with anterior portion nearly twice as long as posterior, very thin and fragile, highly polished, white. Mantle finely fringed; siphons long and slender, upper usually longer, and lower thicker, both papillated, one or both with five to seven short cirri. 10 mm. long, 20 mm. broad. Sandy mud, not common.

A. alba. F. and H. 1853, I, 316 (*Syndosmya*); J. 1863, II, 438 (*Scrobicularia*). Shell oval, rather convex, somewhat inequilateral, thin, opaque, polished, snow white. Mantle fringed with small white papillae; siphons cylindrical, very elastic, covered with a light brown epidermis, apertures plain. 12 mm. long, 20 mm. broad. Estuaries, creeks, etc., in mud from low-water mark.

GARIDAE (ASAPHIDAE).

Shell oblong, nearly equilateral, gaping posteriorly. Two cardinal teeth in each valve; ligament external. Siphons separate, long, with six cirri on inhalant aperture; mantle fringed. Pallial scar well defined, with large sinus.

Gari. Shell coloured, with radiating bands of red or brown alternating with yellow. Sunset shells.

1. Shell sharply angulated posteriorly, with a keel from umbo to angle. **Gari fervensis**
   Shell blunt posteriorly. **G. depressa**

G. fervensis (*Psammobia ferroensis*). F. and H. 1853, I, 274;

J. 1863, II, 396. Shell compressed, rather solid, right valve usually larger than and overlaps left, sculptured by numerous fine ridges which are sharper anteriorly, intersected by numerous slight ones with a few ribs posteriorly, keeled. Pink, with coloured rays; inside shiny, partly iridescent and purple stained. Exhalant siphon curves upwards. Foot very large, long, and bevelled to a sharp edge. 45 mm. long, 20 mm. broad. Rather common in sandy ground.

G. depressa. F. and H. 1853, I, 271 (*Psammobia vespertina*); J. 1863, II, 398. Shell oval or rhomboidal, compressed, rather solid and opaque, somewhat glossy; sculpture as in *fervensis* but longitudinal ribs not persistent in the adult. Yellowish white with rays of purple brown and speckled with white. Exhalant siphonal aperture irregular. Foot very large and rather thick. 32 mm. long, 56 mm. broad. On all coasts, in muddy sand.

SOLENIDAE.

Razor shells. Shell long, narrow, gaping at each end; ligament external. Pallial sinus deep. Two large siphons, partly enclosed in a common sheath, incompletely retractile, fimbriated. Foot very large and tongue-shaped. No byssus. Burrowers in sand.

1.	Shell rectangular with hinge at anterior end.	2
	Shell rectangular, hinge nearly central (i.e. shell almost equilateral).	**Pharus (legumen)**
2.	Shell straight, margins straight.	**Solen marginatus**
	Shell nearly straight or curved.	**Ensis.** 3
3.	Shell curved.	4
	Shell nearly straight. Margins sharply truncated.	**E. siliqua**
4.	Foot cream.	**E. arcuatus**
	Foot reddish brown.	**E. ensis**

Pharus legumen (*Ceratisolen*). F. and H. 1853, I, 256; J. 1865, III, 10. Shell pod-shaped, glossy. Body yellowish white, mantle reddish; siphons rather long, separate, and diverging for most of their length, with cirri on apertures. Foot reddish purple. 102 mm. long, 23 mm. broad. Local, at very low water. See p. 161

Solen marginatus. F. and H. 1853, I, 242; J. 1865, III, 20 (*vagina*). Shell glossy, yellowish brown with dark streaks, margins straight with a deep constriction anteriorly; inside white. Body pale brown; siphons wrinkled, each encircled by a ring of short cirri, inhalant aperture scalloped. Foot dull reddish. 127 mm. long, 20 mm. broad. On sandy shores.

Ensis ensis (*Solen*). F. and H. 1853, I, 250; J. 1865, III, 16. Shell curved, nearly equal in diameter throughout, opaque, glossy, yellowish white with brown streaks. Pallial scar distinct, with shallow sinus, muscle scars of unequal size; body drab, siphons very short, ringed with two rows of cirri. 95 mm. long, 13 mm. broad. Sandy bays. Common.

E. arcuatus. Holme, 1951, *J. mar. biol. Assoc.* XXIX, 639. 150 mm.

E. siliqua (*Solen*). F. and H. 1853, I, 246; J. 1865, III, 18. Resembles *E. ensis* but straighter and more solid, with margins truncated. Body similar but foot yellowish white. 200 mm. long, 25 mm. broad. Common on all sandy shores.

MACTRIDAE.

Trough shells. Shell subtriangular, slightly gaping at the posterior end. Pallial sinus short and tongue-shaped; two cardinal teeth in each valve, four laterals in the right, two in the left. Siphons united, sheathed, of equal length, with cirri. Sand burrowers.

1. Shell rayed from umbo to gape. Lateral teeth and sockets smooth. **Mactra corallina**
   Shell not rayed. Lateral teeth and sockets perpendicularly striated. **Spisula. 2**
2. Shell almost equilateral. **3**
   Shell inequilateral, with prolonged posterior angle. **Spisula subtruncata**
3. Shell heavy, concentrically grooved. To 40 mm. long. **S. solida**
   Shell not very heavy, smooth or nearly so. Up to 30 mm. long. **S. elliptica**

Mactra corallina (*M. stultorum*). F. and H. 1853, I, 362; J. 1863, II, 422. Shell convex, thin and glossy, yellowish white, variegated by longitudinal rays of light reddish brown; growth lines white. Umbones swollen and prominent, curved anteriorly to form a sharp beak. Mantle thickened at edge and fringed with short white filaments. Siphons short, 12 mm. long, covered with a brownish epidermis, apertures with pale red cirri, inhalant with twelve to sixteen, exhalant with sixteen to twenty. 50 mm. long, 38 mm. broad. Not common.

Spisula subtruncata (*Mactra*). F. and H. 1853, I, 358; J. 1863, II, 419. Shell very convex and angular; strongly striated, umbones prominent. Mantle edge crenate; siphons with a white or reddish epidermis, orifices with a double row of white, yellow or pale red cirri. 20 mm. long, 28 mm. broad. Gravel and sand from low-water mark.

S. solida (*Mactra*). F. and H. 1853, I, 351; J. 1863, II, 415. Shell convex, solid and thick, somewhat glossy, numerous irregular striae. Pale yellowish white. Umbones tumid, projecting as blunt beaks. Mantle thickened and fringed but less so on posterior end than elsewhere. Siphons 18 mm. long, each often with a bright yellow line. Foot large, fleshy and white. 44 mm. long, 35 mm. broad. Gravel and sand, common. Low-water mark to a few fathoms.

S. elliptica (*Mactra*). F. and H. 1853, I, 356; J. 1863, II, 417, as *M. solida* var. As *solida* but umbones not prominent and shell smaller, thinner and proportionately longer. 32 mm. long, 20 mm. broad.

LUTRARIIDAE.

Shell elongated, inequilateral, gaping widely at both ends. Ligament partly external. Deep pallial sinus. A spoon-shaped apophysis present for the attachment of the ligament in each valve. Large but not beautiful shells. Siphons long, united almost to their ends. Foot large, compressed. Burrowers in mud.

1. Dorsal margin convex. Shell elliptical. **Lutraria lutraria**
   Dorsal margin concave. Shell almost oblong. **L. magna**

Lutraria lutraria (*L. elliptica*) (Pl. XIII, figs. 53 and 54). F. and H. 1853, I, 370; J. 1863, II, 428. Shell somewhat solid and glossy; irregularly striated, yellowish white. Muscle scars deep, anterior pear-shaped, posterior triangularly oval. Mantle with fine white filaments. Siphons long, sheath with purple-brown blotches and lines, covered by a colourless epidermis; apertures surrounded by white cirri spotted with brown, those of inhalant longer. 140 mm. long, 76 mm. broad. Common in soft sand or mud from low-water mark.

L. magna (*L. oblonga*). F. and H. 1853, I, 374; J. 1863, II, 430. Shell smaller and longer, upturned at its posterior end. Siphons unequal, inhalant much the longer, with thirty to forty fine cirri of two lengths, covered by brown epidermis. 100 mm. long, 50 mm.

broad. Common in soft sand or mud from very low-tide mark.

MYIDAE.

Gapers. Shell oval, slightly inequivalve, gaping. Left valve has spoon-shaped projection (apophysis) for the internal ligament. Pallial sinus deep; mantle largely closed. Siphons well developed, united for most of their length and fringed, enclosed in a sheath. Foot compressed and reduced.

1. Shell truncated posteriorly, about 70 mm. long. **Mya truncata**
   Shell bluntly pointed posteriorly, to 100 or more mm. long. **M. arenaria**

Mya truncata. F. and H. 1853, I, 163; J. 1865, III, 66. Shell rather convex, margin truncated posteriorly, with irregular coarse striae and strong growth lines; greyish white, inside chalky white with a few minute pearls in pallial line. Siphons very long, filaments alternately long and short with a brown spot at base. 65 mm. long, 50 mm. broad. Common in muddy gravel and sand to some depths.

M. arenaria. F. and H. 1853, I, 168; J. 1865, III, 64. Shell oblong, inequivalve, inequilateral; sculpture as in *M. truncata*. Ashy grey, often stained brown, inside chalky white. Apertures of siphons tinged red and fringed with different sized tentacles. 100 mm. long, 65 mm. broad.

HIATELLIDAE.

Shell rhomboidal, inequilateral, inequivalved. Gaping. Ligament external. Siphons elongated, with chitinous sheath, more or less united; cirri around apertures. Mantle extensively closed, with a small pedal orifice. Foot small. Byssus present.

1. Margin of shell rounded, straight posteriorly. Burrowing Molluscs, free when young. **Hiatella gallicana**
   Margin of shell angulated. Mollusc attached by byssus and does not burrow. **H. arctica**

Hiatella gallicana (*Saxicava rugosa*). F. and H. 1853, I, 146; J. 1865, III, 81. Shell with coarse, widely separated wrinkles, slightly ridged, whitish. Mantle thick and wrinkled; siphons very extensile, enclosed within a sheath to a very short distance from extremities, apertures pink with sixteen to twenty short white cirri in a double row. 35 mm. long, 25 mm. broad. Common. Bores into limestone, chalk or New Red Sandstone.

H. arctica (*Saxicava*). F. and H. 1853, I, 141; J. 1865, III, 82, as *S. rugosa* var. Shell oblong, inequivalve, inequilateral. Margins angular. Other features as in *H. gallicana*. 20 × 10 mm. Attached by a byssus to rocks or shells and does not burrow.

PHOLADIDAE.

Piddocks. Shell equivalve, inequilateral, gaping anteriorly and usually posteriorly; no ligament or hinge, apophyses in both valves. Largely closed mantle, very long united siphons. Foot short and truncated. No byssus. Dorsal region above ligament has accessory shelly plates or shields.

1. Shell with one accessory dorsal plate. 2
   Shell with two accessory dorsal plates. **Pholadidea (loscombiana)**
   Shell with four accessory dorsal plates. **Pholas (dactylus)**
2. Dorsal plate very small. Valves divided into an anterior and a posterior portion by a ridge radiating from the umbo. **Zirfaea (crispata)**
   Dorsal plate fairly large. Valves not ridged. **Barnea.** 3
3. Valves beaked. Siphonal apertures scalloped. **B. parva**
   Valves not beaked. Inhalant aperture with about twelve papillae, exhalant either plain, or with a few papillae. **B. candida**

Pholas dactylus (Pl. XIV, fig. 60). Common piddock. F. and H. 1853, I, 108; J. 1865, III, 104. Shell elongate, moderately solid, with forty to fifty longitudinal rows of prickles formed by the intersection of ribs and striae in anterior part of the shell, and gradually thinning out posteriorly; apophysis spoon-shaped. Four accessory dorsal shields, two anterior (often cracked), one in the middle between these, and one posterior which is long and narrow. Siphons papillate, exhalant shorter, aperture plain or with a few cirri, inhalant with about twelve fringed tentacles. 128 mm. long, 50 mm. broad. Burrows into slate, shale, New Red Sandstone, chalk, marl, peat and submerged wood.

Barnea candida (*Pholas*). White piddock. F. and H. 1853, I, 117; J. 1865, III, 107. Shell elongated, tumid and thin, with twenty-five to thirty longitudinal rows of sharp prickles except at the ends. Chalky white. One dorsal shield. Body dirty white, tinged brown.

Siphons more slender and elongated than in *P. dactylus*. Foot small, oval. 75 mm. long, 25 mm. broad. Burrows as *P. dactylus*.

B. parva (*Pholas*). F. and H. 1853, I, 111; J. 1865, III, 109. Shell oblong, rather compressed and solid, sculpture of very numerous transverse rows of imbricating and flexuous ridges, puckered on the crests, due to the intersection of slight, less numerous, longitudinal ones. White, often stained brick red. One dorsal shield. Body milk-white, siphons striped brown and white inside, apertures scalloped only. 46 mm. long, 20 mm. broad. Burrows.

Zirfaea crispata (*Pholas*). F. and H. 1853, I, 114; J. 1865, III, 112. Shell oval, convex, with a depression in the middle, solid, coarse and rugged. About twenty rows of prickles anterior to the furrow. Dorsal shield small. Body reddish brown; siphons long and bearing branched papillae. Sheath velvety. 70 mm. long, 40 mm. broad. Burrows.

Pholadidea loscombiana (*P. papyracea*). F. and H. 1853, I, 123; J. 1865, III, 116. Shell convex, thin, delicate, with numerous transverse scalloped ridges on the upper half. Hinge plate broad with two dorsal shields, often united. Body bluish white; siphons long, aperture of larger with 20 white cirri; sheath pale reddish brown. 20 mm. long, 40 mm. broad. Burrows.

TEREDINIDAE.

Ship-worms (Pl. XIV, fig. 61). Shell globular, covering only a part of the long, vermiform animal. No true hinge. Ligament obsolete. Curved apophysis in each valve. Siphons very long, united, but free at the tips, with two posterior calcareous "pallets". Foot reduced, sucker-shaped. Wood borers. Inhabit a calcareous tube within wood.

1. Shell white. Tube not chambered at the narrow end. Pallets small, forked at the base. **Teredo navalis**
   Shell greenish brown. Tube with septa across the narrow end. Pallets not forked at the base. **T. norvegica**

Teredo norvegica. F. and H. 1853, I, 66; J. 1865, III, 168. Shell convex, solid and opaque, sculpture lines in three directions. Colour whitish, stained brown. Pallets large oval blades with cylindrical stalks. Thick tubular sheath, divided by a ridge near the opening which separates siphons. Body white or pale grey.

Apertures with fine cirri. Valves 15 mm. long, 16 mm. broad. Pallets 20 × 8 mm. Sheath 300 × 20 mm. Young develop in a brood pouch to veliger stage. Local.

T. navalis. F. and H. 1853, I, 74; J. 1865, III, 171. Shell smaller, thinner and more finely sculptured; white. Sheath less solid. Valves 8 mm. long, 8 mm. broad. Pallets 5 × 2·5 mm. Sheath 150 × 8 mm. Eggs discharged into the sea.

PANDORIDAE.

Hinge with a primary tooth in each valve. Ligament internal. Ascending lamella of ctenidium reduced. Muscle impressions weak. Pallial sinus small. Siphons united below, with short, fringed diverging orifices.

Pandora albida. F. and H. 1853, I, 207 (*rostrata*); J. 1865, III, 24 (*inaequivalvis*). Shell elongated, attenuated posteriorly. 28 × 12 mm. The only littoral species. Not common.

EXPLANATION OF PLATE XIII

Lamellibranchiata (i), Figs. 53–57

Fig. 53. *Lutraria lutraria.* Left valve of the shell to show general structure and muscle scars. The shell is equivalve and inequilateral, has a spoon-shaped apophysis for the ligament and a single hinge tooth. Note the large pallial sinus and forward position of the protractor muscle of the foot. × ½.

Fig. 54. *Lutraria lutraria.* Right valve of the shell removed to show the body of the Mollusc lying in the left valve, the organs being viewed through the thin mantle, whose rim is thickened along the pallial line of attachment to the shell. The siphons are united throughout, with two sutures. There is a large pallial sinus. × ½.

Fig. 55. *Anomia ephippium.* Under view of a specimen detached from rock. The shell is inequivalve, the upper and larger convex valve being the left, the under and smaller flat valve the right. A small ligament connects the two. The right valve is perforated for the passage of a calcareous byssus. × ½.

Fig. 56. Left valve of four members of the Anomiidae to show muscle scars. After Winckworth. *a*, *Anomia ephippium*. *b*, *Heteranomia squamula*. *c*, *Monia patelliformis*. *d*, *Monia squama*. The larger scar is that of the byssus muscle, the smaller that of the single adductor muscle. In *Anomia*, the byssus muscle scar is divided into two parts.

XIII. MOLLUSCA (LAMELLIBRANCHIATA)

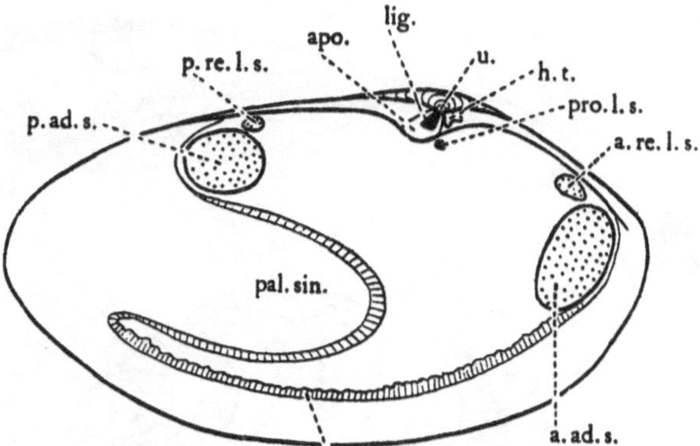

53

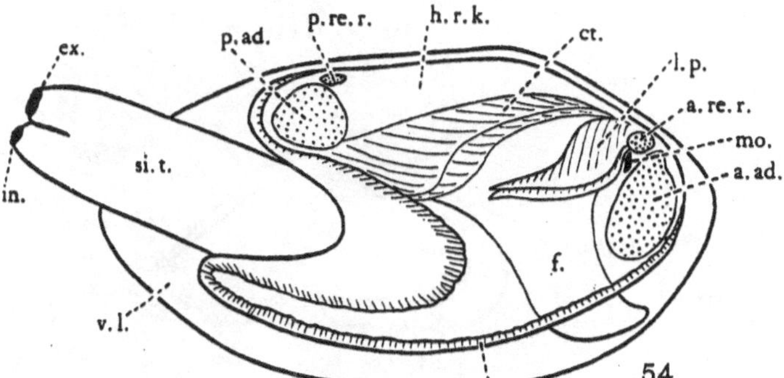

54

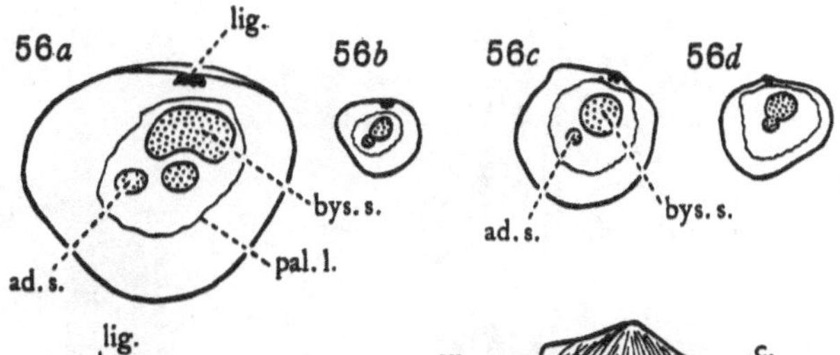

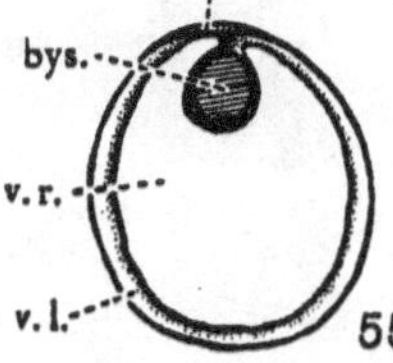

55

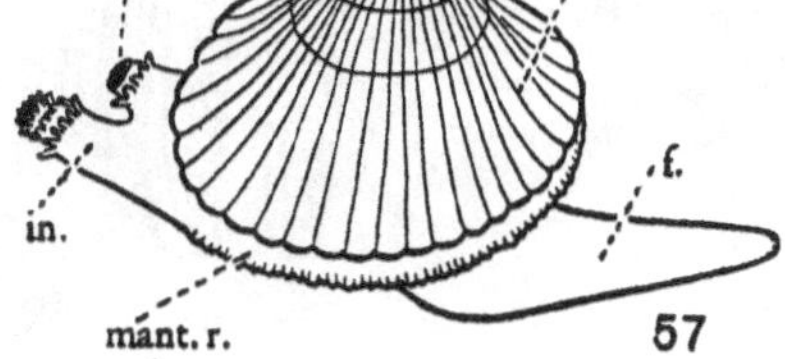

57

XIV. MOLLUSCA (LAMELLIBRANCHIATA)

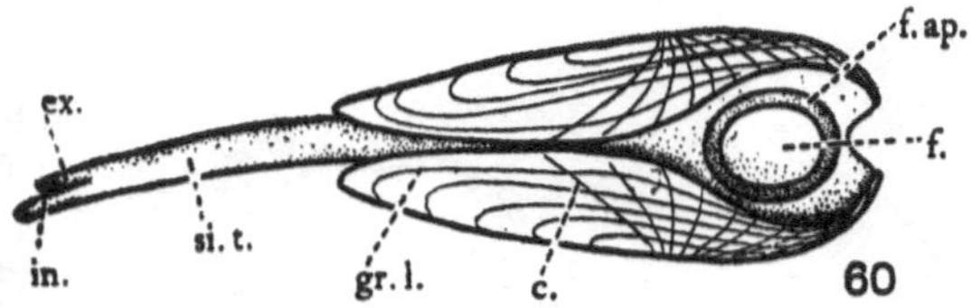

60

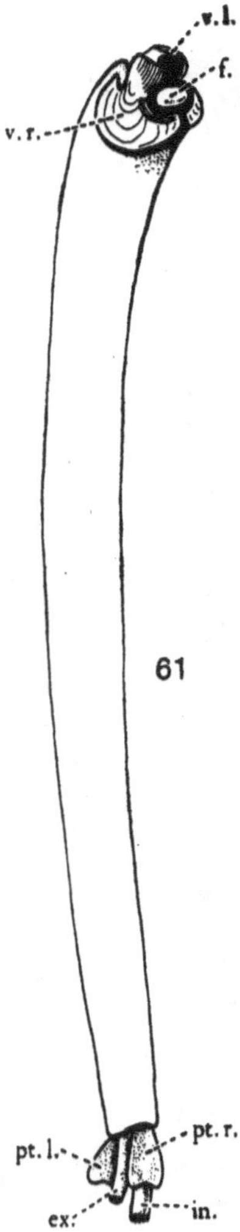

61

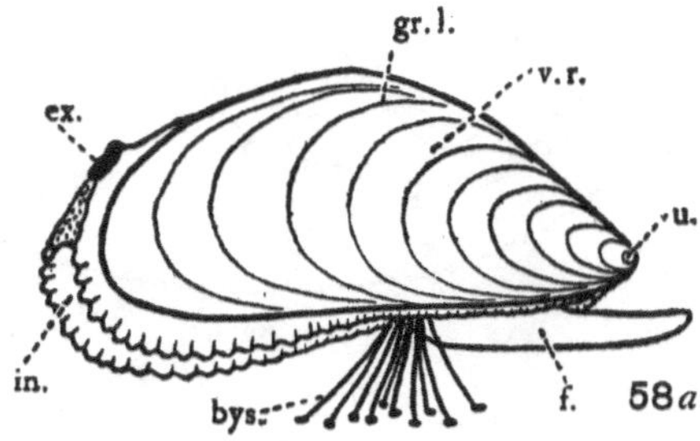

58*a*

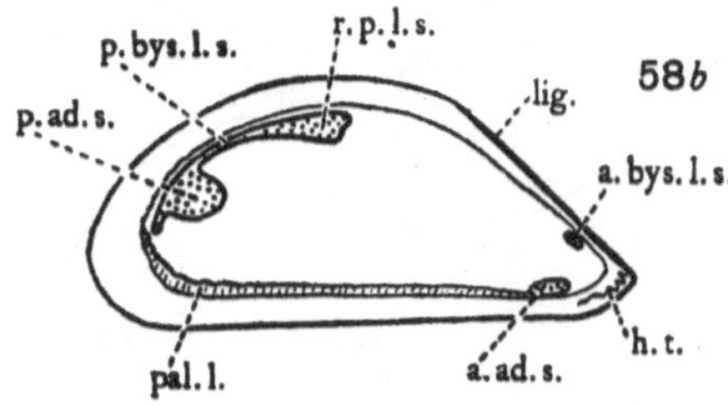

58*b*

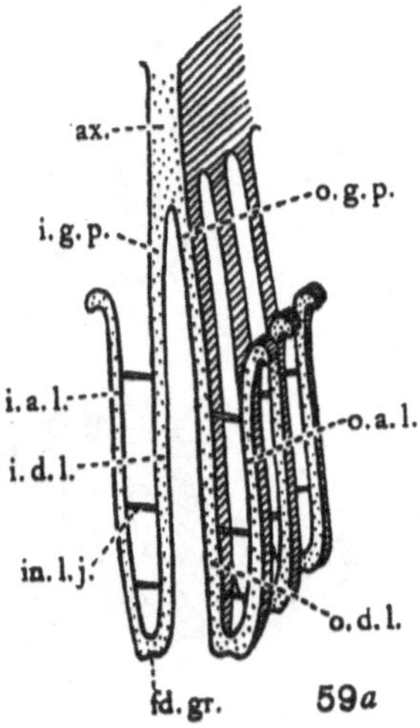

59*a*

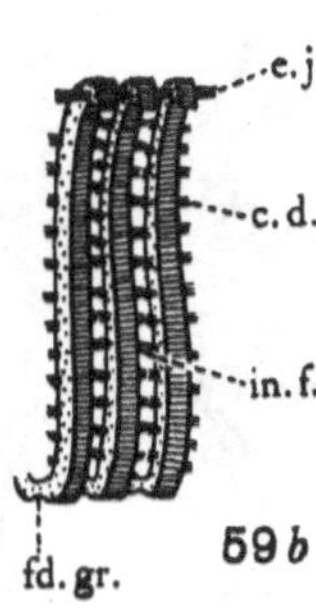

59*b*

(*facing p.* 171)

Fig. 57. *Cardium edule.* Right lateral view, with foot, mantle and siphons protruding. The shell has radiating ribs or costae, and a crenate margin. There are two sutures and short siphons. The foot is bent (geniculate). × 1.

a.ad.	anterior adductor muscle	l.p.	labial palp
		lig.	ligament
a.ad.s.	anterior adductor muscle scar	mant.r.	right fold of mantle
		mo.	mouth
a.re.l.s.	left anterior retractor pedis muscle scar	p.ad.	posterior adductor muscle
a.re.r.	right anterior retractor pedis muscle	p.ad.s.	posterior adductor muscle scar
ad.s.	adductor muscle scar	p.re.l.s.	left posterior retractor pedis muscle scar
apo.	apophysis		
bys.	byssus	p.re.r.	right posterior retractor pedis muscle
bys.s.	byssal muscle scar		
c.	costa (rib)	pal.l.	pallial line
ct.	ctenidium	pal.sin.	pallial sinus
ex.	exhalant aperture	pro.l.s.	left protractor pedis muscle scar
f.	foot		
h.r.k.	position of heart, rectum and kidney	si.t.	siphonal tube
		u.	umbo
h.t.	hinge tooth	v.l.	left valve of shell
in.	inhalant aperture	v.r.	right valve of shell

EXPLANATION OF PLATE XIV

Lamellibranchiata (ii), Figs. 58–61

Fig. 58. *Mytilus edulis.* *a*, The animal seen from the right side, with foot, thread-like byssus and mantle protruding from the shell. Note the almost terminal, anterior position of the umbo. There is one suture, and the exhalant aperture is completely separated from the widely open inhalant. × ½. *b*, The inside of the left valve to show muscle scars. × ½.

Fig. 59. *Mytilus edulis.* *a*, A portion of the right ctenidium to show the ctenidial axis forking into outer and inner gill plates, each of which is bent back upon itself to form descending and ascending lamellae, linked by vascular interlamellar junctions. On the right of the figure three filaments of the outer gill plate are shown, but the ciliated interfilamentar junctions have not been inserted. × 20. *b*, Surface view of three filaments of the ascending lamella to show a few of the ciliated disks locking adjacent filaments together, and the single edge junction joining the hooked ends of the filaments. × 20.

Fig. 60. *Pholas dactylus.* Ventral view. Shell equivalve, inequilateral, gaping. Three sutures, viz. those forming inhalant and exhalant apertures, and the hole for the foot. The siphons have been slightly twisted in order to show both. × 1.

Fig. 61. *Teredo navalis.* Half-lateral view of specimen extracted from its tube. The pallets have been somewhat displaced. × 1½.

a.ad.s.	anterior adductor muscle scar
a.bys.l.s.	left anterior byssal retractor muscle scar
ax.	ctenidial axis
bys	byssus
c.	costa (rib)
c.d.	ciliated disc
e.j.	edge junction (vascular)
ex.	exhalant aperture
f.	foot
f.ap.	aperture in mantle for foot
fd.gr.	food groove
gr.l.	growth line of shell
h.t.	hinge tooth
i.a.l.	inner ascending lamella
i.d.l.	inner descending lamella
i.g.p.	inner gill plate
in.	inhalant aperture
in.f.j.	interfilamentar junction
in.l.j.	interlamellar junction
lig.	ligament
o.a.l.	outer ascending lamella
o.d.l.	outer descending lamella
o.g.p.	outer gill plate
p.ad.s.	posterior adductor muscle scar
p.bys.l.s.	left posterior byssal retractor muscle scar
pal.l.	pallial line
pt.l.	left pallet
pt.r.	right pallet
r.p.l.s.	left retractor pedis muscle scar
si.t.	siphonal tube
u.	umbo
v.l.	left valve of shell
v.r.	right valve of shell

AMPHINEURA

(Plate XV, figs. 62 and 63)

The littoral representatives of this small class are the eight-valved Chitons. Chitons are bilaterally symmetrical Molluscs, with mouth and anus at opposite ends of the flat oval body, whose ventral surface is the broad flat foot. The mantle is dorsal in position and has spicules embedded in it; it forms a *girdle* round the edge of the body surrounding the eight transverse calcareous *shell plates* or *valves* situated mid-dorsally. The pallial cavity is a shallow elliptical space between the mantle and foot, in this cavity lie the numerous gills—typical ctenidia.

The *head* is distinct and is separated from the body by a short neck. It bears a pair of labial palps and a snout-like buccal region, but no eyes or tentacles. A radula or rasping tooth ribbon is present. The eight shell valves are articulated in a longitudinal series, with sufficient overlapping to enable the animal to roll itself up into a ball by curvature of the foot. Each shell plate is formed of two layers, an upper tegumentum and a lower articulamentum, and since the former may be the smaller of the two the articulamentum may project laterally and anteriorly. In the overlapping of the plates the anterior projection is covered by the plate in front. A mesial thickening of a plate is called a *beak*. Chitons feed on seaweeds and are slow moving Molluscs.

POLYPLACOPHORA (LORICATA)

1. Girdle with tufts of bristles. **Cryptoplacidae**, p. 174
   Girdle granular or smooth. 2
2. Notches in margin of shell valves, beaks prominent. Girdle rather broad. 3
   Margin of shell valves slightly wavy but not notched. Girdle narrow. **Lepidopleuridae**, p. 173
3. Girdle beaded with fringe of fine spines on margin. Colour pale. **Ischnochitonidae**, p. 175
   Girdle finely granular. Colour red, green or brown. **Lepidochitonidae**, p. 173

LEPIDOPLEURIDAE.

Body oval or oblong. Margin of valves slightly crenulated, not notched. Mantle girdle narrow, granular, with flat scales.

Lepidopleurus asellus (*Chiton*). F. and H. 1853, II, 407; J. 1865, III, 218 (*C. cinereus*). Body broad, shell flattened, valves oblong and short with small beaks, covered with small shining granules. Girdle less than a quarter the width of a valve, scales overlapping, longitudinally ridged. 13 mm. long, 9 mm. broad. Very common from extreme low water to considerable depths.

LEPIDOCHITONIDAE.

Girdle broad, covered with fine granules; valves broad with prominent beaks and lateral margins notched.

1. Shell dull; girdle with granules of irregular size and colour. **Lepidochitona**
   Shell glossy; valves smooth to the naked eye. 2
2. Girdle broad, with elongated scales and groups of curved spines. Fifteen notches on head valve. **Callochiton**
   Girdle finely granulated, marginal spines spatulate. Eight notches on first (head) valve. **Tonicella.** 3

3. Girdle granular. Eight notches on tail valve. Outside red, inside of valves rose-coloured in centres and greenish laterally. **T. rubra**
Girdle membranous, minutely pustulated. Reddish brown, variegated on outside, inside yellow. **T. marmorea**

Tonicella rubra (*Chiton ruber*). F. and H. 1853, II, 399; J. 1865, III, 224. Glossy. Valves with a longitudinal keel which is not prominent. Reddish brown outside, rose-coloured in the centre and greenish on the lateral margin. Eight notches on the head valve, eight to ten on the tail valve, and one each side on the others. 17 mm. long, 9 mm. wide. Common between low-water mark and shallow water.

T. marmorea (*Chiton marmoreus*). F. and H. 1853, II, 414; J. 1865, III, 227. Glossy. Keel indistinct, covered with minute tubercles. Girdle very broad, membranous, minutely pustulated. Reddish brown, variegated white or yellow, inside yellowish. Eight notches on the head valve, ten on the tail valve, and one each side on the others. 25 mm. long, 15 mm. broad. The largest British Chiton. A rare northern form. Lower shore.

Lepidochitona cinereus (*Chiton marginatus*). F. and H. 1853, II, 402; J. 1865, III, 221. Valves dull, with prominent keel, a row of granules across valves. Girdle moderately broad, covered with minute granules and cigar-shaped marginal spines. Colour varied, in reds, browns, and greens. Eight notches on head valve, ten on tail valve and one on each side of the others. 15 mm. long, 7 mm. broad. Very common below high-water mark.

Callochiton achatinus (*Chiton laevis*). F. and H. 1853, II, 411; J. 1865, III, 226. Valves smooth, with conspicuous "eyes" on the surface, keel more or less prominent. Girdle broad. Reddish brown, variegated with white or green, inside flesh coloured. Fifteen deep notches on head valve, fifteen on tail valve, three on each side of the others. 19 mm. long, 10 mm. broad. Between low-water mark and shallow water.

CRYPTOPLACIDAE.

Body oval or elliptical, shell somewhat compressed and dull, with a broad keel down the centre, granulated. Girdle fairly broad, covered with spines and bearing tufts of bristles. Five notches on the head valve, two on the tail valve and one on each side of the other valves, but these are not easily seen.

1. Shell valves with pear-shaped granules. **Acanthochitona crinitus**
Shell finely granular. Twenty bristle tufts. **A. discrepans**
Shell granules minute. Eighteen bristle tufts. **A. communis**

Acanthochitona crinitus (*Chiton fascicularis*) (Pl. XV, figs. 62 and 63). F. and H. 1853, II, 393; J. 1865, III, 211. Body elliptical. Girdle rather narrower than in the next species, covered with fine spines and bearing eighteen tufts of bristles. Colour brown, yellow or red, underside greenish. 15 mm. long, 8 mm. wide. Local, south and west coasts, littoral and sub-littoral.

A. discrepans. J. 1865, III, 212 (*Chiton fascicularis* var. *gracilis*). Body relatively longer than *A. crinitus*, with broader girdle. Spines on girdle few and broad, eighteen tufts of bristles and often two extra ones on the tail. 20 mm. long, 6 mm. broad. Local.

A. communis. Matthews, 1953, pp. 245–6. Large, with broad girdle. 30 mm. long.

ISCHNOCHITONIDAE.

Body oval; shell with narrow, somewhat glossy valves, bearing numerous small granules in irregular lines which converge to the beaks, valves sharply keeled, beaks small and prominent. Girdle rather broad, covered with bright granules, fringed with small spines on the margin. Valve edges notched, thirteen notches on head valve, eleven on the tail valve and one on each side of the others.

Ischnochiton albus (*Chiton*). F. and H. 1853, II, 405; J. 1865, III, 220. Shell yellowish white, porcellanous inside. Girdle with bead-like granulations and small spines round the edges. 9 mm. long, 5 mm. broad. From low-water mark to shallow water. Local.

GASTROPODA

(Plates XV–XVII, figs. 64–81)

This very large class contains the one-shelled asymmetrical Molluscs like the limpets and winkles. Great variation occurs between the more primitive and the more specialised members.

The head is well developed, and bears the mouth and one or two pairs of retractile tentacles. Eyes are usually present. The *foot* is a flat creeping sole as in the Amphineura, but it may be modified for swimming, digging, floating, etc., and it may develop an anterior *propodium*, dorso-lateral *epipodium* or lateral *parapodium*. In some Gastropods the postero-dorsal region of the foot secretes a horny plate for closing the shell, known as the *operculum*. The spiral growth of the operculum is always inverse from that of the shell. The *mantle* covers the protruding visceral mass but rarely the head. There is a mantle cavity, usually asymmetrically placed, containing a single plume-shaped *ctenidium* (or occasionally a pair of ctenidia), an *osphradium* near the ctenidium, and the external apertures of the gut and renal organs. The edge of the mantle is in some rolled up and produced to form a *pallial siphon* for the intake of water. Such a siphon is not a complete tube, but is open along the ventral side. The mantle secretes the shell, which is typically rolled in a spiral; even those Gastropods which appear to have symmetrical conical shells (e.g. *Patella*) pass through a stage with a spiral shell. The shell may be *dextral* or *sinistral*. To determine this, place the shell on its base, mouth downwards, with the summit towards you. Trace the course of the whorls of the shell from the apex to the outer lip of the aperture. A dextral shell, the commoner type, will show a clockwise spiral, a sinistral shell an anti-clockwise spiral. Most species have a dextral shell, a sinistral form being a rare abnormality, but a few have a normally sinistral coil. The line of junction of the whorls is called a *suture*, the axis of the shell around which the spiral is built is the *columella*, and to this central pillar the body of the animal is attached by a columellar muscle, arising from the foot. The columella may be solid, or it may be tubular and open on the inner side of the aperture by a fissure, the *umbilicus*. The last and widest region of the shell is called the *body whorl*. The aperture of the shell is said to be *entire* or *siphonate*, the latter type indicating the presence of a siphon on the mantle, since the siphon affects the shape of the aperture by forming a siphonal groove. The free side of the aperture forms the *outer lip*, the fixed side is the *inner* or *columellar lip*. When the lip margin is thickened it is called the *peristome*. The lines of growth on the shell will be transverse, new calcareous matter being added to the outer lip. Projecting ridges are *ribs*, smaller lines are *striae*. Either may run transversely or longitudinally.

Many Gastropods possess *jaw plates*, usually composed of rodlets of horny matter arranged in closely packed rows. There is a strong muscular buccal mass containing a characteristic tooth-bearing apparatus, the *odontophore*. The odontophore consists of a pulley-shaped cushion or rotella, with movable "cartilages" and muscles, on the surface of which lies a strong membrane bearing a number of teeth in regular transverse rows. This "tooth" ribbon is called the *radula*. So characteristic is the radula of each species that it is often possible to identify the animal from its radular pattern. In carnivorous Gastropods the whole buccal region may be protrusible as a *proboscis* during feeding.

Gastropods are divided into two very large sub-classes—the Streptoneura (or Prosobranchs) and the Euthyneura. The Prosobranchiate Gastropods can be recognised by their single pair of tentacles and well-developed shell. There is usually a ctenidium anterior to the heart, and the body is in nearly all cases asymmetrical and spirally coiled. Prosobranchs are dioecious (sexes separate) though there may be a reversal of sex during the life of the individual (usually with protandry, i.e. functional male sex organs preceding a change over to femaleness). The external asymmetry is even more pronounced internally and affects the nervous system, the visceral connectives being always crossed over in a figure of 8 (streptoneury).

The Euthyneura have been derived from Streptoneurous forms by a process of detorsion or untwisting of the spirally coiled body, whereby the visceral nerve connectives have in many become uncrossed. The head typically bears two pairs of tentacles, known as *oral tentacles* (anterior) and *rhinophores* or olfactory tentacles (posterior). A shell may or may not be present, an operculum is rarely found. A ctenidium is present in many forms, and lies posterior to the heart (Opisthobranchiata); in others true ctenidia are absent, but adaptive gills, such as *branchial plumes*, *branchial lamellae*, *cerata*, etc., may take their place. Both male and female gonads occur in the same individual (monoecious), and there may be one genital aperture from a common male and female duct (the *monaulic* condition, e.g. *Aplysia*); or the male and female ducts may be separate and open to the exterior by two apertures, either distant or contiguous (the *diaulic* condition, e.g. *Aeolidia*); or there may be three openings, a male, a female and a copulatory aperture (the *triaulic* condition, e.g. *Archidoris*). The Euthyneura are divided into marine and air-breathing forms, known as Opisthobranchs and Pulmonates respectively. Opisthobranchs have either ctenidia or adaptive gills, the former situated posterior to the heart. There is usually a pallial cavity between the mantle and the body wall, and in this cavity lie the anus, the renal and genital apertures, the ctenidium and osphradium, and sometimes special glands of a defensive character. The Opisthobranchs are divided into Tectibranchs and Nudibranchs. The Tectibranchs mostly retain their shell, mantle, ctenidium and osphradium, while the Nudibranchs have lost all of these except the mantle, which forms a dorsal cloak to the body. If gills are present they are of a new type, such as a circlet of *branchial plumes* around the dorsal anus, papillose processes of the mantle roof known as *cerata* (singular *ceras*), or lamellate processes known as *branchial lamellae* pendent from the under side of the mantle skirt. Cerata, except in the families Tritoniidae and Dendronotidae, contain diverticula of the digestive gland.

The Pulmonates live on land or in fresh water, but breathe atmospheric oxygen. A few inhabit coastal areas. The mantle cavity is closed except for a pulmonary aperture, and its vascular walls act as a breathing organ. A shell is present in some.

1. One pair of cephalic tentacles. Shell well developed. An operculum usually present. Ctenidium anterior to the heart. Typically dioecious and asymmetrical. **Prosobranchiata (Streptoneura)** 2
   Two pairs of cephalic tentacles (rarely one pair). Detorsion frequent in adult, with secondarily acquired external symmetry. Monoecious. **Euthyneura.** 20
2. Shell external. 3
   Shell wholly or partly internal. Spire concealed. **Lamellariidae**, p. 195
3. Shell raised. 4
   Shell flat, but spirally coiled, minute. **Skeneopsidae**, p. 192
4. Shell conical, with or without traces of a spire. Shell aperture occupies the whole base of the shell. No operculum. 5
   Shell spiral, though the spire may be concealed. 7
5. No spire in the adult shell. 6
   Shell of adult with small, apical remains of spire. Flat "diaphragm" partly across aperture. **Calyptraeidae**, p. 193
6. No true ctenidia, but pallial tentacular branchiae. Pallial scar indistinct or else away from margin of shell. **Patellidae**, p. 181
   One bipectinate ctenidium on left side. Pallial scar distinct and near margin. **Acmaeidae**, p. 183
   Two symmetrical ctenidia. Shell with an apical or marginal slit. **Fissurellidae**, p. 183

7. Spire visible. 8
Spire of shell partly or wholly concealed by the last or body whorl. Aperture of shell an elongated slit. Mantle capable of being reflected over whole of shell. Cowries. **Cypraeidae**, p. 196
8. Siphon on mantle and siphonal canal on shell. 9
No siphon or siphonal canal. 13
9. Shell turreted; no operculum. **Turridae**, p. 198
Shell not turreted, dextral. 10
Shell not turreted, sinistral. **Triphoridae**, p. 192
10. Eyes at base of tentacles laterally. Siphonal canal a deep but short groove. **Cerithiidae**, p. 191
Eyes at base of tentacles anteriorly. **Cerithiopsidae**, p. 192
Eyes on swellings joined to tentacles one-half to one-third way up tentacles. 11
11. Columella ridged. **Nassariidae**, p. 197
Columella smooth. 12
12. Siphon short. **Muricidae**, p. 196
Siphon long. **Buccinidae**, p. 197
13. No operculum. Parasitic on Echinoderms. **Styliferidae**, p. 193
Operculum circular. Three or four epipodial tentacles. 14
Operculum ear-shaped. 15
14. Operculum horny. **Trochidae**, p. 184
Operculum calcareous. **Turbinidae**, p. 186
15. Shell globular, spire blunt and usually very short. Foot large with a propodium which is reflected over head. **Naticidae**, p. 194
Shell spirally conical. No propodium. 16
16. Tentacles wholly or partly ciliated. 17
Tentacles smooth. 18
17. Foot with two wing-like expansions to the opercular lobe, beneath which may be one or more tentacular appendages (caudal cirri). Minute tentacle at upper corner of shell mouth as a rule. **Rissoidae**, p. 189
No opercular lobe expansions or caudal cirri. **Hydrobiidae**, p. 191
18. Mentum (a shelf projecting anteriorly between head and foot) present. Eyes mesial to tentacles. **Pyramidellidae**, p. 193
No mentum. Eyes lateral to tentacles. 19
19. Shell turreted, beautifully ribbed. Aperture nearly round, with complete peristome. Retractile proboscis. **Epitonidae**, p. 192
Shell conical. Aperture obliquely square or oval. Head snout-shaped. **Lacunidae**, p. 186

EUTHYNEURA

20. Marine, with aquatic respiration. Pallial cavity usually widely open. Shell internal or external, sometimes absent. Ctenidium posterior to the heart, or, if absent, adaptive gills (pallial branchiae) usually present. **Opisthobranchiata.** 21
Pallial cavity usually adapted for aerial respiration. No ctenidium. Mantle fused with the neck except for a small space which forms

the pulmonary aperture. Shell external, internal or absent. No operculum in adult. **Pulmonata.** 40

21. Mantle and shell usually present. Ctenidium and osphradium in mantle cavity. **Tectibranchiata.** 22
    No shell, ctenidium or osphradium. Usually with external symmetry and adaptive gills. Mantle forming a dorsal cloak. **Nudibranchiata.** 27

TECTIBRANCHIATA

22. Operculum present. **Acteonidae**, p. 199
    No operculum. 23
23. One pair of tentacles which are fused to form a cephalic shield. 24
    Two pairs of tentacles, no cephalic shield. 26
24. Shell external. Foot with lateral swimming lobes reflected over the shell. 25
    Shell internal. **Philinidae**, p. 200
    Shell absent in British forms. **Peltidae**, p. 200
25. Head flattened, bifid posteriorly. **Atyidae**, p. 199
    Head elongated, snout-like, without posterior processes. **Akeratidae**, p. 200
26. Pedal sole long and narrow. Foot with upstanding lobes (parapodia) which shield mantle and internal shell and are used for swimming. Purple dye exuded by glands in the mantle. **Aplysiidae**, p. 201
    Foot broader than body. No parapodia. Mantle projecting laterally, overhanging but not enclosing the ctenidium. **Pleurobranchidae**, p. 201

NUDIBRANCHIATA

27. Anus median, dorsal. 28
    Anus lateral. 32
28. Branchial plumes surround anus. 31
    Back covered with cerata (branchial papillae) which contain diverticula of digestive gland. 30
    No branchial plumes or cerata. 29
29. Body with latero-dorsal fin-like expansions for swimming. **Elysiidae**, p. 203
    No lateral expansions. **Limapontiidae**, p. 203
30. Cerata numerous, some anterior to rhinophores. **Zephyrinidae**, p. 209
    Cerata numerous but none in front of rhinophores. **Stiligeridae**, p. 202
31. Branchial plumes retractile. **Dorididae**, p. 208
    Branchial plumes non-retractile. **Polyceridae**, p. 204
32. Cerata present, containing diverticula of the digestive gland. 34
    Cerata present, arborescent, but not containing diverticula of the digestive gland. Rhinophores retractile within sheaths. 33

33. Rhinophore much branched, with one longer, smooth truncate process. Sheath short and smooth. **Tritoniidae**, p. 204
Rhinophore lamellate, sheath long and much branched. **Dendronotidae**, p. 209

34. Cerata branched. **Heroidae**, p. 210
Cerata unbranched. **35**

35. Oral tentacles represented by a veil. Cerata few, tuberculate, shaped like fir cones. **Dotoidae**, p. 210
Oral tentacles linear. Cerata simple, varied in shape but not like fir cones. **36**

36. Body densely covered with cerata. Foot with pointed antero-lateral processes, usually broad, with short tail. **Aeolidiidae**, p. 211
Cerata less numerous. Foot slender, rounded anteriorly or with short lateral processes, tail long and slender. **37**

(The remaining four genera belong to the two families *Tergipedidae* and *Flabellinidae*, in which the points needed for classification are not distinguishable without dissection. The following key, therefore, separates into *genera*.)

37. Anterior foot margin with lateral processes. Cerata reddish or orange. **Coryphella**, p. 213
Anterior foot margin rounded. **38**

38. Cerata numerous, slender, not swollen. **Cratena**, p. 212
Cerata club-shaped, few in number. **39**

39. Cerata four pairs only. Much swollen. Body very slender, with tapering foot. **Tergipes (despectus)**, p. 212
Cerata more than 8. **Eubranchus**, p. 214

PULMONATA

40. Shell external. One pair of non-invaginable tentacles. **Basommatophora. 41**
No shell. Two pairs of invaginable tentacles, eyes on the tips of the posterior pair. **Stylommatophora. 42**

41. Shell with short spire and wide oval aperture. Tentacles short, knob-like, bearing eyes. **Otinidae**, p. 215
Shell with prominent spire and denticulate aperture. **Ellobiidae**, p. 214

42. Male and female apertures distinct, the former nearer the head, the latter postero-mesial and close to the mesial anus. Pulmonary cavity reduced. Slug-like forms. **Onchidiidae**, p. 215

PROSOBRANCHIATA

PATELLIDAE.

Shell conical, aperture wide, forming the complete base of the cone. Mantle fringed at the edges. Eyes prominent, on eye cushions at the base of the tentacles. Pallial branchiae forming a circle between the mantle and foot. Foot very large and elliptical, occupying the whole of the lower side of the animal.

1. **Margin of shell crenate. Mantle scar distinct and away from the margin. Tentacles short. Patella. 3**
   Margin of shell smooth. Mantle scar indistinct. Tentacles long. Patina. 2
2. **Shell semi-transparent and glossy, with bright blue longitudinal stripes. Patina pellucida** (young)
   Shell solid and opaque. P. pellucida (old)
3. **Shell a regular cone; inside glossy white or yellow, to blue and brown in centre. Marginal tentacles transparent. Patella vulgata**
   Shell rather flat with fine ribs. Marginal tentacles white or cream. 4
4. **Shell oval, inside dark, with cream head scar. Foot dark. P. depressa**
   Shell oblong, inside pale, with orange head scar. Foot cream. P. aspera (athletica)

Patella. (Limpets.) Shell opaque and dull, ribbed. See Fig. 111.

P. vulgata. Common limpet. F. and H. 1853, II, 421; J. 1865, III, 236. Shell a rather tall cone; ribs from apex to margin, where they are stronger, smaller ones between the larger, margin scalloped by ribs. Inside yellow or white with central body scar varying from white to dark brown. Body brownish yellow. Base 40–65 mm. long, 55 mm. broad. Very common between tide marks on all shores where there are rocks. Spawning October to November.

P. depressa. J. 1865, III, 237, as *P. vulgata* var. *intermedia.* Shell smaller, flatter and more oval than above, with finer ribs and strongly rayed margin. Base 45 mm. long, 35 mm. broad. South west coasts, to Anglesey only. Middle shore. Summer breeder.

P. aspera (*athletica*). F. and H. 1853, II, 425; J. 1865, III, 237, as *P. vulgata* var. *depressa.* Shell flatter still, oblong; ribs fine and short, apex nearer anterior end. Inside porcellanous with a pale orange head scar. Body cream-coloured. Base 45 mm. long, 35 mm. broad. On all coasts. Below L.W.N. or in pools.

Patina. Shell conical, convex and smooth. Mantle with thirty to sixty fine white cirri alternately long and short.

P. pellucida. Blue-veined limpet. F. and H. 1853, II, 429, as *Patella*; J. 1865, III, 242, as *Helcion.* A beautiful shell, transparent, yellowish brown with Kingfisher-blue stripes. Inside highly polished. Base 20 mm. long, 15 mm. broad. Foot oval, yellowish white, often sunk in an excavated hollow in a *Laminaria* stipe or frond. Common at low water. Older specimens, formerly called *P. laevis*, have solid shells, lose their stripes, and live in cavities excavated in the holdfasts of *Laminaria*. Graham & Fretter, *J. mar. biol. Ass. U.K.* 1947, XXVI, no. 4.

FISSURELLIDAE.

Shell ovately conical, ribbed, solid and opaque, with a notch or hole at the apex or in the margin through which a tubular process of the mantle emerges. Two symmetrical ctenidia.

1. Shell notch marginal. **Emarginula**
   Shell with apical hole. **Diodora**

Emarginula reticulata. Slit limpet. F. and H. 1853, II, 477; J. 1865, III, 259, as *fissura*. Shell with a notch in the anterior margin; twenty-five to thirty-five strong ribs from apex to margin, crossed by twenty to thirty concentric striae. Mantle finely fringed. Base 11 mm. long, 9 mm. broad. Widespread on stones, etc., at low water.

Diodora apertura (Pl. XV, fig. 64). F. and H. 1853, II, 469, as *Fissurella reticulata*; J. 1865, III, 266, as *F. graeca*. Shell with an apical hole; about twenty-five ribs crossed by about thirty concentric ones. Mantle with stout cirri which correspond with the ribs of the shell. 34 × 20 mm. Fairly widespread from low-water mark.

ACMAEIDAE.

Shell conical, usually flattened, with striations from the apex, which is nearer the anterior margin. Pallial scar nearly marginal. Mantle fringed. A short plume-like ctenidium.

1. Shell with dark red irregular streaks on a grey ground. Inside with a dark brown central scar. **Acmaea (Patelloida) tessulata**
   Shell white or pink, with sixteen to twenty pink rays; inside white or pink. **A. virginea**

Acmaea tessulata. The tortoiseshell limpet. F. and H. 1853, II, 434, as *Acmaea testudinalis*; J. 1865, III, 246, as *Tectura testudinalis*. Shell rather thin, dull, greyish, with dark red irregular streaks, or sometimes reddish with white rays or spots, central scar chocolate coloured. 22 mm. long, 18 mm. broad. Beneath stones at low water on northern coasts.

A. virginea. White tortoiseshell limpet. F. and H. 1853, II, 437; J. 1865, III, 248, as *Tectura.* Shell more solid than the above. Yellowish white or pale pink, with sixteen to twenty broad pink or brown rays which sometimes have white spots, apex nearly overhangs the anterior margin, inside white or pink, pallial scar with a row of white dots. 10 mm. long, 8 mm. wide. Common on lower shore.

TROCHIDAE.

Top shells. Shell roundish or conical, spirally coiled, somewhat nacreous; umbilicus usually present; aperture roundish. Operculum horny, thin, circular, multispiral, with a central nucleus. Mantle with a lappet on each side of the probosciform head. Tentacles long, ringed purple or black, with eyes on peduncles at their bases. Single plume-like ctenidium, three or four epipodial tentacles.

1. Shell a sharply pointed cone without umbilicus. **2**
   Shell apex blunt, whorls usually tumid, with or without umbilicus. **6**
2. Shell large with ten to twelve beautifully marked whorls. Four pairs of epipodial tentacles. **Calliostoma (zizyphinum)**
   Shell small, usually with six to eight whorls. Three pairs of epipodial tentacles. **Cantharidus. 3**
3. Shell ridged. **4**
   Shell finely striated only. **5**
4. Shell ridges six to eight. Base very broad. **C. clelandi**
   Shell ridges fewer. Base moderately broad. **C. exasperatus**
5. Shell with flat whorls. **C. striatus**
   Shell with slightly rounded whorls. **C. montagui**
6. Shell without umbilicus, except in young stages. **Monodonta (lineata)**
   Shell with umbilicus. **Gibbula. 7**
7. Shell round and ridged. Umbilicus very wide. **G. magus**
   Shell smoother. Umbilicus narrow, sometimes partly covered by inner lip. **8**
8. Shell grey with purplish grey markings and spots. **G. cineraria**
   Shell grey with a few reddish or purple rays, not spotted. **G. umbilicalis**

Calliostoma zizyphinum (*Trochus zizyphinus*) (Pl. XV, fig. 65). F. and H. 1853, II, 491; J. 1865, III, 330. Shell a regular pyramid with a flattened base, slightly glossy, spire pointed, pale yellow or pink with longitudinal streaks of red. Aperture rhomboidal, no umbilicus. Small intertentacular lobes present. Eyes large and prominent. Four pairs of epipodial tentacles. A white variety is fairly common. 25 mm. high, 25 mm. broad. Common from low-water mark on rocky shores.

Cantharidus. Shell a sharply pointed cone without umbilicus. Intertentacular lobes very small. Three pairs of epipodial tentacles.

C. striatus (*Trochus striatus*). F. and H. 1853, II, 508; J. 1865, III, 322. Shell solid and rather glossy, pale yellow or white with streaks which vary from dark red or brown to nearly black; seven flat whorls. Inner lip with a small blunt tooth. 9 mm. high, 8 mm. broad. From low-water mark, especially on *Zostera*.

C. montagui (*Trochus*). F. and H. 1853, II, 511; J. 1865, III, 320, as *montacuti*. Shell narrow at base, moderately solid and opaque, six or seven fine spiral ridges on each whorl; seven whorls gradually enlarging. Yellowish white with red-brown spots on each ridge or with longitudinal streaks on last whorl. Body yellowish white, speckled with pale brown. 8 mm. high, 6 mm. broad. General, from low-water mark.

C. exasperatus (*Trochus*). F. and H. 1853, II, 505 (*exiguus*); J. 1865, III, 324. Shell resembles *C. striatus* but is more solid, with a few well-marked spiral ridges. Pink and red in colour. Animal tinged red, with white eye stalks. 9 mm. high, 8 mm. broad. South and west coasts.

C. clelandi (*Trochus millegranus*). F. and H. 1853, II, 502; J. 1865, III, 325. Shell with six to eight concentric ridges, greatly expanded and flattened at base, upper whorls somewhat keeled, ten rather flattened whorls, last very large. Yellowish white, speckled or blotched with red. Tentacles with a reddish brown line down the middle. 10 mm. high, 10 mm. broad. From low-water mark. Not common, but generally distributed.

Monodonta. (*Osilinus*). Shell with tumid whorls, umbilicus overgrown by inner lip in adult.

M. lineata (*Trochus*). F. and H. 1853, II, 525; J. 1865, III, 317. Shell regularly conical, very thick and dull, smooth in adult; spire bluntly pointed and slightly raised. Six whorls, yellow or grey with a greenish tinge and numerous purple zigzag streaks and markings; apex eroded, yellow. Aperture covering umbilicus and with a large tubercle. Body dark coloured. Intertentacular veil fringed. Tentacles slender, with alternately large and small annulations. 24 mm. high, 25 mm. broad. Local on west coasts, below high-water mark. Northern range Cardigan Bay.

Gibbula. Cone-shaped with an umbilicus. Two large fringed intertentacular lobes and three pairs of epipodial tentacles.

G. magus (*Trochus*). F. and H. 1853, II, 522; J. 1865, III, 305. Shell somewhat scalariform, solid, rough, rather dull, with short, oblique longitudinal ribs and spiral ridges, the base encircled with a stronger ridge. Eight whorls. Yellowish white with streaks of pinkish red, umbilicus very wide. Body papillose, yellow, blotched with purple or purple and white. Tentacles long and slender, usually partly ringed with black. 22 mm. high, 27 mm. broad. Fairly common from low-water mark.

G. cineraria (*Trochus cinerarius*). F. and H. 1853, II, 516; J. 1865, III, 309. Shell variable in height, solid and rather dull, with seven or eight spiral ridges, light grey with purple brown streaks and spots. Six or seven whorls, umbilicus small and narrow. Body purplish grey, spotted or barred with white, yellow or purple. Some smaller processes are present as well as the three pairs of epipodial tentacles. Cephalic tentacles ringed with purple or alternately purple and white. 12 mm. high, 13 mm. broad. Abundant on all rocky shores on lower middle shore.

G. umbilicalis (*Trochus umbilicatus*). F. and H. 1853, II, 519; J. 1865, III, 312. Shell flat, with a flatter base than the other species. Longitudinal rays red or purplish, broad and few, sometimes zigzag but never spotted. Body light brown. Tentacles marked with alternately large and small, often incomplete, rings (resembling the plant *Equisetum*). 13 mm. high, 17 mm. broad. On upper middle shore.

TURBINIDAE.

Shell conical and spiral. A calcareous white operculum which is convex on outer side, concave on inner. Body as in Trochidae.

Tricolia pullus (*Phasianella pulla*). The pheasant shell. F. and H. 1853, II, 538; J. 1865, III, 338. Shell conical, semi-transparent and glossy, with short spire; five to six whorls, last longer than all the rest together. Colour varies, but usually yellowish with reddish zigzag lines or spots. Three pairs of epipodial tentacles. 9 mm. long, 6 mm. broad. Common.

LACUNIDAE.

Winkles, etc. Shell conical and spiral, not nacreous. Operculum horny, thin and ear-shaped, with a few whorls and a lateral nucleus.

Head snout-shaped, tentacles long, eyes on very short stalks at outer base of tentacles. Single plume-like gill.

1. Shell without umbilicus; columella smooth. No caudal appendages. **Littorina.** 2
   Shell with umbilicus; columella grooved. Two caudal appendages. **Lacuna.** 5
2. Shell obtuse, yellow, brown or reddish; on *Fucus* between tide marks. **Littorina littoralis**
   Shell conical. 3
3. Shell thick, with a moderately tall cone. Tentacles annulated. Foot striped. At low-water mark. The largest species. **L. littorea**
   Shell a small cone. Tentacles not annulated. Foot self-coloured. At high-water mark. 4
4. Shell a pointed cone, smooth except for growth lines. Tentacles broad at base, slender above. Eyes in middle of tentacle base. The smallest species. **L. neritoides**
   Shell a short cone; surface with several spirals and wrinkles; tentacles slender. Eyes at sides of tentacles. **L. saxatilis**
5. Shell much longer than broad. 6
   Shell very little longer, or shorter than broad. 7
6. Shell with turreted whorls, deep suture. **Lacuna crassior**
   Shell with smoothly enlarging whorls. **L. vincta**
7. Shell globular with expanded aperture. Usually with three red bands. **L. parva**
   Shell triangular with very wide aperture. Pale coloured. **L. pallidula**

Littorina. Winkles. Shell solid with a short spire, no umbilicus, aperture oval, with lips usually not meeting, columella smooth, operculum with a process inside. No caudal appendages.

L. neritoides. J. 1865, III, 361. The smallest species of the genus. Shell a pointed cone, glossy when young, nearly smooth except for the lines of growth. Dark coloured. Eggs pelagic, in square capsules. 7 mm. long, 6 mm. broad. Above high-water mark, in clefts of rock.

L. saxatilis (*rudis*). F. and H. 1853, III, 32; J. 1865, III, 364. Shell a short, sharply pointed cone, dull, with several spiral ridges crossed by irregular growth lines. Surface always covered with wrinkles. Colour variable, brown, red, grey to black. Six to nine whorls, sutures deep. Tentacles slender, with two pale stripes down length as a rule. Foot pale. Viviparous. A very variable species, with seven named subspecies (see Winckworth, 1932, and

J. 1865, III, 365). 16 mm. high, 12 mm. broad. Common on stony and rocky beaches in barnacle zone, often in rock clefts.

L. littoralis (*obtusata*). F. and H. 1853, III, 45; J. 1865, III, 356. Shell obtuse, with a blunt spire, thick, sculptured with very fine striae; colour very variable, being yellow, green, brown, red, orange, purple or self-streaked. Aperture large, occupying half the lower part of the shell. Tentacles tapering, with white tips, eyes small, with white rings. 17 mm. long, 13 mm. broad. Common below high-water neap tide level. Egg masses on *Fucus*.

L. littorea. Common winkle. F. and H. 1853, III, 29; J. 1865, III, 368. The largest species of the genus. Shell a moderately tall cone, thick and dull, with fine spiral ridges and wrinkles as in the following species, but more marked. Spire sharp, seven or eight whorls. Tentacles strongly contractile, annulated. Black eyes with lighter rings, on very short stalks. Foot striped. 30 mm. long, 25 mm. broad. Common on weed near low-water mark. Eggs pelagic, usually two in a flattened capsule.

Lacuna. Chink shells. Shell with an umbilicus.

L. vincta. F. and H. 1853, III, 62; J. 1865, III, 346, as *divaricata*. Shell conical, expanded, somewhat thin, transparent and glossy. Sculpture of numerous slight spiral lines; six whorls. Colour from white to yellowish, usually with reddish brown spiral bands. 12 × 8 mm. Common on weed. Eggs in rings, cream, yellow or green, on *Fucus*.

L. crassior. F. and H. 1853, III, 67; J. 1865, III, 344. Shell turreted, angular at base, solid, opaque and dull. Sculpture like *vincta*, six or seven whorls, suture deep. Yellowish coloured. 13 mm. long, 8 mm. broad. Local, on stones and old shells in sand and mud, from low-water mark. Eggs in solid masses.

L. parva (*puteolus*). F. and H. 1853, III, 58; J. 1865, III, 348. Shell globular, expanded at the base, solid, opaque, glossy, sculpture as above, but often striae in growth lines. Colour yellowish, sometimes purple or reddish or with three brownish red bands. Three or four whorls. Aperture expanded. 7 mm. broad, 7 mm. high. On weed at low-water mark. Eggs in solid masses.

L. pallidula. F. and H. 1853, III, 56; J. 1865, III, 351. Shell triangular, rather thin, opaque and glossy, with irregular striae in growth lines, and a very few spiral striae. 11 mm. high, 9 mm. broad. On *Laminaria*, etc. Eggs in solid masses.

RISSOIDAE.

Shell very small, spiral, oblong, or oval, usually without an umbilicus, spire typically elongated. Head with a long snout, tentacles wholly or partly ciliated, eyes on small tubercles at outer base of tentacles. Mantle with a minute tentacle at the outer corner of the shell aperture (except in *Barleeia*). Opercular lobes of foot large with two wing-like expansions, beneath which is a short tentacular appendage (except in *Barleeia*). A large family, only the commonest littoral species being described below.

1. No tentacular appendage on mantle or foot. **Barleeia (unifasciata)**
   Tentacular and mantle appendage present. 2
2. Strong spiral striations on shell, outer lip strengthened by a rib. 4
   Faint spiral striations or none, outer lip plain. **Cingula.** 3
   Smooth outer lip thin. **Cingulopsis fulgida**
3. Six slight spiral ridges, twelve distinct ones below the periphery. Streaked with reddish brown. **C. semistriata**
   Twelve to twenty slight spiral ridges. Two reddish bands on the last whorl, one on the others. **C. cingillus**
4. Strong longitudinal ribs. **Rissoa.** 5
   Faint longitudinal ribs. Lips thickened to form a peristome. **Onoba (semicostata)**
5. Fifteen to eighteen ribs on penultimate whorl, on last whorl do not extend all the way. **Rissoa membranacea**
   Eight ribs on last, twelve on penultimate whorl. A brown streak from outer lip. **R. parva**
   Fifteen ribs on penultimate whorl, on under side of last only. **R. lilacina**

Rissoa parva. F. and H. 1853, III, 98; J. 1867, IV, 23. Shell somewhat glossy. Ribs strong and curved, spire short, six or seven convex whorls. Yellowish white, occasionally with brown streaks and a band round the base, always a brown streak from the outer lip. Opercular lobe large and dark purple, appendage cylindrical and rather long, whitish. 4 mm. long, 2–3 mm. broad. On weed at low-water mark, abundant.

R. membranacea. F. and H. 1853, III, 109, as *labiosa*; J. 1867, IV, 30. Shell semi-transparent and rather glossy. Ribs prominent. Whitish with sometimes a brown tinge; aperture occasionally violet or edged brown. Spire short, seven somewhat compressed whorls. Opercular lobe large and dark purple. Appendage white and conspicuous. 8 mm. long, 2 mm. broad. On weed, from low-water mark.

R. lilacina. F. and H. 1853, III, 106 (*rufilabrum*); J. 1867, IV, 33 (*violacea*). Shell oval, solid, glossy. The aperture has a large labial rib which is tawny on the outside and violet within. Colour whitish or yellowish. Tentacles slender with an orange line; a single caudal appendage. 5 mm. high, 2 mm. broad. Lower shore.

Cingulopsis fulgida (*Rissoa*). F. and H. 1853, III, 128; J. 1867, IV, 43. Shell rather thin, semi-transparent and glossy, smooth, pale yellow, with two reddish brown bands on the body whorl. 1·5 mm. high. On weed, *Zostera*, etc., common on southern coasts.

Cingula semistriata (*Rissoa*). F. and H. 1853, III, 117; J. 1867, IV, 46. Semi-transparent, somewhat solid and glossy, spire rather short and pointed. Six rounded whorls, last nearly two-thirds of the spire, sutures slight. Pale yellowish white with a row of reddish brown streaks on upper part of each whorl. 0·5 mm. high. Lower shore.

C. cingillus (*Rissoa*). F. and H. 1853, III, 122; J. 1867, IV, 48. Shell rather solid and glossy, semi-transparent, buff or yellowish white. Snout semi-transparent; tentacles flexible with somewhat bulbous tips; eyes black and conspicuous; foot narrow and short. About 8 mm. high.

Barleeia unifasciata. F. and H. 1853, III, 120, as *Rissoa rubra*; J. 1867, IV, 56, as *Barleeia rubra*. Shell conical, semi-transparent and solid, glossy and smooth. Dark red or yellowish coloured, operculum dark crimson. Tentacles club-shaped, with a brownish line, or speckled with yellow. 15 mm. high, 3 mm. broad. Common on weed in the south, from low-water mark.

Onoba semicostata (*Rissoa striata*). F. and H. 1853, III, 94; J. 1867, IV, 37. Shell oblong, solid, nearly opaque and usually dull, spire elongated with a blunt point; six somewhat flattened whorls. Aperture small. Outer lip thin, with a thick rib; inner lip reflected and forming with the other a complete peristome. White or yellowish. Snout long and narrow. Appendage short, white and pointed. About 2 mm. high. Common on all coasts at low-water on lower shore.

HYDROBIIDAE.

Shell oblong, rather solid and smooth; spire yellowish, long and tapering. Tentacles thread-like, ciliated, with eyes on tubercles arising from their bases. The Molluscs of this family resemble the Rissoidae in structure except that they have no appendages to the opercular lobe. Estuarine.

Hydrobia.

1. Whorls flattened, spire blunt. **H. ulvae**
   Whorls rounded, spire pointed. **2**
2. Whorls well rounded with deep suture. **H. (Potamopyrgus) ventrosa**
   Whorls rounded, last much the largest. **H. (Potamopyrgus) jenkinsi**

H. ventrosa. F. and H. 1853, III, 138, as *Rissoa*; J. 1862, I, 66. Shell glossy, semi-transparent and horn-coloured when the animal has been removed. Six well-rounded whorls with a pointed apex. 6 mm. high. In estuaries, abundant.

H. jenkinsi. Ellis, 1926, p. 78. Shell tapering, glossy, not very thin. Five or six whorls, last one large and sometimes with a keel, apex pointed, peristome pronounced. Umbilicus slit-like and small. No male has yet been found in this species. Female viviparous. 4–5 mm. high, 2–3 mm. broad. Brackish water.

H. ulvae. F. and H. 1853, III, 141, as *Rissoa*; J. 1867, IV, 52; Ellis, 1926, p. 76, as *Sabanea ulvae.* Shell conical, elongated, with a blunt spire, thick and smooth. Reddish brown, four to six flattened whorls, last equalling half the shell. 5–10 mm. high, 3–5 mm. broad. Common in estuaries.

CERITHIIDAE.

Shell pyramidal, tuberculated; no umbilicus, spire tapering, whorls numerous, aperture small with a short siphonal groove; nearly circular horny operculum with a not quite central nucleus. Foot notched; head snout-like with awl-shaped tentacles connected by a membrane. Eyes on protuberances lateral to the tentacles.

Bittium reticulatum **(*Cerithium*).** F. and H. 1853, III, 192; J. 1867, IV, 258. Shell solid, not very glossy, marked by thin spiral ribs which are intersected by coarser longitudinal ones to form nodules. Reddish brown colour, nodules owing to rubbing may be white. Fifteen or sixteen rather convex whorls. Aperture triangular, outer lip rounded, indented by the spiral ridges. 13 mm. high, 3 mm. broad. West and south coasts chiefly.

SKENEOPSIDAE.

Shell minute, umbilicated; circular and flat, with a depressed or even involute spire. Tentacles long and divergent. Eyes large on broad protuberances; foot short.

Skeneopsis planorbis (*Skenea*). F. and H. 1853, III, 156; J. 1867, IV, 65. Four whorls, thin and opaque, smooth, pale brown or greenish. Very small, about 2 mm. in diameter. Under stones and on weed.

CERITHIOPSIDAE.

Shell cylindrical, finely tuberculated; no umbilicus, spire elongated, whorls numerous with narrow and deep sutures, short siphonal canal; operculum horny with terminal nucleus. Eyes on swellings anterior to the base of the tentacles.

Cerithiopsis tubercularis. F. and H. 1853, III, 365; J. 1867, IV, 266. Dark chestnut, with thirteen or fourteen whorls and a turreted spire. 6 mm. high, 2 mm. broad. Under stones on lower shore.

TRIPHORIDAE.

Shell sinistral, an elongated pyramid, solid and opaque, slightly glossy, with tubercles due to spiral and longitudinal striae crossing one another. Tentacles long and slender, connected by a veil; eyes on short stalks at their bases.

Triphora perversa (*Cerithium*). F. and H. 1853, III, 195, as *C. adversum*; J. 1867, IV, 261, as *C. perversum*. Dark chestnut to yellowish brown. 8 mm. high, 2 mm. broad. Lower littoral zone.

EPITONIDAE.

Shell turreted, beautifully ridged; spire elongated. Aperture nearly circular with a complete peristome which is angular below. Operculum ear-shaped.

Clathrus clathrus (*Scalaria communis*) (Pl. XV, fig. 69). F. and H. 1853, III, 206; J. 1867, IV, 91. One of the most beautiful Gastropods. Shell slender, conical, solid, opaque and not glossy, with curved reflected ribs; colour varies from cream to fawn, sometimes with bands or mottling; fifteen or sixteen gradually enlarging whorls. Deep suture. Body white with black streaks. Animal discharges a purple dye. 38 mm. long, 15 mm. broad. Extreme low water.

STYLIFERIDAE.

Shell oval or elongated, apex styliform and excentric; aperture round or oval; no operculum. Head suctorial; eyes sessile behind the tentacles. Parasitic on Echinoderms.

Pelseneeria stylifera (*Stylifer turtoni*). F. and H. 1853, III, 226; J. 1867, IV, 195. 3 × 2 mm. Not common.

PYRAMIDELLIDAE.

Shell smooth, conical or pyramidal with a pointed spire, aperture oval. Columella with a tooth. Operculum ear-shaped. Long retractile proboscis; mentum present, eyes sessile, mesial to the tentacles.

1. Shell an elongated cone. Tooth minute or invisible. Ribs curved and oblique. **Turbonilla elegantissima**
   Shell nearly cylindrical. Tooth visible. Ribs curved but not oblique. **T. innovata**

Turbonilla elegantissima. F. and H. 1853, III, 242 (*Chemnitzia elegantissima*); J. 1867, IV, 164 (*Odostomia lactea*). Shell rather solid and glossy, white. Twenty to twenty-five curved longitudinal oblique ribs. Twelve moderately convex whorls which enlarge gradually. Suture deep. Body clear white; tentacles short and leaf-like. 9 mm. high, 2·5 mm. broad. Common at low-water mark. Parasitic. (Fretter, *J. mar. biol. Assoc. U.K.* 1948.)

T. innovata. F. and H. 1853, III, 244 (*C. pusilla*); J. 1867, IV, 167 (*O. pusilla*). Shell cylindrical, longitudinal ribs curved, not oblique. Twelve rounded whorls. Snout and tentacles with a pale purple streak. White. 7 mm. high, 2 mm. broad. Low water.

CALYPTRAEIDAE.

Shell depressed, partly spiral, with a very wide base; circular or oval aperture occupying the whole of the base of the shell; inside with a partition or diaphragm representing a specialised columella. Eyes on swellings half-way up tentacles. Foot broad, and separated from the lower part of the body.

1. Shell and body round. Apex central. **Calyptraea**
Shell and body oval. Apex posterolateral. **Crepidula**

Calyptraea chinensis. F. and H. 1853, II, 463 (*sinensis*); J. 1865, III, 273. Shell thin, conical, glossy at the central apex, with numerous fine circular striae. Colour white with a yellow tinge. Inside white. Diaphragm oblique, occupying nearly half the posterior side. 20 mm. in diameter. South coasts. At low water.

Crepidula fornicata (Pl. XV, fig. 66). Orton, 1909, *Proc. Roy. Soc.* B, LXXXI, p. 468. Shell solid, apex spirally curved towards the right and projecting beyond the right posterior part of the base. Growth lines distinct, at first spiral then gradually flattening out. Diaphragm deep set, convex towards the inside, equal to half of the shell base. Brownish grey, with numerous narrow brown streaks arranged in longitudinal lines, diaphragm white, remainder of inside streaked with brown, but variable. Animal irregularly pigmented with black. Gill filaments very long. Young stages motile, but adults sedentary, fixing themselves one above the other in a permanent chain of from two to twelve individuals. Protandrous hermaphrodites, the youngest being always functional males, older ones hermaphrodites and oldest females; all stages occur in the chain, whose base carries the oldest member. Shell 40 mm. long, 25 mm. broad. On oyster beds. Introduced with American oysters about 1880, now spreading along south and east coasts. A detritus feeder, using up the oysters' food.

NATICIDAE.

Carnivorous Molluscs living in sand. Shell opaque and globular, umbilicate, aperture large, semi-circular. Proboscis long and retractile; tentacles widely separated with an eye at the base situated laterally. Foot large with propodium reflected over head and behind over shell. Operculum ear-shaped. Eggs in stiff, sandy ribbons, 25 mm. wide.

1. Spire short; umbilicus large and deep. **Natica catena**
Spire short, but more prominent; umbilicus partly covered by the lip. **N. poliana**

Natica catena (Pl. XV, fig. 67 and Pl. XVI, fig. 71). F. and H. 1853, III, 326, as *monilifera*; J. 1867, IV, 220. Shell glossy, pale yellow with reddish brown streaks or zigzag markings at the top of each whorl, with seven swollen, rapidly enlarging whorls. Aperture equals one-third of the shell in area and two-thirds of the spire in length; expanded and angulated at the base, umbilicus large and deep. 36 mm. square. Sandy bays, local.

N. poliana. F. and H. 1853, III, 330, as *nitida*; J. 1867, IV, 224, as *alderi.* Shell buff with spiral rows of reddish brown or light chocolate streaks and spots, with six convex and compressed whorls; aperture three-quarters of the spire in length; inner lip forms a very thick ridge at the upper angle and a large callosity which projects over the umbilicus. 18 mm. high, 17 mm. broad. Abundant in sand from low-water mark, but local.

LAMELLARIIDAE.

Shell internal, ear-shaped, with a very short lateral spire. Aperture forms the base of the shell. Body flattened; mantle, which is folded to form a short anterior siphon, covers the shell. Tentacles separated by a membrane, bearing eyes on tubercles at their outer bases.

1. Shell completely covered, white and thin. **2**
   Shell partly uncovered, yellowish brown, not very thin. **Velutina (velutina)**
2. Mantle thick; body convex. **Lamellaria perspicua**
   Mantle thinner, more opaque; body flatter. **L. latens**

Velutina velutina (*laevigata*). F. and H. 1853, III, 347; J. 1867, IV, 240. Shell oval, yellowish brown, semi-transparent, with regular but not deep spiral ridges; 3½ whorls; aperture roundish oval, equalling seven-eighths of the shell. Body with fine opaque lines; mantle pale yellow, thick, partly covering shell. 20 mm. high, 18 mm. broad. From low-water mark, sometimes found floating.

Lamellaria perspicua (Pl. XVI, fig. 72). F. and H. 1853, III, 355; J. 1867, IV, 235. Shell slightly convex, white and very thin, transparent when fresh, with 2½–3 tumid whorls. The animal varies in colour, being lemon with clear spots, yellow mottled with red and white, reddish brown, ash grey, etc.; mantle thick, outer epidermis thick in the centre and thinner at the edges. 17 mm. high, 13 mm. broad. Feeds on *Trididemnum* and other Ascidians, which it resembles closely. Lays its eggs in the Ascidian's cloaca.

L. latens. F. and H. 1853, III, 358 (*tentaculata*); J. 1867, IV, 235 (*perspicua*, male). Shell flatter and smaller, with less convex whorls. Body flatter, mantle thinner. Yellowish or white with yellow and black spots. Tentacles long and slender. Feeds on ***Trididemnum*** and other Ascidians.

CYPRAEIDAE.

Shell oval, smooth and polished, with the spire hidden. A narrow slit-like aperture which extends the whole length of the shell; short, straight siphonal canal. Mantle capable of being expanded over shell; short anterior siphon. Tentacles cylindrical with eyes on short stalks which arise laterally at the base.

1.	Shell with three purplish brown spots.	**Trivia monacha**
	Shell without spots.	**T. arctica**

Trivia (Cowrie). Shell with twenty to twenty-five ribs, colour pale. Mantle very large, brightly spotted. Vase-shaped stalked egg cases attached to the jelly of compound Ascidians.

T. monacha (*Cypraea europaea*). F. and H. 1853, III, 495. Shell with three purplish brown spots. 12 mm. × 9 mm. Lower shore, common.

T. arctica. In literature as a variety of above sp. Shell without the spots. Not so common; deeper water form but occurs on lower shore also.

MURICIDAE.

Shell oval, or oblong with a well-developed spire, sculptured. Aperture oval, columella straight; siphonal canal in the shell; horny, non-spiral operculum. Mantle large, with a siphon; head small, with a retractile proboscis; eyes on stalks which are joined along their length to the tentacles.

1.	Shell with strong longitudinal ribs.	**2**
	Shell without longitudinal ribs.	**Nucella**
2.	Ribs rugged and wavy. Siphonal canal often closed.	**Ocenebra**
	Ribs more regular. Siphonal canal open.	**Urosalpinx**

Nucella lapillus (*Purpura*). Dog Winkle. F. and H. 1853, III, 380; J. 1867, IV, 276. Shell a short cone, thick and opaque, with numerous flat, spiral ridges and close-set striae in the growth lines. Colour variable, usually white but often banded; six or seven convex whorls. Mantle thick, siphon short. Foot oblong. Egg capsules like little vases, tinged with purple, attached to rocks in spring. 45 mm.

high, 26 mm. broad. Very common. Gregarious on rocks. Middle shore.

Ocenebra erinacea (*Murex*). Sting Winkle. F. and H. 1853, III, 370; J. 1867, IV, 306. Shell conical, solid, rather dull, with strong longitudinal ribs and some spiral ones which alternate in size; close-set wavy striae present. Siphonal canal closed in older specimens. Yellowish white with red and brown streaks, eight to ten convex angulated whorls. Foot small and narrow. Shell 50 mm. high, 25 mm. broad. Common at low-water mark and on lower shore.

Urosalpinx cinerea. Oyster-drill. Orton, *Essex Naturalist*, XXII, 298, 1930. Shell somewhat resembling that of *Ocenebra* but with smaller, more fluted and more regular longitudinal ribs. Spiral ribs less strong, alternately large and small. Six or seven whorls. Siphonal canal open, curved or obliquely twisted. Brown or greyish and rough outside, inside lighter, shiny and porcellanous. 30 mm. high, 17 mm. broad. Introduced from America with the oyster trade and now spreading round the south and east coasts. Like *Crepidula fornicata*, it is a serious pest in oyster beds.

Buccinidae.

Shell oval, aperture oval, whorls tumid. Outer lip slightly grooved, siphonal canal wide. Operculum ringed, non-spiral, columella smooth. Eyes small, one-third of the way up the tentacles. Pallial siphon extensile. Proboscis long and powerful. Penis in male. Foot large. Predaceous Molluscs.

Buccinum undatum (Pls. XV, fig. 68 and XVI, fig. 70). Common Whelk. F. and H. 1853, III, 401; J. 1867, IV, 285. Shell opaque and dull, sculptured with numerous fine ridges, crossed by fine longitudinal ones, colour yellowish white, seven to eight rounded whorls. Animal greyish white with black spots. 83 mm. high, 50 mm. broad. Low-water mark to deep water. Egg capsules fastened together in sponge-like masses, October to May.

Nassariidae.

Shell small and solid, with an elongated spire, sculptured; columella ridged, short siphonal canal, operculum horny with a terminal nucleus. Tentacles moderately long, with eyes on stalks which are joined to them about half-way up their length. Foot large, bilobed posteriorly, with two tentacle-like processes.

1. Shell with brown blotch on base. **Nassarius incrassatus**
Shell without brown blotch on base. **N. reticulatus**

Nassarius reticulatus (*Nassa reticulata*). F. and H. 1853, III, 388; J. 1867, IV, 346. Ribs of shell strong, slightly curved, crossed by spiral striae which give a tubercular appearance. Buff coloured with a white aperture and purple-brown band at base of the whorls. Ten whorls; outer lip with eight to twelve crenations. Long slender tentacles widely separated by a mentum. 32 mm. high, 17 mm. broad. Egg capsules flat, bottle-shaped, in rows on *Fucus*, 5 mm. high. Lower shore.

N. incrassatus (*Nassa incrassata*). F. and H. 1853, III, 391; J. 1867, IV, 351. Shell somewhat glossy; ribs strong, obliquely curved, crossed by inconspicuous striae forming tubercles. Pale buff, to brown and purple, with three bands of reddish brown. Brown blotch on base, near siphonal canal. Eight or nine whorls; outer lip with eight to ten folds. 15 mm. high, 8 mm. broad. Abundant on rocky ground from low-water mark. Egg capsules solitary.

Turridae.

Shell spindle-shaped or conical, turreted, aperture oblong, outer lip notched; columella smooth, siphonal canal nearly straight, no operculum. Siphon hardly protrudes from shell; tentacles pointed, far apart, eyes on stalks which are joined laterally to them; foot double-edged in front.

1. Shell spindle-shaped; aperture grooved. **Philbertia. 2**
Shell conical; aperture narrow, smooth. **Mangelia (attenuata)**
2. Ribs and striae fine and prominent; tentacles extensible. Body white. **P. linearis**
Ribs and striae numerous, not prominent, giving a granular appearance. Tentacles very short; body tinged purple. **P. purpurea**

Mangelia attenuata (*Pleurotoma*). F. and H. 1853, III, 488; J. 1867, IV, 377. Shell elongated, pointed at each extremity, thin, glossy, with fine sharp ribs which extend from base and across suture, giving the appearance of continuous ribs from apex to base. Pale tawny coloured, with a few reddish brown bands, nine or ten whorls. 15 mm. high, 5 mm. broad. Gravelly and muddy sand from low-water mark. Commonest in the south.

Philbertia purpurea. F. and H. 1853, III, 465, as *Mangelia*; J. 1867, IV, 373, as *Defrancia*. Shell solid and dull, with numerous fine ribs (longitudinal and spiral) which produce a granular appearance; suture deep. Purplish or reddish brown; spire long and tapering. 26 mm. long, 10 mm. broad. From low-water mark, common in the south.

P. linearis. F. and H. 1853, III, 470 (*Mangelia*); J. 1867, IV, 368 (*Defrancia*). Shell fairly broad, solid and opaque. Ribs strong, buttress-like, crossed by fine striae. Yellowish white with reddish brown lines; spire rather long. Tentacles extensile, saffron coloured, with bulbous tips. 9 mm. high, 4 mm. broad. Widely distributed on lower shore.

TECTIBRANCHIATA

Acteonidae.

Shell well developed, ovate, thick, external, with prominent spire and elongated aperture, finely striated and spirally banded; columellar lip spirally coiled; operculum present. Mantle slightly reflected over shell. Tentacles modified to form a bifid cephalic shield, with which the edges of the foot are continuous. This is used for digging. Foot oblong, truncated anteriorly. Penis long, reflected.

Acteon tornatilis (Pl. XVI, fig. 73). F. and H. 1853, III, 523 (*Tornatella fasciata*); J. 1867, IV, 433. Shell pale red, with spire of seven whorls, last large; aperture occupies two-thirds of the shell. Animal yellowish white. Cephalic shield prolonged posteriorly to form two leaf-like processes. 20 mm. high, 10 mm. broad. On sandy coasts, often abundant locally, to shallow water.

Atyidae.

Shell external, oval, spire reduced, usually concealed; aperture extending whole length of shell; no operculum. Body thick, not all enclosed within the shell. Tentacles partly distinct but forming a continuation of the head and usually projecting backwards as a pair of flattened processes. Foot large, with broad lateral lobes which cover part of the shell and are used for swimming. Plume-shaped tentacle (Hancock's organ) dorsal to the spermatic groove.

1. Shell oval, outer lip projecting. **Haminoea navicula**
   Shell oblong, outer lip not projecting. **H. hydatis**

Haminoea navicula (*Bulla hydatis*). F. and H. 1853, III, 530;

J. 1867, IV, 437. Shell fragile and thin, oval, with outer lip projecting. 25 mm. high, 20 mm. broad. Local, in mud.

H. hydatis. J. 1867, IV, 439, as *Bulla elegans.* Smaller oblong shell which is fairly solid and narrower than that of *navicula.* Outer lip does not project beyond the shell. Local, in mud.

AKERATIDAE.

Shell resembling that of *Haminoea*, thin and elastic, with short spire and deeply sunk sutures. Aperture auriculate, occupying nearly the whole length of the shell. No operculum. Body not wholly retractile within the shell. Head snout-like, elongated, cephalic shield flat without posterior processes. Foot very large, its sides prolonged into flexible swimming lobes shaped like those of a Pteropod. These lobes are wrapped round the shell when not in use. Mantle with a tentacular process which protrudes through a notch in the shell near the spire.

Akera bullata. F. and H. 1853, III, 527; J. 1867, IV, 430. Shell semi-transparent, glossy, greenish, elastic. To 30 mm. long, 16 mm. broad, but usually smaller. Body speckled and streaked with dark colours on a varying background of grey, white or orange. Gregarious. On mud flats at extreme low-water mark on south and west coasts.

PHILINIDAE.

Shell thin, wholly internal, with small, loosely coiled spire. Aperture very large; tentacles form a cephalic shield. No eyes in British form. Foot broad and folded on each side to form a pair of flaps.

Philine aperta (Pl. XVI, fig. 75). F. and H. 1853, III, 539; J. 1867, IV, 457. Shell squarish oval, 20 mm. high, 18 mm. broad, semi-transparent and glossy, white. Aperture forms seven-eighths of the lower surface. Animal white, cut dorsally into four lobes:—cephalic shield, two lateral foot lobes and a mantle lobe covering the shell posteriorly. To 50 mm. long, 30 mm. broad. From low-water mark on all coasts.

PELTIDAE.

No shell. Body small, slug-like, depressed; mantle separated from the foot by a deep groove; eyes sessile, far apart. Cephalic shield continuous with dorsal integument of body.

Pelta coronata (*Runcina hancocki*). F. and H. 1853, III, 612; J. 1869, v, 15. Elongated, smooth. Mantle nearly black, merging into buff or brown. Ctenidium of three plumes only. Foot yellow with reflexed sides, extending backwards beyond the mantle, and pointed. Minute animal, 6 mm. long, 3 mm. broad. In rock pools, between tide marks, often present in abundance.

APLYSIIDAE.

Shell reduced and internal, with very little trace of a spire, flat, enclosed by the mantle, which may have an aperture. Margins of foot upstanding as swimming lobes (parapodia). Pedal sole long, used for crawling. Seminal groove present.

Aplysia punctata (Pl. XVI, fig. 74). Sea hare. F. and H. 1853, III, 554 (*A. hybrida*); J. 1869, v, 5. Anterior tentacles forming a broad oral veil. Posterior tentacles or rhinophores ear-shaped, with eyes at their bases. Shell placed centrally on the back. 30 mm. long, 20 mm. broad. Animal olive green, spotted with white and black, red when young; ejects purple from pallial glands, also acrid substance from the subpallial glands. Large animals may attain a length of 120 mm. Lower shore and shallow water, amongst weed, on all coasts. At times very abundant. It comes inshore to spawn in spring. Egg strings very long, orange or pink, coiled round weed, gelatinous.

PLEUROBRANCHIDAE.

Shell absent or covered by mantle, which is very large, flabby, and may project over the foot. Two well-developed rhinophores with eyes at bases. Oral tentacles form a broad veil over mouth. Foot large, flattened, without parapodia. No pallial cavity. Ctenidium on right side between mantle and foot, often hanging out.

1. Animal yellow with white spots and smooth mantle; shell apex coiled. Ctenidium free at apex. **Berthella (plumula)**
   Animal yellow to brown, with tuberculated mantle. Spire of shell vestigial. Ctenidium long and slender. **Pleurobranchus (membranaceus)**

Berthella plumula (*Pleurobranchus*) (Pl. XVI, fig. 76). F. and H. 1853, III, 559; J. 1869, v, 11. The body is flabby, ovate or oblong in outline, and raised or humped in the centre, where the shell lies, covered by the voluminous overhanging mantle. When the animal is creeping on the broad flat foot, the short blunt diverging rhinophores and truncated oral veil often project beyond the mantle rim, and on the right side the ctenidium may hang out postero-laterally. 40 mm. long, 25 mm. wide. Common locally at low-water mark, under rocks. Feeds on Tunicates. Its egg capsules are buried in the jelly of compound forms.

Pleurobranchus membranaceus. F. and H. 1853, III, 558; J. 1869, V, 10. Body oval, rather flattened, with rugged nodular mantle of a brown or rusty red colour. Mantle relatively smaller than in *Berthella* and not overhanging the foot. Foot, rhinophores, oral veil and ctenidium paler, usually yellow. To 100 mm. long. From low-water mark, but less common than *Berthella*.

NUDIBRANCHIATA

Note. Some authors place the Stiligeridae, Elysiidae and Limapontiidae in a separate order, the Ascoglossa.

STILIGERIDAE.

Anus dorsal. Cerata which contain diverticula of the digestive gland are present. No jaws.

1. Anus postero-dorsal. Rhinophores reduced or absent. **Alderia (modesta)**

   Anus antero-dorsal. Rhinophores folded longitudinally. **Hermaea. 2**

2. Rhinophore tips unequal, appearing bifid. Animal yellow and red. **H. bifida**

   Rhinophore tips not bifid. Animal green and white. **H. dendritica**

Alderia modesta. F. and H. 1853, III, 611; J. 1869, V, 33; A. and H. 1845, Fam. 3, pl. 41. Anus postero-dorsal. Cerata obtuse, coloured like the body, in six or seven diagonal rows. Head small, lobed, but without definite tentacles. Greenish yellow or yellowish fawn with dark blotches and opaque white spots. 12 mm. long. Local, in brackish water. East coast of England and Scotland, Irish coasts. Feeds on green weed *Vaucheria*.

Hermaea. Active Molluscs. Anus antero-dorsal. Cerata long and simple. Rhinophores auriculate, oral tentacles indistinct.

H. bifida. F. and H. 1853, III, 610; J. 1869, V, 33; A. and H. 1845, Fam. 3, pl. 39. Cerata numerous, transparent, pointed, unequal in size, slightly tuberculated towards the tip. Very slender, pale yellow and pellucid; hepatic diverticula appear bright red through the skin of the cerata. 25 mm. long. On the weed, *Griffithsia flosculosa*, between tide marks and in shallow water.

H. dendritica. F. and H. 1853, III, 609; J. 1869, v, 34; A. and H. 1845, Fam. 3, pl. 40. Anus on a short conspicuous tube. Cerata long, in eight rows of three or four. Greenish white with green dendritic markings. Hepatic diverticula bright green. 7 mm. long. On the green weed *Codium tomentosum*.

ELYSIIDAE.

No cerata or branchiae. Anus dorsal. Two auriculate tentacles. No jaws. Body with latero-dorsal lobes for swimming.

Elysia viridis. F. and H. 1853, III, 614; J. 1869, v, 31; Eliot, 1910, pl. vii. Body oval, grass green to bluish green, with light green spots. Lateral lobes with white margins and white round the eyes. Resembles *Aplysia* but flatter. 20 mm. long. On green weed, middle to lower shore. Annual.

LIMAPONTIIDAE.

Body small, smooth, slug-like, without cerata or branchial plumes. Anus dorsal and posterior. No jaws.

1. Rhinophores well developed. **Actaeonia**
   Rhinophores represented by crests. **Limapontia. 2**
2. Body linear. Up to 3 mm. **L. capitata**
   Body depressed, expanded posteriorly. Up to 10 mm. **L. depressa**

Acteonia cocksi (*Cenia cocksii*). F. and H. 1853, III, 616; J. 1869, v, 30; Eliot, 1910, pl. vii. Body rounded. Head with black line down centre and continuous as a broad band along the body. Rhinophores, rings round eyes, and the sides of the body pale coloured, green, yellow or cream. Foot yellow. 5 mm. long. On *Cladophora* spp. In shallow rock pools. Annual.

Limapontia capitata (*nigra*). F. and H. 1853, III, 614; J. 1869, v, 28. Body slender, tentacles represented by pale lateral crests. Eyes on bright patches. Anus dorsal. Colour brownish green or black; 3 mm. Gregarious, on green weeds between tide marks, or in rock pools. Sometimes very abundant. Active. Annual.

L. depressa. J. 1869, v, 29; Eliot, 1910, pl. vii. Body oblong, depressed, swelling posteriorly. Dark coloured, with yellowish spots. Eyes on oblong light patches. Anus posterior. Length 8 mm. Occurs in brackish water. Sluggish.

TRITONIIDAE.

Anus lateral. Rhinophores retractile within simple, undulating sheaths, consisting of a bundle of equal branched processes and one long, truncated, smooth process. Foot rounded anteriorly. Cerata plumose, arranged on a ridge of the mantle (pallial crest) on side of back; without ramifications of the digestive diverticula. Oral veil and oral tentacles present.

Tritonia lineata. F. and H. 1853, III, 583; J. 1869, v, 68; A. and H. 1845, Fam. 2, pl. 4. Slender, transparent white, oral veil narrow, with four long filaments, the two central ones the longest, rhinophore branches delicate. Sheaths long, with undulating margin. Pallial ridges with about four pairs of cerata, which are slender, imperfectly pinnate, white, with an opaque white line running out from the longitudinal line which follows the pallial ridge. Cerata all small but the first the largest. 18 mm. long. On *Alcyonium*.

POLYCERIDAE.

No cerata, though lateral processes may be present on the mantle. Non-retractile branchial plumes surround a dorsally placed anus. Mantle usually rather small, sometimes much smaller than the body, or a little larger than the body.

1. Rhinophores retractile, with distinct sheaths. 2
   Sheath absent or indistinct. 3
2. Pair of processes behind branchial circlet. **Thecacera pennigera**
   No processes, but a ring of tubercles. **T. virescens**
   Processes present but neither tubercles nor processes round branchial circlet. **Euphurus claviger**
3. Lateral processes near branchial circlet. 4
   No lateral processes. 5
4. One pair of smooth conical lateral processes behind branchial circlet. **Polycera quadrilineata**
   Five pairs of smooth conical processes beside branchial circlet. **Ancula cristata**
   Three branched mammillate processes anterior to branchial circlet. **Aegires punctilucens**
5. Mantle small, with a distinct rim, not overhanging the foot. 6
   Mantle overhanging the foot laterally. Animal shaped like a Dorid. 9

6. Branchial plumes more than five. **Goniodoris.** 7
Branchial plumes five or fewer. **Palio.** 8
7. Thirteen large pinnate plumes. Body white. **Goniodoris nodosa**
Seven to nine tripinnate plumes. Body red-brown. **G. castanea**
8. Three branchial plumes. Body yellowish green. **Palio dubia**
Five branchial plumes. Body dark, brownish green. **P. nothus**
9. Mantle papillose. Seven to nine branchial plumes with white lines on the stems. **Acanthodoris pilosa**
Mantle tuberculate. 10
10. Branchial plumes nine large and two small, distant from the anus. **Adalaria (proxima)**
Branchial plumes more than nine, equal. **Onchidoris.** 11
11. Eleven small branchial plumes. 12
Twenty or more branchial plumes. **O. fusca**
12. Eyes visible. Mantle covering the tail. **O. depressa**
Eyes not visible. Tail exposed. **O. muricata**

Thecacera. Body smooth, mantle not clearly defined. No oral tentacles, but an oral veil may be present. Rhinophores lamellate, retractile into sheaths. Branchiae plumose, non-retractile, surrounding the dorsal anus.

T. pennigera. F. and H. 1853, III, 575; J. 1869, v, 72; A. and H. 1845, Fam. 1, pl. 21*a*. Body smooth, elongated and humped, white, but heavily marked with orange blotches and black spots. A pair of club-shaped processes behind the branchial circlet. Three short, white, tripinnate branchiae arise from a single stem. 10 mm. long. A southern species, not common. Feeds on *Bugula*.

T. virescens. F. and H. 1853, III, 576; J. 1869, v, 73. Body somewhat humped, smooth, pale coloured with green blotches. Rhinophores green, laminated, with simple sheaths. Five green branchiae edged with white and surrounded by a row of tubercles. 5 mm. long. Rare.

Euphurus claviger (*Triopa*). F. and H. 1853, III, 573; J. 1869, v, 71; A. and H. 1845, Fam. 1, pl. 20. Body elongated, depressed, white, with a row of ten to twelve orange-tipped processes on each side of the mantle edge and round its anterior border, twenty to twenty-four in all, and orange tubercles on back. Rhinophores with fourteen or fifteen yellow laminae and small plain sheaths. Three rather long bipinnate branchial plumes. 20 mm. long. Local. Feeds on Bryozoa. Probably an annual.

Polycera quadrilineata. F. and H. 1853, III, 576; J. 1869, v, 75; A. and H. 1845, Fam. 1, pl. 22. Body smooth, white, with rows of yellow tubercles. Mantle veil with four yellow digitations, and continuous as a slight pallial ridge. Seven large pinnate branchiae, and occasionally two small ones as well; on each side of circle a single, large, yellow-tipped process directed posteriorly. 20 mm. long. On weed between tide marks and in shallow water. Very dark sooty specimens occur locally. Egg ribbon a broad white strap, 3–4 mm. wide, forming one coil. Feeds on Bryozoa.

Ancula cristata (Pl. XVII, fig. 79). F. and H. 1853, III, 580; J. 1869, V, 77; A. and H. 1845, Fam. 1, pl. 25. Body somewhat linear, with a slender tail, transparent, white. Mantle represented by ten processes tipped with orange, around the branchial circlet. Rhinophores large, each with two yellow-tipped basal appendages. 13 mm. long. Intertidal. Feeds on Bryozoa and Tunicates.

Aegires punctilucens (*Aegirus*). F. and H. 1853, III, 574; J. 1869, V, 70; A. and H. 1845, Fam. 1, pl. 21. Body stout, humped, purplish fawn coloured, with minute white spots and large dark blotches with a brilliant turquoise blue centre. Covered all over with regular lines of tubercles which are mammillated at the top and have a bright red spot; three large branched ones in front of branchial circlet. Rhinophores stout, smooth, with a blunt five-lobed mammillated sheath and bifid apex. Three broad tripinnate branchial plumes. 20 mm. long. Local. Feeds on *Membranipora*.

Goniodoris. Mantle smaller than the body, with a sinuous margin. Foot broad. Branchial plumes numerous.

G. nodosa (Pl. XVII, fig. 78). F. and H. 1853, III, 572; J. 1869, V, 81; A. and H. 1845, Fam. 1, pl. 18. Body transparent, white. Mantle small, exposing head and foo , one or two rows of tubercles down centre, margin reflected and usually wavy. Rhinophores yellow. Head produced laterally into tentacular processes. Thirteen large pinnate branchial plumes, and sometimes three smaller ones nearly surrounding the anus. 25 mm. long. Fairly common between tide marks on all coasts. Feeds on Bryozoa and Tunicates.

G. castanea. F. and H. 1853, III, 572; J. 1869, V, 82; A. and H. 1845, Fam. 1, pl. 19. Body oval, reddish brown, with opaque white spots. Mantle small with a reflected margin, a pale-coloured ridge down the centre. Branchial plumes meet round anus. 25 mm. long. Feeds on *Botryllus* and *Botrylloides*.

Acanthodoris pilosa (*Doris pilosa*). F. and H. 1853, III, 570; J. 1869, V, 93; A. and H. 1845, Fam. 1, pl. 15. Body oval, very convex, semi-pellucid, colour ranging from white to purple-brown Rhinophores long, curved in a posterior direction for about half their length, with nineteen or twenty delicate laminae; small denticulate sheaths. Oral tentacles form a veil. 13–32 mm. long. Between tide marks or in shallow water. Feeds on Bryozoa.

Palio. Mantle small, with a lobed or wavy margin. Branchial plumes few, not forming a complete circlet. Body tuberculated.

P. dubia (*Polycera lessonii*). F. and H. 1853, III, 577; J. 1869, V, 76; A. and H. 1845, Fam. 1, pl. 24. Body yellowish green, wrinkled. A row of yellow tubercles on slightly projecting mantle margin. Three bipinnate branchial plumes, lateral ones with large branches. 10 mm. long. East and west coasts. Low-water mark to shallow water. Common on north-east coast, less common on south and west coasts. Feeds on Bryozoa.

P. nothus (*Polycera ocellata*). F. and H. 1853, III, 578; J. 1869, V, 76; A. and H. 1845, Fam. 1, pl. 23. Body greenish black, with large tubercular pale-coloured spots. Mantle-edge wavy, margined with white. Three long and two short imperfectly tripinnate plumes. 13 mm. long. Under stones between tide marks.

Adalaria proxima (*Doris*). J. 1869, V, 88; A. and H. 1845, Fam. 1, pl. 9. Body elliptical, orange yellow. Rhinophores linear, lamellate for nearly half their length. Mantle large, with dorsal tubercles, loaded with spicules. Branchial plumes nine large and two small, simply pinnate, set away from the anus. 12 mm. long. Between tide marks. Feeds on the Bryozoan *Electra pilosa*.

Onchidoris. Mantle large, covering all the body except the tail, tuberculate, loaded with spicules. Rhinophores lamellate. Oral tentacles not defined but fused to form a veil over the mouth. Branchial plumes simply pinnate. Feeds on Bryozoa and *Balanus*.

O. fusca (*Doris bilamellata*). F. and H. 1853, III, 567; J. 1869, V, 90; A. and H. 1845, Fam. 1, pl. 11. Body elliptical, dull white, mottled with brown. Branchial plumes variable in number, pinnate (twenty to twenty-nine in larger individuals), forming a circle which is incomplete posteriorly around the anus. 13–32 mm. Common in spring on all coasts. Feeds mostly on Barnacles.

O. depressa (*Doris*). F. and H. 1853, III, 568; J. 1869, V, 91; A. and H. 1845, Fam. 1, pl. 12. Body oval, thin, transparent and very flat. Yellowish, spotted with red or brown. Spicules visible through the skin, in radiating lines. Ten or eleven very small pinnate plumes.

Eyes visible at base of tentacles. 3–9 mm. long. Under stones between tide marks and in shallow water.

Onchidoris muricata (*Doris aspera*). F. and H. 1853, III, 567; J. 1869, V, 87; A. and H. 1845, Fam. 1, pl. 9. Body elliptical, white, semi-transparent, with spicules shining through. Mantle with stout flattened tubercles. Liver shows through as a reddish patch dorsally, body of a pink tint generally. Eleven small branchial plumes. 8 mm. long. An active northern species, rare on the south coast. Under stones between tide marks. Diet Bryozoa.

DORIDIDAE.

Anus median dorsal, surrounded by *retractile* branchial plumes. Mantle large, without appendages, usually overhanging the foot on all sides, or at least laterally.

1. Oral tentacles linear. 2
   Oral tentacles short. Branchial plumes tripinnate. 3
2. Up to fifteen large white tripinnate plumes. Anus brown. **Jorunna (tomentosa)**
   Ten small, simply pinnate plumes. Mantle red. **Rostanga (rufescens)**
3. Five small branchiae. **Cadlina (laevis)**
   Nine branchiae. **Archidoris.** 4
4. Rhinophores with ten or eleven laminae. Mantle red. **A. flammea**
   Rhinophores with about twenty alternately large and small laminae. Mantle buff, mottled. **A. britannica**

Archidoris britannica (*Doris tuberculata*). The sea lemon. F. and H. 1853, III, 563; J. 1869, V, 83; A. and H. 1845, Fam. 1, pl. 3. The largest British Nudibranch. Body elliptical, yellow or buff coloured, blotched with pink, brown, or green, rarely white. Tuberculated. Rhinophores conical with about twenty alternately large and small laminae. Nine tripinnate branchiae. 50–75 mm. Common in spring on all coasts. Egg ribbons 25 mm. wide, 40 cm. long, white, 1–3 turns. March. Feeds on *Halichondria*. Lives two years.

A. flammea (*Doris*). F. and H. 1853, III, 564; J. 1869, V, 83; A. and H. 1845, Fam. 1, pl. 4. Body red, mantle covered with flattish tubercles. Rhinophores orange, large, with ten or eleven scarlet laminae. Nine tripinnate branchiae. 20–30 mm. Fairly common, often on red sponges.

Cadlina laevis (*Doris repanda*). F. and H. 1853, III, 566; J. 1869, V, 86; A. and H. 1845, Fam. 1, pl. 6. Semi-transparent, white, mantle large with opaque white tubercles and an irregular row of white or yellow ones down the sides. White or yellowish rhinophores with twelve or thirteen laminae, sheath tuberculated. Five small tripinnate branchiae. 25 mm. long. Northern. Amongst rocks between tide marks. Feeds on sponges.

Rostanga rufescens (*Doris coccinea*). F. and H. 1853, III, 565; J. 1869, v, 86; A. and H. 1845, Fam. 1, pl. 7. Mantle minutely tuberculated, scarlet, with small black spots. Rhinophores stout, yellow, with tubercles but no sheath at base, and ten deeply cleft laminae. Oral tentacles long and slender. Ten small, simply pinnate branchiae forming a small circle. 13 mm. Southern. Under stones, between tide marks, often on red sponges.

Jorunna tomentosa (*Doris johnstoni*). F. and H. 1853, III, 564; J. 1869, v, 85; A. and H. 1845, Fam. 1, pl. 5. Animal ovate, flattish. Yellowish white or buff with a few brown spots. Mantle large with fine tubercles. Up to 50 mm. long. Common on rocky shores and in shallow water. Egg ribbons of four coils, with scalloped margin, buff. May–June. Feeds on *Halichondria*.

DENDRONOTIDAE.

Rhinophores laminated, retractile into much branched sheaths. Head with a branched veil. Cerata also arborescent, in two rows along the body. Anus lateral. Feeds on hydroids.

Dendronotus frondosus (*D. arborescens*). F. and H. 1853, III, 586; J. 1869, v, 62; A. and H. 1845, Fam. 3, pl. 3. Body elongated, reddish or yellowish, mottled with brown and white. Cerata very much branched, six or seven in a row on each side. Foot narrow, rounded in front. Up to 25 mm. long. From low-water mark.

ZEPHYRINIDAE.

Rhinophores non-retractile, without sheaths. Oral tentacles small, with a veil. Cerata irregular in number, round mantle edge and in front of rhinophores, containing diverticula of digestive gland. Anus dorsal.

1. Cerata smooth. **Janolus cristatus**
   Cerata mucronate or tuberculate. **J. hyalinus**

Janolus cristatus. F. and H. 1853, III, 609 (*Antiopa splendida*); J. 1869, v, 58 and A. and H. 1845, Fam. 3, pl. 44 (*Antiopa cristata*).

Body ovate, flattened, pale buff coloured; rhinophores yellow with white tips. Very numerous large smooth cerata, with white tips, arranged laterally on the body, leaving a bare median strip when the cerata are depressed. Foot deeply grooved in front. 40 mm. long. On south and west coasts. Feeds on Bryozoa.

J. hyalinus (*Antiopa hyalina*). J. 1869, v, 59; A. and H. 1845, Fam. 3, pl. 44. Elliptical, transparent, yellowish white, with red-brown markings. Rhinophores short, united at the base by a crest. Cerata numerous, densely and irregularly arranged on the ridge-like mantle edge, tuberculated, pellucid, sprinkled brown and opaque white. 10 mm. long. South coasts. On *Tubularia*.

DOTOIDAE

Body slender. Rhinophores smooth, slender, retractile into plain sheaths. Cerata elliptical, shaped like fir-cones, containing diverticula of the digestive gland, in a single series on each side of the body. Anus lateral.

1. Rhinophores with short sheaths. Nine pairs of close set cerata with about seven to nine rows of tubercles. **Doto (Idulia) fragilis**
   Rhinophores with fairly long sheaths. Five to seven pairs of cerata with three to five rows of tubercles. **D. coronata**

Doto fragilis. F. and H. 1853, III, 588; J. 1869, v, 60; A. and H. 1845, Fam. 3, pl. 5. Body oblong, yellowish brown or green. Nine close-set stout cerata on each side of the body, with ten rows of tubercles. 26 mm. long. On Plumularian hydroids.

D. coronata. F. and H. 1853, III, 587; J. 1869, v, 61; A. and H. 1845, Fam. 3, pl. 6. Slender, yellowish white with rose-coloured spots, rhinophores fairly long, sheaths trumpet-shaped. Head veil slight, produced into processes laterally, five to seven cerata, elliptical, stalked, with three to five rows of tubercles, each with a purple spot at the tip. Foot white, linear. 13 mm. long. On hydroids.

HEROIDAE.

Cerata branched, containing diverticula of the digestive gland, in a single series on each side of the body. Oral tentacles joined to form a head veil, which is produced at the sides. Rhinophores smooth, with sheaths. Anus lateral. On hydroids.

Hero formosa. J. 1869, v, 63; Eliot, 1910, pl. iv. Body slender, transparent white or rose-coloured, with three lines of opaque white. Two rows of six cerata, which are stalked and branched at the tip; first pair in front of rhinophores. 25 mm. long.

AEOLIDIIDAE.

Cerata very numerous, arranged in clusters, forming a dense covering to the body and usually extending forwards anterior to the rhinophores. Antero-lateral borders of the foot prolonged into short processes; posterior end (tail) relatively short. Anus lateral.

1. Rhinophores smooth. **3**
   Rhinophores lamellate. **Facelina. 2**
2. Seven or eight laminae on rhinophores. **F. longicornis**
   Twenty to thirty laminae on rhinophores. **F. drummondi**
3. Antero-lateral processes of foot long. **Favorinus (branchialis)**
   Antero-lateral processes of foot short. **4**
4. Body broad, eighteen to twenty rows of conical cerata. **Aeolidia (papillosa)**
   Body elongated, about fourteen rows of tapering cerata. **Eolidina (glauca)**

Facelina. Rhinophores lamellate or ringed.

F. longicornis (*Eolis coronata*). F. and H. 1853, III, 592; J. 1869, v, 39; A. and H. 1845, Fam. 3, pl. 12. Slender, white, tinged rose. Rhinophores yellow with seven or eight strong laminae. Oral tentacles very long and tapering. Cerata in six or seven clusters on each side of the back, rather long, crimson with bright blue tips. To 33 mm. Under stones, between tide marks.

F. drummondi (*Eolis*). A. and H. 1853, III, 593; J. 1869, v, 39; A. and H. 1845, Fam. 3, pl. 13. Body transparent, flesh-coloured, rather broad at head and tapering posteriorly. Rhinophores fawn-coloured, rather long. Oral tentacles very long. Cerata long, numerous, in four to six clusters, first with thirty to forty cerata. To 35 mm. long. An allied species, *F. punctata*, has opaque white spots. On hydroids.

Favorinus. Rhinophores smooth. Antero-lateral processes of foot long and slender.

F. branchialis (*Eolis alba*). F. and H. 1853, III, 597; J. 1869, v, 43; A. and H. 1845, Fam. 3, pl. 21. Body slender, white. Rhinophores dark coloured, bulbous near the tips, with white apices, oral tentacles long. Cerata linear, white with a green band near the tips, in five or six clusters. 20 mm. long. On shore sponges and hydroids.

Aeolidia. Rhinophores smooth. Foot broad, with short antero-lateral processes.

A. papillosa (*Eolis*). F. and H. 1853, III, 590; J. 1869, v, 37; A. and H. 1845, Fam. 3, pl. 9. Broadly ovate, brownish grey spotted with lilac, brown and white. Rhinophores simple, short; oral tentacles a little longer. Cerata conical, brown and white, twelve to twenty-four in a row. Up to 9 cm. Under stones between tide marks. Common. The largest Eolid. Egg ribbons gelatinous, white, spiral. On *Tubularia* and Anemones.

Eolidina. Rhinophores smooth. Body slender.

E. glauca (*Eolis*). F. and H. 1853, III, 591; J. 1869, v, 38; A. and H. 1845, Fam. 3, pl. 11. Body elongated, reddish. Rhinophores red, tipped white, slightly shorter than oral tentacles. About fourteen rows of tapering cerata. Foot with small antero-lateral processes. 45 mm. long. On Anemones.

TERGIPEDIDAE.

Cerata usually few in number, arranged in a single row on each side of the body with the members of the two rows alternating, or in transverse rows. All the cerata posterior to the smooth rhinophores. Foot very narrow, rounded anteriorly, prolonged into a long slender tail posteriorly.

1. Cerata, in two longitudinal rows, alternating. **Tergipes**
   Cerata in transverse rows. **Cratena. 2**
2. Colour predominantly green with a reddish streak between the tentacles. Six to eight rows of three or four cerata. **C. foliata**
   Colour predominantly red with orange rhinophores. Cerata in ten or eleven close-set rows. **C. aurantia**

Tergipes despectus (*Eolis despecta*) (Pl. XVII, fig. 80). F. and H. 1853, III, 605; J. 1869, v, 56; A. and H. 1845, Fam. 3, pl. 36. Slender, transparent, white, with greenish viscera. Cerata large, club-shaped, in two rows of four down the back, those of each side alternating. 6 mm. long. Lower shore. On hydroids.

Cratena. Cerata in transverse rows.

C. aurantia (*Eolis aurantiaca*). F. and H. 1853, III, 600; J. 1869, v, 48; F. and H. 1845, Fam. 3, pl. 27. Rather stout, pale buff. Rhinophores fairly long, orange. Cerata in ten or eleven close transverse rows, purplish red, with white bands and orange tips. 12 mm. long. Between tide marks. On hydroids.

C. foliata (*Eolis olivacea*). F. and H. 1853, III, 599; J. 1869, v, 47; A. and H. 1845, Fam. 3, pl. 26. Pale yellow with opaque white spots and greenish cerata. Rhinophores short. A reddish streak between tentacles and down the head. A blotch sometimes on the back of head. Six to eight rows of three or four stout cerata, olive coloured with paler tips. 12 mm. long. Under stones.

FLABELLINIDAE.

Cerata fairly numerous though fewer than in the Aeolidiidae, none anterior to the smooth or faintly ringed rhinophores. Foot elongated, with blunt projections, or rounded anteriorly. Tail rather long and slender. Anus lateral.

Coryphella. Rhinophores finely ridged or simple but not lamellate. Feeds on hydroids.

1. Oral tentacles nearly as long as the rhinophores. **C. verrucosa rufibranchialis**
   Oral tentacles longer than or equal in length to the rhinophores. 2
2. Body white with orange or scarlet cerata, tipped white. **C. verrucosa pellucida**
   Body violet with orange cerata. **C. pedata**
   Body white with three opaque white lines. Cerata rose coloured. **C. lineata**

C. verrucosa rufibranchialis (*Eolis rufibranchialis*) (Pl. XVII, fig. 81). F. and H. 1853, III, 593; J. 1869, v, 41; A. and H. 1845, Fam. 3, pl. 14. Slender and tapering, white. Rhinophores slender, slightly ridged. Cerata rose-coloured with white tips, in six or seven groups on each side. 25 mm. long. Low-water mark.

C. verrucosa pellucida (*Eolis gracilis*, *E. pellucida*). F. and H. 1853, III, 595, 596; J. 1869, v, 42; A. and H. 1845, Fam. 3, pls. 18 and 19. Body slender, white. Rhinophores long and white. Cerata scarlet or orange with white tips. 20 mm. long.

C. pedata (*Eolis landsburgi*). F. and H. 1853, III, 596; J. 1869, v, 43; A. and H. 1845, Fam. 3, pl. 20. Slender, violet. Rhinophores long and tubular, with white tips. Oral tentacles longer, also tipped white. Five or six clusters of linear cerata, orange red with white tips. 8 mm. long.

C. lineata (*Eolis*). F. and H. 1853, III, 594; J. 1869, v, 41; A. and

H. 1845, Fam. 3, pl. 16. Four or five clusters of cerata, rose-coloured, with white tips and white lines. Foot linear, produced at sides. 25 mm. long. Intertidal and in shallow water.

Eubranchus. Cerata club-shaped. Rhinophores simple.

1. Cerata few, in five or six rows. **E. exiguus**
   Cerata numerous, in thirteen or fourteen rows. **E. tricolor**

E. exiguus (*Eolis*). F. and H. 1853, III, 606; J. 1869, v, 56; A. and H. 1845, Fam. 3, pl. 37. Yellowish white, mottled green and brown. Rhinophores long with two brown bands, oral tentacles short with one brown band. Five or six rows of oval, white cerata with two or three brownish green bands, first row of six, third of four, then only two. Foot short. Very small, only 5 mm. long. On hydroids.

E. tricolor (*Eolis*). F. and H. 1853, III, 604; J. 1869, v, 54; A. and H. 1845, Fam. 3, pl. 34. Body somewhat broad, buff coloured to yellow. Rhinophores stout, fairly long. Oral tentacles about half as long. Thirteen or fourteen rows of three to five large, ovate purple-streaked cerata which taper abruptly to the apex where they are encircled by a ring of golden yellow. Foot short. 25 mm. long. Feeds on hydroids.

PULMONATA

BASOMMATOPHORA

ELLOBIIDAE.

Shell oval or spindle-shaped, glossy, spire conical, aperture narrow, oblong, with flutings. Columella with folds, two or three of which are enlarged to form denticulations. No umbilicus, no operculum. Head with a snout, eyes sessile at inner base of the club-shaped tentacles. Respiratory aperture on the right.

1. Shell solid, white. Tentacles flattened. **Leucophytia bidentata**
   Shell thin, red, brown or yellow. Tentacles thick, streaked with purple. **Phytia myosotis**

Leucophytia bidentata. F. and H. 1853, IV, 191, as *Conovulus*; J. 1869, v, 104, as *Melampus*. Shell somewhat solid, nearly opaque, white, with six or seven rounded whorls. Tentacles very short and compressed. 6 mm. long, 3 mm. broad. Under stones between tide marks.

Phytia myosotis. F. and H. 1853, IV, 194, as *C. denticulatus* var.

myosotis; J. 1869, v, 106, as *Melampus*. Shell thin, semi-transparent, reddish, brown or yellowish, of seven or eight whorls; tentacles thick, annulated in the middle, tipped and streaked with purple. Respiratory aperture very small. 8 mm. long, 3 mm. broad. Mud flats and salt marshes.

OTINIDAE.

Shell ear-shaped, not covering the whole animal. Spire short, lateral, with introverted apex. Aperture very large. Tentacles very short with sessile eyes in the centre of each. Respiratory aperture a narrow slit.

Otina ovata (*otis*). F. and H. 1853, III, 321; J. 1869, v, 110. Shell rather thin, semi-transparent, dull coloured, reddish brown or purplish, with two convex whorls, last occupying nearly all the shell. Body white, eyes distinct. 3 mm. long, 2 mm. broad. On rocks between tide marks, in caves, and inside dead Barnacle shells, etc.

STYLOMMATOPHORA

ONCHIDIIDAE.

No shell. Body slug-like, head with a veil formed of two flattened lobes. Two tentacles, invaginable, with eyes at their tips. Respiratory aperture under hinder edge of mantle.

Onchidella celtica (*Onchidium celticum*). F. and H. 1853, IV, 3; J. 1869, v, 95. Body thick, greyish green, covered with close-set irregular tubercles, mantle thick, greyish white and covering the head. Head lobes broad, tentacles short and conical, with bulbous tips. 12 mm. long, 9 mm. broad. Local, on rocks between tide marks. South-west coasts.

EXPLANATION OF PLATE XV

Amphineura, Figs. 62 and 63; *Gastropoda* (i), Figs. 64–69

Fig. 62. *Acanthochitona crinitus*. Dorsal view. Six of the eight valves are keeled and beaked. Note the pear-shaped markings on the valves. × 4.

Fig. 63. *Acanthochitona crinitus*. Ventral view. The pinnate ctenidia are more numerous than in the figure. × 4.

Fig. 64. *Diodora apertura*. Right lateral view. The head has been twisted to the right to show the eyes. Note the absence of spire on the conical ribbed shell. × 1½.

Fig. 65. *Calliostoma zizyphinum*. Right lateral view. To show a spiral shell, with operculum, and a foot bearing lateral epipodia. × 1.

Fig. 66. *Crepidula fornicata*. Right lateral view to show the depressed shell with traces of the spire on the right side and the diaphragm or shelf set within the large oval aperture. × ¾.

Fig. 67. *Natica catena*: Non-siphonate type of shell with large umbilicus. × 1.

Fig. 68. *Buccinum undatum*. Siphonate type of shell without umbilicus. Ribs and ridges (striae) present. × ½.

Fig. 69. *Clathrus clathrus*. Turreted, non-siphonate type of shell, without umbilicus, but with complete peristome and longitudinal ribs. × 1½.

a.	apex of shell	i.t.l.	intertentacular lobe
an.	anus	mant.	mantle
ap.	aperture of shell	mant.t.	mantle tube
bk.	beak	mo.	mouth
b.t.	bristle tuft	o.l.	outer lip
b.w.	body whorl	operc.	operculum
col.	columella	p.	palp
col.l.	columellar lip	pst.	peristome
ct.	ctenidium	rib	rib
dia.	diaphragm	si.gr.	siphonal groove
e.	eye	su.	suture
epi.	epipodium	t.	tentacle
epi.t.	epipodial tentacle	umb.	umbilicus
f.	foot	v.1.	first valve
gir.	girdle	v.8.	eighth valve
gr.l.	growth line of shell		
ho.	hole in shell for mantle tube		

XV. MOLLUSCA (AMPHINEURA AND GASTROPODA)

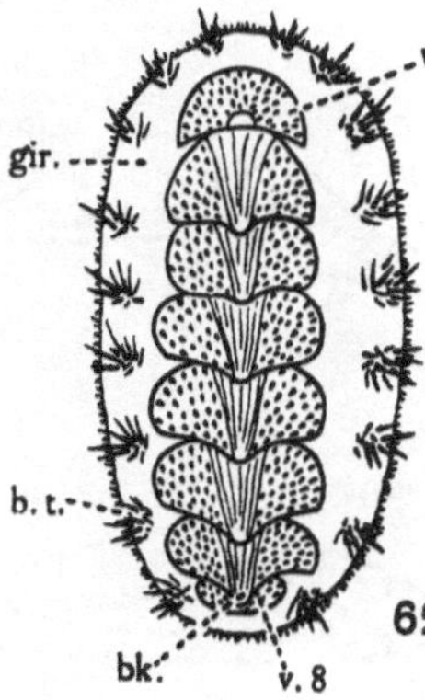

62

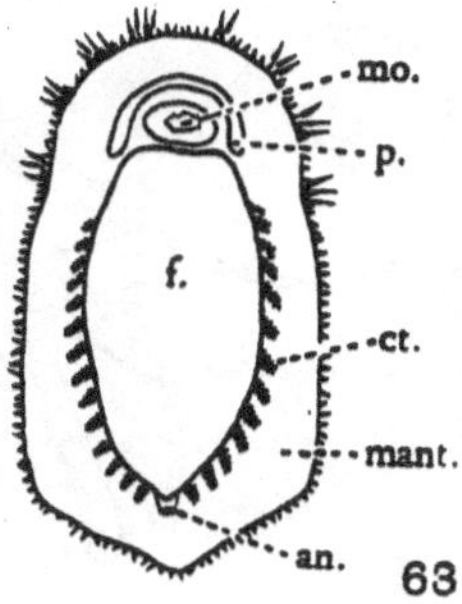

63

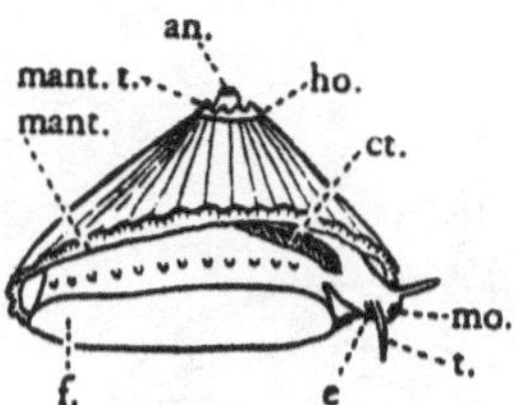

64

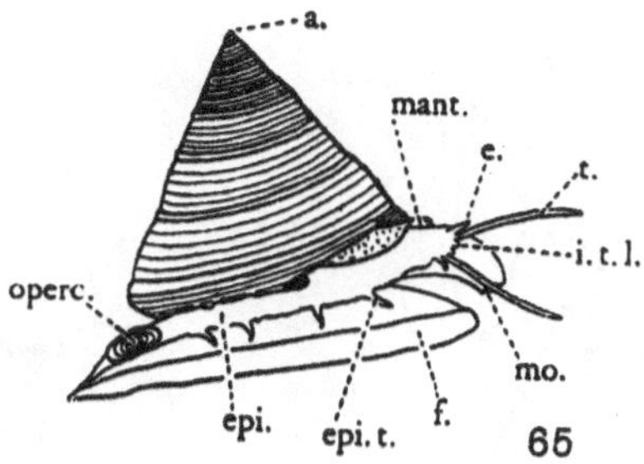

65

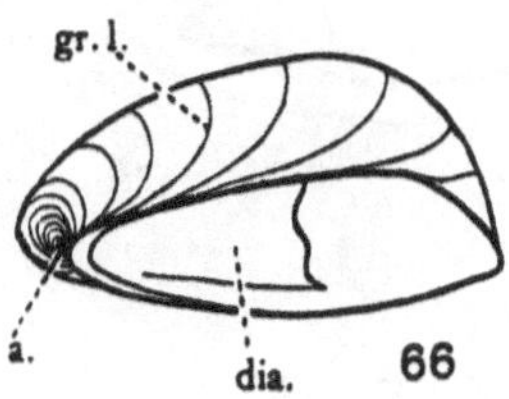

66

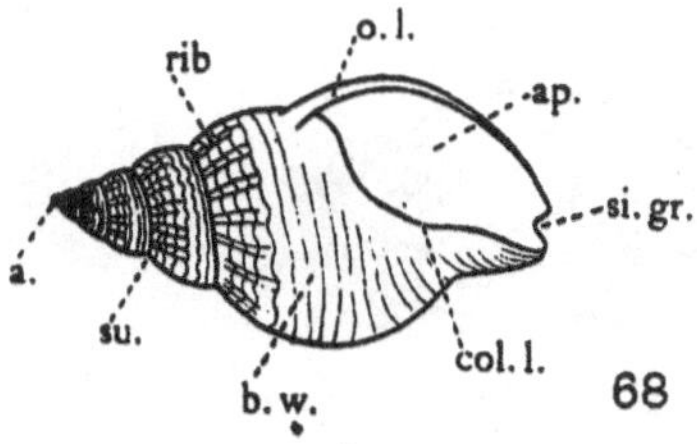

68

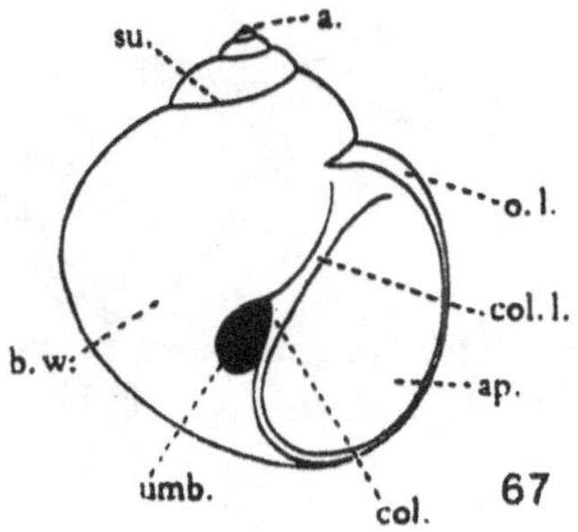

67

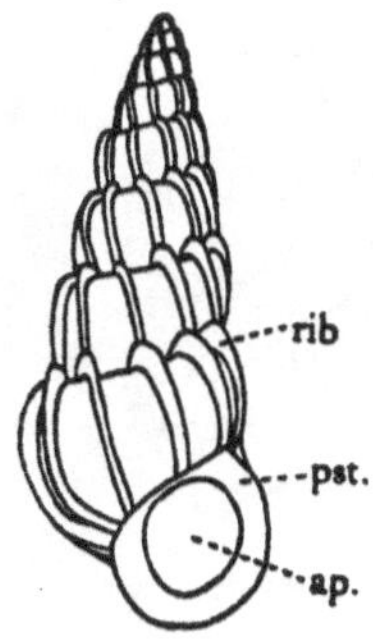

69

XVI. MOLLUSCA (GASTROPODA)

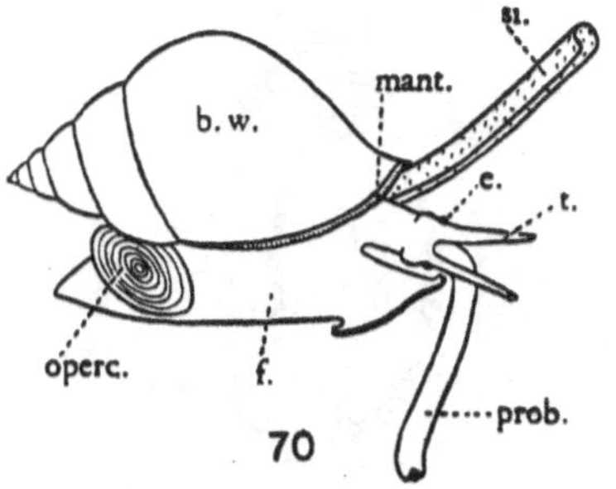

70

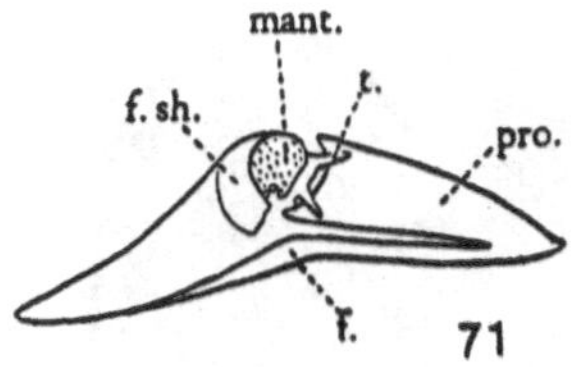

71

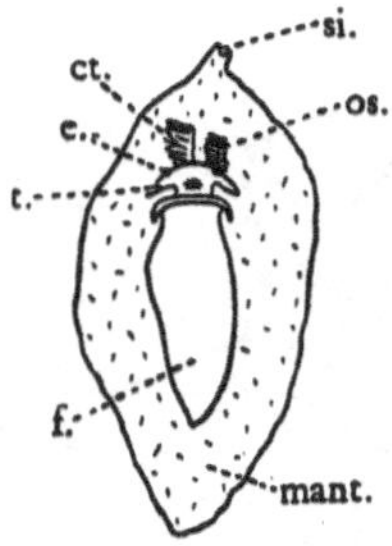

72*a*

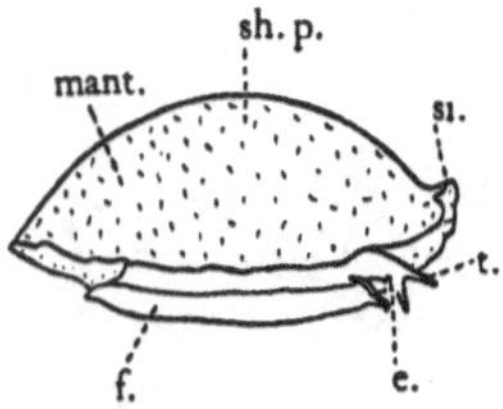

72*b*

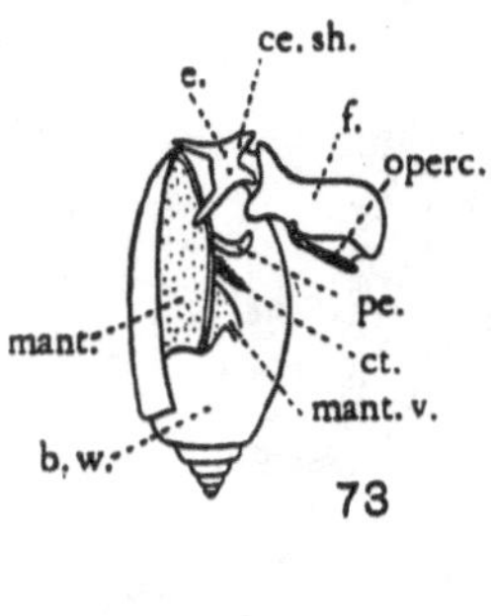

73

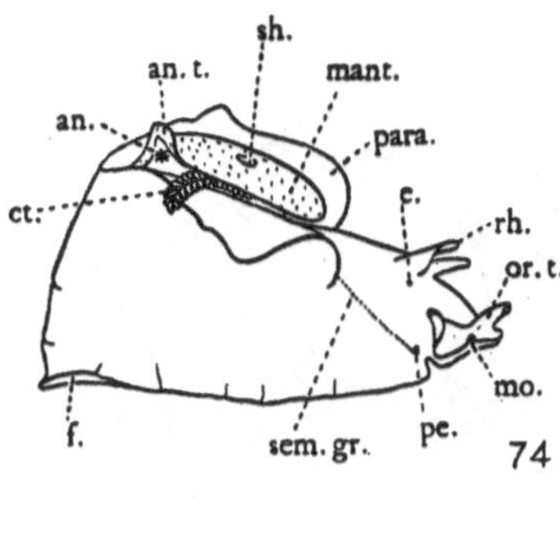

74

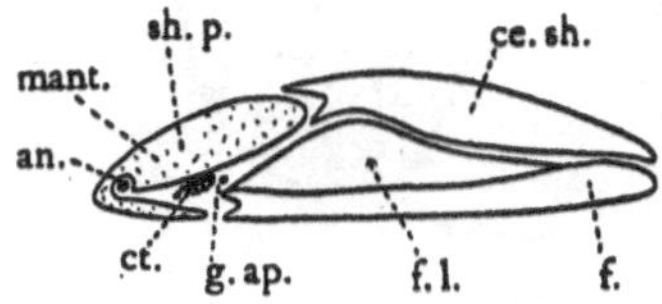

75

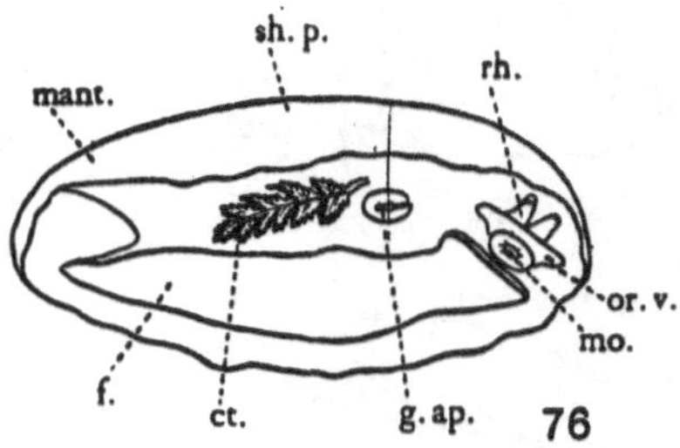

76

(*facing p.* 217)

EXPLANATION OF PLATE XVI

Gastropoda (ii), Figs. 70–76

Fig. 70. ***Buccinum undatum.*** Right lateral view of the animal in movement, with proboscis and siphon extended. In this and subsequent figures the mantle has been marked with short lines. After Meyer and Möbius. × ½.

Fig. 71. *Natica catena.* Right lateral view of the animal removed from its shell to show the small extent of the mantle and the large foot, with a propodium anteriorly, and a flat projection posteriorly partly covering the shell. The small eyes are not figured. × ½.

Fig. 72. ***Lamellaria perspicua.*** A prosobranch with shell enclosed by the mantle so that it bears a superficial resemblance to a Nudibranch. *a*, ventral, *b*, right lateral view. × 1½.

Fig. 73. ***Acteon tornatilis.*** Right lateral view. Cephalic shield and operculum present. × 1½.

Fig. 74. ***Aplysia punctata.*** Right lateral view of a contracted specimen in the "Sitting Hare" position. The right parapodium has been turned back to expose the ctenidium. × ½.

Fig. 75. ***Philine aperta.*** Right lateral view to show the large cephalic shield, large foot and posteriorly placed mantle. × 1.

Fig. 76. ***Berthella plumula (Pleurobranchus).*** Ventro-lateral view to show the overhanging mantle skirt, head and pallial organs. The anus is hidden beneath the ctenidium. In this figure the mantle has been left plain. × 1.

an.	anus	or.t.	oral tentacle
an.t.	anal tube of mantle	or.v.	oral veil
b.w.	body whorl	os.	osphradium
ce.sh.	cephalic shield	para.	parapodium
ct.	ctenidium	pe.	penis
e.	eye	pro.	propodium
f.	foot	prob.	proboscis
f.l.	lateral lobe of foot	rh.	rhinophore
f.sh.	part of foot enveloping shell	sem.gr.	seminal groove
		sh.	shell
g.ap	genital aperture	sh.p.	position of shell within mantle
mant.	mantle		
mant.v.	ventral lobe of mantle	si.	siphon
mo.	mouth	t.	tentacle
operc.	operculum		

EXPLANATION OF PLATE XVII

Gastropoda (iii), Figs. 77–81

Fig. 77. Types of Rhinophore in Nudibranchs. After Alder and Hancock. *a*, ***Archidoris britannica***. Laminated, with short, sunken, smooth sheath. × 4. *b*, ***Dendronotus frondosus***. Laminated, with tall sheath bearing arborescent processes. × 3. *c*, ***Tritonia lineata***. Branched, with terminal truncate process. Sheath with wavy edge. × 15.

Fig. 78. ***Goniodoris nodosa***. Right lateral view, after Alder and Hancock. Note the sheathless rhinophores, short mantle skirt and pinnate branchial plumes. × 2½.

Fig. 79. ***Ancula cristata***. Right lateral view, after Alder and Hancock. Both rhinophores and mantle bear lateral processes, tipped with orange. The branchial plumes are bipinnate. × 2½.

Fig. 80. ***Tergipes despectus***. Right lateral view, after Alder and Hancock. Rhinophores and oral tentacles simple. Cerata inflated. × 10.

Fig. 81. ***Coryphella verrucosa rufibranchialis***. Right lateral view, after Alder and Hancock. Rhinophores simple, rough. Cerata in about eight bundles on each side, not inflated. × 3.

an.	anus	or.t.	oral tentacle
br.pl.	branchial plume	rh.	rhinophore
cer.	ceras	rh.pr.	rhinophore process
f.	foot	rh.sh.	rhinophore sheath
lam.	lamella	t.p.	truncated process of rhinophore
lat.pr.	lateral process		
mant.	mantle		

XVII. MOLLUSCA (GASTROPODA)

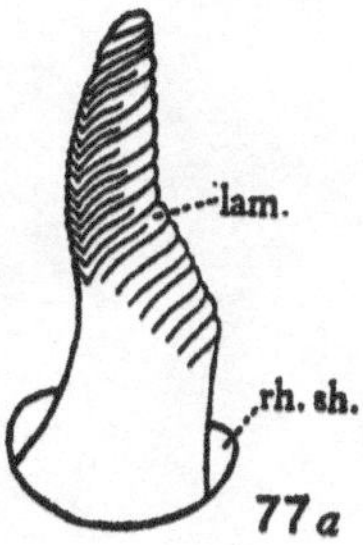

77a

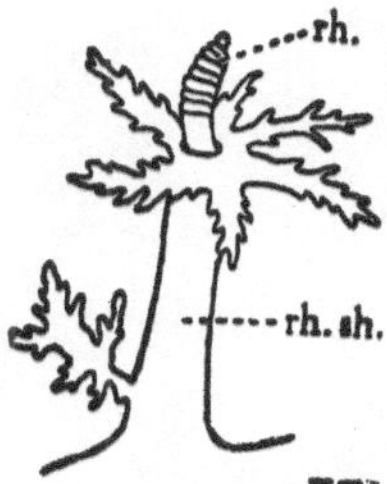

77b

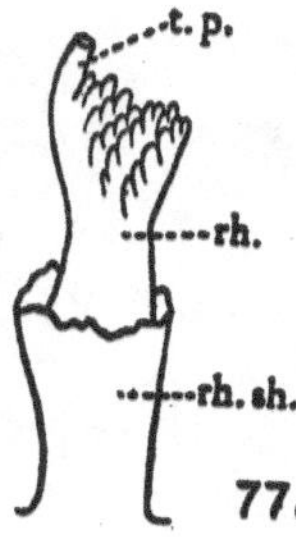

77c

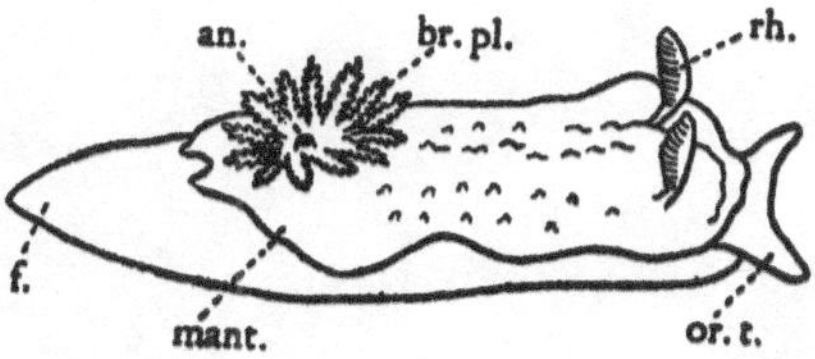

78

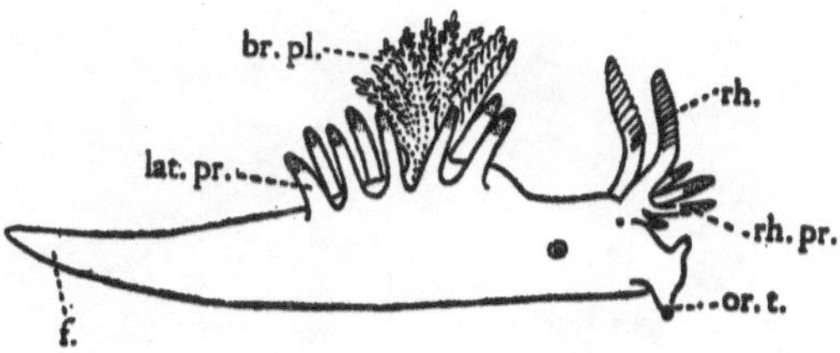

79

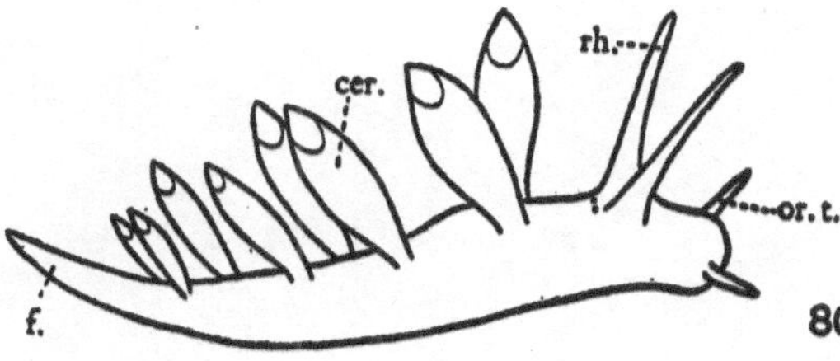

80

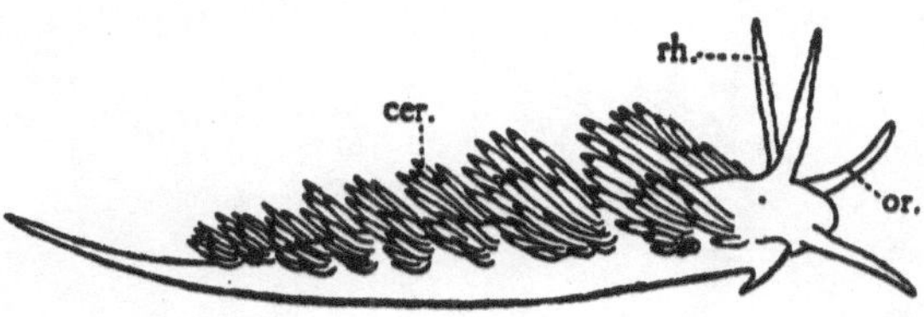

81

(*facing p.* 218)

BRYOZOA

(Plate XVIII)

AUTHORITIES

HINCKS, T. A. 1880. *A History of the British Marine Polyzoa.* London: Van Voorst.

HARMER, S. F. 1891. "On the British species of *Crisia.*" *Quart. J. micr. Sci.* XXXII, 127–81.

The Bryozoa (or Polyzoa) form a small phylum of doubtful affinities but ancient lineage.

Most are small, plant-like or zoophyte-like colonies, inhabiting shallow or deep water. A few live within the tidal zone, or are washed up on the shore after storms. The unit of the colony, or *zooarium*, is the polyp or zooid, part of whose body is protrusible as the *polypide*. The protective house for the zooid is the *zooecium*, which is attached to the body of the zooid. The zooecium may be horny, or calcareous, permanently open (Cyclostomata) with toothed aperture (Ctenostomata), or with hinged lid (Cheilostomata). On some, modified zooecia are set apart for the development of the ova, and are called *ooecia* or ovicells. In others (Cheilostomata) certain zooecia are modified for defence as strong, specialised muscular zooecia opening and closing like birds' beaks (*avicularia*); or elongated whip-like structures (*vibracula*).

The polypide has a superficial resemblance to a hydroid polyp because of the circlet or horseshoe of filiform tentacles surrounding the mouth. But these tentacles in the Bryozoa are ciliated. They spring from an expanded base, the *lophophore*. The alimentary canal is U-shaped, with a dorsally situated anus, which in some is within, in others without the lophophore. Most Bryozoa are hermaphrodite, the genital elements being discharged into a spacious coelom.

1. Anus outside the lophophore. **Ectoprocta.** 2
   Anus inside the lophophore. **Endoprocta.** 10
2. Throat not protected by a lip or epistome. Lophophore circular. **Gymnolaemata.** 3

3. Orifice of zooecium protected by an operculum. 4
Orifice of zooecium circular and incapable of being closed. No avicularia or vibracula. **Cyclostomata.** 8
4. Operculum a hinged lid. Avicularia or vibracula and ooecia often present. **Cheilostomata.** 5
Operculum a circlet of teeth. No ooecia, avicularia or vibracula. Gelatinous. **Ctenostomata.** 9
5. Colony incrusting, or at least with basal crust. 6
Colony erect. 7
6. Zooecium partly calcareous, but with membranous unprotected front wall. **Membraniporidae**, p. 221
Front wall of zooecium protected by radiating calcareous ribs. **Cribrilinidae**, p. 222
Zooecium calcareous, without membrane or ribs. **Escharidae**, p. 222
7. Branches flat, zooids back to back, avicularia sessile, simple. **Flustridae**, p. 221
Zooarium dichotomously branched, zooids on one side only. Avicularia sessile, vibracula sometimes present. **Cellulariidae**, p. 220
Plant-like, often spiral in growth. Avicularia pedunculate, jointed. **Bicellariidae**, p. 221
8. Zooarium erect, jointed, attached by rooting fibres. **Crisiidae**, p. 223
9. Zooarium erect. **Alcyonidiidae**, p. 223
Zooarium incrusting. **Flustrellidae**, p. 224
10. Solitary or colonial. Zooids on retractile peduncles. **Pedicellinidae**, p. 224

ECTOPROCTA

CELLULARIIDAE.

Colony erect, compact. Branches flat, zooids in two or more series, closely united and arranged side by side in the same plane, so that there is a barren side. Sessile avicularia usually present, and sometimes vibracula. Aperture of zooecium often spiny.

Scrupocellaria. Long whip-like vibracula and usually avicularia present, the latter on the edges of the fertile sides of the fronds and down their centre.

S. scruposa (Pl. XVIII, fig. 83). Hincks, p. 45, pl. 7. Erect, white, bristly, common between tide marks. The only avicularia are the large lateral ones. Often found attached to worm tubes, shells or weeds.

S. reptans. Hincks, p. 52, pl. 7. Creeping, with short, flat branches not more than 1·5 cm. in length. The zooecium has an antler-like operculum. Lateral avicularia very small, but there are large pedunculate central (ventral) avicularia. On sandy shores.

BICELLARIIDAE.

Zooarium erect, plant-like and unjointed. Zooecia in two or more series, boat-shaped, with large apertures. Avicularia pedunculate and jointed. The ooecia are rounded and are distinct from the zooecia.

Bugula. Avicularia usually large, shaped like birds' heads, and typically there is one to each zooecium.

B. turbinata (Pl. XVIII, fig. 82). H. p. 77, pl. 10. From 3–5 cm. high, orange, with the branches spirally arranged in corkscrew fashion. The zooecium has a spine at each upper angle. Under overhanging rock.

B. plumosa. H. p. 84, pl. 12. Tall, plumose, tapering to a fine point at the apex of the colony. More delicate and feathery than the former species and usually reaches a greater height (5–7 cm.). Avicularia very small. The zooecium has one short spine. Common on rocks, shells, etc.

FLUSTRIDAE.

Zooarium horny and flexible, with little calcareous matter. The fronds are usually flat, broad, leaf-like and dichotomously branched, covered all over with rows of zooecia, on one or on both sides of the frond. Avicularia simple, rounded, slightly modified zooecia.

Flustra. Typically in shallow water, but colonies are frequently thrown up on the shore after storms in great quantity. It is known as the sea mat.

F. foliacea. H. p. 115, pl. 16. Brown or yellow in colour, attaining a height of 10–15 cm. Branches flattened, broadened at the extremities. Zooecia in alternate rows and on both sides of the frond, each with four to five spines on its anterior rim. Simple ooecia are present, as outgrowths of the zooecia.

MEMBRANIPORIDAE.

Zooarium incrusting, the zooids lying flat on stones, shells or algae. The zooecia may be membranous, calcareous or semi-calcareous, arranged regularly or irregularly, with raised margins, and often closed by membrane only. Larva a *Cyphonautes*.

Membranipora membranacea. H. p. 140, pl. 18. Large colonies form a lace-like network on *Fucus*, *Laminaria*, etc. Zooecia entirely decumbent, quadrangular, and arranged in fairly regular alternating lines. Two stout hollow spines at the anterior angles beside the small hinged lid. No avicularia. Tall, closed, tubular, tower-like processes arise from certain of the zooecia.

Electra pilosa. H. p. 137, as *Membranipora*, pl. 23. Zooecia ovate, spines numerous, variable.

CRIBRILINIDAE.

Zooarium frequently incrusting, calcareous, white or silvery. The zooecia have the front wall fissured with radiating ribs of calcareous matter on a membranous foundation. Avicularia present.

Cribrilina punctata. H. p. 190, pl. 26. In some varieties of this species the radiating ribs are not present, but the zooecia are strongly punctate and there are typically four short spines at the anterior end and a small avicularium at each anterior corner. The aperture is mucronate (i.e. notched). The ooecia are rounded and hood-shaped. Inter-tidal to shallow water.

ESCHARIDAE.

Incrusting or erect. Zooecia calcareous, without membranous area or raised margin. In some the aperture opens directly to the exterior, in others there is a secondary, elevated orifice enclosing an avicularium.

1. Front of zooecium radiate, a median avicularium. **Umbonula**
   Front of zooecium not radiate, a pair of avicularia. **Mucronella**

Umbonula verrucosa. H. p. 317, pl. 39. The colony forms rose-red incrustations on stones, shells or *Laminaria* roots. The primary orifice of the zooecium is rounded or quadrangular and just below the mouth is a prominent umbo or boss, supporting an avicularium. The front is traversed by radiating ridges and the anterior border may bear short spines.

Mucronella coccinea. H. p. 371, pl. 34. This species also forms calcareous crusts of orange red colour on weed, stones, etc. The primary orifice of the zooecium is, however, obscured by a mucronate elevation of the peristomial area, which covers one at least of the three denticles of the lower lip. There are six marginal spines anterior to the aperture. The surface is pitted but not punctate or radiate. An avicularium lies on either side.

CRISIIDAE (see Harmer, S. F.).

Zooarium erect, jointed, and attached by rooting fibres, calcareous, but membranous at the joints. Zooecia tubular, without operculum, and arranged in one or two series. Neither avicularia nor vibracula are present, but simple ooecia closed by membrane, occur. The polypide is small and simple, with eight tentacles.

1.	Zooecia in a single series.	**Crisidia**
	Zooecia in two series.	**Crisia**
2.	Aperture of ooecium circular.	**C. ramosa**
	Aperture of ooecium transversely elongated.	**C. eburnea**

Crisidia cornuta. H. p. 419, pl. 56. Zooecia in a single series on the slender branches, and curved. A long jointed spine beside each. Ooecia in axils of branches, oval, speckled, with tubular aperture. 6–12 mm. high. On red weed, rocks and shells, from low-tide mark to shallow water.

Crisia eburnea (Pl. XVIII, fig. 84). H. p. 420, pl. 56; Harmer, p. 131. Dense white tufts attached to algae, zoophytes or stones. The branches are curved inwards. The zooecia alternate, in two rows, slightly curved outwards, with a long lateral spine, and transversely elongated aperture. The spines get broken off in dead colonies. Joints yellow or brown. 10–20 mm. high.

C. ramosa. Harmer, p. 134. A straggling species with fan-shaped branches. Ooecia the largest of the British species, pear-shaped, with apertures on a funnel. 20 mm. On stones, shells, red weeds or other Bryozoa.

ALCYONIDIIDAE.

Colony gelatinous, erect or incrusting. Zooecia wholly or partly immersed, orifice with membranous papillae (teeth). No ooecia, avicularia, or vibracula.

Alcyonidium.

A. gelatinosum (Pl. XVIII, fig. 85). H. p. 491, pl. 69. Erect, gelatinous, greenish yellow, translucent branches which end bluntly and are round in section. Zooecia small, closely packed together, with twelve or thirteen blunt teeth round the orifice. It may reach a height of 60 cm. or more. Lower shore, attached to rocks and weeds, but commoner in shallow water.

A. hirsutum. H. p. 493, pl. 70. Either incrusting or with erect, compressed, palmate branches, surface covered with tall imperforate conical papillae. The zooecial aperture has four to eight minute teeth. 12–15 cm. On shore algae.

FLUSTRELLIDAE.

Zooecia immersed in a gelatinous crust, orifice with two lips or a toothed edge. The polypide has numerous tentacles (thirty or more).

Flustrella hispida. H. p. 506, pl. 72. Forms a thick brown crust, with reddish brown horny spines, on *Fucus*. The orifice of the ovate zooecium is raised and toothed and tall spines surround it. Polypides relatively large and when expanded form a milky or pale blue film on the colony. Very common in littoral zone. Shelled larva.

ENDOPROCTA*

PEDICELLINIDAE.

Colony never calcareous. Small, solitary or stolonic forms with non-retractile tentacles. The zooids have a quick, jerky movement on their retractile peduncles. Most are hermaphrodite.

Pedicellina cernua (Pl. XVIII, fig. 86). H. p. 565, pl. 81. The stolon buds repeatedly, each bud giving rise to a single erect zooid, whose webbed tentacles close in over the mouth and anus, but do not contract much. A simple epistome is formed by the raised mouth edge. The peduncle bears short obtuse spines and tapers as it nears the head. A temporary brood pouch is formed within the vestibule and may be so crowded with embryos that the polypide is distorted. About 2 mm. high. Attached to zoophytes or algae.

Barentsia gracilis. H. p. 570, pl. 81, as *Pedicellina.* Peduncle not spiny, otherwise this species closely resembles *P. cernua* in form and habitat. Stolon and peduncles slender.

* Modern systematists regard this group as a phylum, using the original spelling **Entoprocta**.

XVIII. BRYOZOA

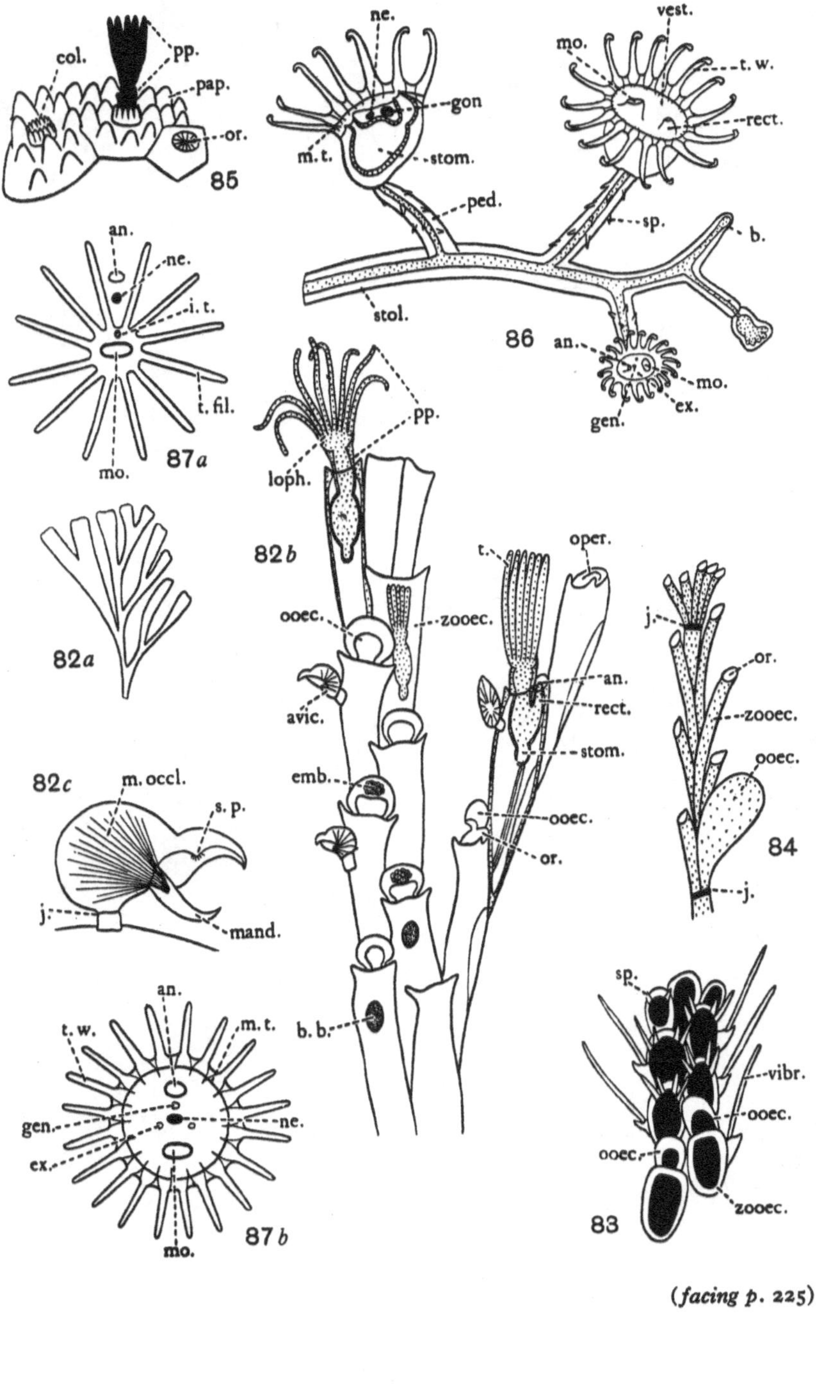

(*facing p.* 225)

EXPLANATION OF PLATE XVIII

Bryozoa, Figs. 82–87

Fig. 82. *Bugula turbinata*. Cheilostomata. *a*, Part of the zoarium to show form and branching. ×3. *b*, Two branches in ventral and lateral views. Two expanded polypides and one retracted individual have been figured. Their lower portions are seen through the wall of the zooecium, but further details of the interior have not been shown. Some of the avicularia have been omitted. ×35. *c*, An avicularium. Its occlusor but not its divaricator muscle is figured. ×140.

Fig. 83. *Scrupocellaria scruposa*. Cheilostomata. A portion of a dead zoarium to show vibracula. The details of the zooecium are not shown. The spines have been omitted from the two lowest zooecia. ×15.

Fig. 84. *Crisia eburnea*. Cyclostomata. Part of a dead branch of the zooarium. The ooecium is young and has not yet acquired its aperture. ×15.

Fig. 85. *Alcyonidium gelatinosum*. Ctenostomata. Surface of colony to show retracted and expanded condition of the polypides. The operculum takes the form of a toothed collar and is shown open on the left and closed on the right. Centrally is an expanded polypide with upstanding collar. ×60.

Fig. 86. *Pedicellina cernua*. Portion of a stolon with two buds and three polypides. The polypide on the left is shown in sagittal section. ×15.

Fig. 87. Schemes to show arrangement of the chief organs in (*a*) an Ectoproctous (Gymnolaematous) and (*b*) an Endoproctous Bryozoan.

an. anus
avic. avicularium
b. bud
b.b. brown body
col. collar
emb. embryo
ex. excretory aperture
gen. genital aperture
gon. gonad (hermaphrodite gland)
i.t. intertentacular organ (acting as excretory and genital duct)
j. joint
loph. lophophore
m.occl. occlusor muscle
m.t. circular muscle ring for tentacles
mand. mandible
mo. mouth
ne. nerve ganglion
ooec. ooecium
oper. operculum
or. orifice
pap. papilla
pp. polypide
ped. peduncle
rect. rectum
s.p. sensory patch
sp. spine
stol. stolon
stom. stomach
t. tentacle
t.fil. filiform tentacle
t.w. webbed tentacle, with bulbous base
vest. vestibule
vibr. vibraculum
zooec. zooecium

PHORONIDEA

This small group contains a single genus, *Phoronis*, whose distribution is local despite the wide dissemination of the larva (the Actinotrocha) in the plankton.

Phoronis is a tubiculous sedentary animal, living gregariously but not colonially, i.e. the individuals of a group are not joined together. The tube is a secretion, and to it sand grains and other marine substances are adherent. The animal living in the tube is very shy, withdrawing itself instantly if even a shadow falls across it. It is very small in the British species, *P. hippocrepia*, which is less than a centimetre in length. The body is soft and worm-like, and the upper end bears a horseshoe-shaped, double crown of twenty to eighty ciliated tentacles (lophophore). Within the horseshoe lies the mouth; outside it, but within the gap at the ends of the horseshoe, is the anus. Mouth and anus, are however, separated by a lip or epistome. The alimentary canal, as in the Bryozoa, is U-shaped, the anus being dorsal in position. Phoronis is hermaphrodite, and the tentacles protect the young during the early stages of their development. The red blood may show through the skin.

This species has been found on the south-west coast (Devon), the west coast (Menai Straits) and the north-east coast (Firth of Forth). Its membranous tubes are usually embedded in limestone, shells, etc.

ECHINODERMATA

(Pls. XIX–XXI)

AUTHORITIES

FORBES, E. 1841. *A History of British Starfishes.*
KOEHLER, R. 1921. *Faune de France.* I. *Echinodermes.*
MORTENSEN, T. 1927. *Handbook of the Echinoderms of the British Isles.*

The Echinoderms form a distinct and readily recognisable group of animals. Not all are spiny skinned, but all possess some degree of radial symmetry, usually pentaradiate. Locomotion is usually by means of extensile *tube feet* or *podia*, which are arranged typically in a double series along each ray, radius or *ambulacrum*, the intervening portion of the body being devoid of tube feet and known as the interradius or *interambulacrum*. The tube feet are connected with an internal water vascular system, consisting of ampullae, radial canals, a ring canal around the mouth and usually a communication with the exterior called the *madreporite*. The tube feet may be adhesive or merely tentacle-like. Most Echinoderms have a well-developed skeleton composed of numerous calcareous plates embedded in the skin. To these plates may be attached movable spines, and in some, modified spines form the *pedicellariae*, or snapping organs. The body may be star-shaped, with disk and arms, as in the sea stars, feather stars, and brittle stars, globular as in sea urchins, or soft and sausage-shaped with reduced plating, as in the sea cucumbers. In the feather stars the arms are feather-like, owing to repeated branching to form pinnules.

The length of the arm of a starfish relatively to the radius of the disk is expressed by a formula. Thus $R=5r$ (where $r=$ the radius of the disk, $R=$ distance from centre of disk to apex of a ray) means that the total distance from centre of disk to apex of an arm is five times that of the radius of the disk. In the Ophiuroids, as the arms are distinctly marked off from the disk, it is usual to measure the diameter of the disk and the length of the arm separately.

1. Body with disk or calyx, and arms. — 2
   Body globular or sausage-shaped, without arms. — 4
2. Arms feather-like with ambulacral grooves on the upper oral side. **Crinoidea** (feather stars)
   Arms not feather-like. Mouth on the lower side. — 3

3. Ambulacral grooves on under side. Arms not sharply delimited from the disk. Tube feet used for attachment. **Asteroidea** (starfish)
   No ambulacral grooves. Tube feet tactile. Arms slender, brittle, sharply marked off from the disk. **Ophiuroidea** (brittle stars)
4. Body globular or egg-shaped, with movable spines set on a firm skeleton of plates. **Echinoidea** (sea urchins)
   Body sausage-shaped, without spines or regularly arranged calcareous skeleton. **Holothurioidea** (sea cucumbers)

CRINOIDEA

The dying remnant of the class Pelmatozoa (feather stars and sea lilies). The body forms a cup-shaped *calyx*, the *arms* are typically ten in number, and feather-like owing to the presence of *pinnules*. The animal is attached permanently by a stalk in the sea lilies, but in the feather stars there is a temporary attachment by means of *cirri* in the adult, although the larva has a long stalk for fixation. The mouth and anus are on the upper side, where the open *ambulacral grooves* from the arms extend across the calyx to the edge of the mouth, acting as ciliated food grooves. The pinnules of the arms bear gonads, sensory podia without fixation disks, and rounded bodies called *sacculi*. There is no definite madreporite, but pores are present on the calyx.

Antedonidae.

Animal stalked in the larval stage only. Adult attachment temporary by means of aboral, jointed and clawed cirri. The five original arms are divided to form ten at the second joint. Mouth central, anus excentric on a spout.

Antedon bifida (*rosacea*) (Pl. XIX, fig. 88). Rosy feather star. M. p. 28; Forbes, p. 5, as *Comatula*. Rosy red to deep purple, yellow or orange. Twenty-five cirri. Diameter about 10–15 cm., when the arms are spread, but these are deeper water specimens, occurring in about 5 fathoms. In calm weather on sheltered coasts young individuals are found on the shore, e.g. on the South Wales coast.

ASTEROIDEA

Free-living, flattened, star-shaped Echinoderms. The *arms* or rays are usually not sharply marked off from the central disk, and in some species are extremely short, so that the body may be pentagonal in outline, as in the bird's foot star. The tube feet (*podia*) lie in an *ambulacral* groove on the oral (under) surface of each arm. The *mouth*, which is soft-walled and devoid of jaws, lies in the centre of the under side, the anus and madreporite are on the upper or aboral side. The skeleton consists of overlapping calcareous plates bearing *spines* and *pedicellariae*. There are dermal gills (*papulae*) in the areas of skin forming the meshes of the skeleton. The spines may be single (*Asterias*), grouped in sheaves (*Solaster*) or arranged in brush-like masses called *paxillae*. Some authors call all grouped spines paxillae, and this is followed here.

Starfishes are carnivorous, feeding on bivalve Molluscs, worms, Crustacea, fish and other Echinoderms.

1. Marginal plates of skeleton large and conspicuous. Papulae on aboral side. Podia in two series. **Phanerozonia.** 3
   Marginal plates usually inconspicuous. Papulae on both sides. 2
2. Paxillae usually present. Podia in two series. Pedicellariae simple or absent. **Spinulosa.** 4
   Rarely with paxillae. Podia in four series. Straight and crossed pedicellariae present, the latter pedunculate. **Forcipulata.** 6
3. Podia with disks. No pedicellariae. Aboral spines minute or absent. **Poraniidae**, p. 231
4. Body flattened, with a distinct edge limiting oral and aboral sides. Paxillae small. **Asterinidae**, p. 230
   Body without a sharp limit between upper and lower sides. 5
5. Disk fairly large. Arms usually more than five. Paxillae conspicuous. **Solasteridae**, p. 230
   Disk small. Arms five, long, slender, cylindrical. Paxillae small. **Echinasteridae**, p. 231
6. Aboral skeleton irregularly reticulate. Spines single. No paxillae. **Asteriidae**, p. 229

ASTERIIDAE.

Rays five, rarely six. Aboral skeleton reticulate. Spines single Podia in four rows with terminal disks. Straight and crossed pedicellariae present.

Asterias.

Asterias rubens. Common starfish. M. p. 139; F. p. 83, as *Uraster*. Reddish brown to pale red or yellow. There is usually a distinct line of spines along the middle of each ray. Skeleton not very strong so that the skin of the aboral side is somewhat flabby. $R=5r$. A red "eye-spot" at the tip of each ray. Shore specimens are mostly small; those in deeper water may reach 26 cm. from the centre of the disk to the tip of an arm. Common all round the coast on lower shore.

A. (*Marthasterias*) *glacialis*. Spiny starfish. M. p. 143; F. p. 78, as *Uraster*. Larger, more transparent, more spiny and less common than *A. rubens*. Usually of a grey-green or reddish colour. Disk smaller relatively, arms longer. $R=5$ to $9r$. West and south coasts.

Asterinidae.

Body flattened, border with concealed marginal plates. Aboral side with imbricating scale-like plates and irregular groups of spines or low paxillae. Arms very short and disk large.

Asterina gibbosa (Pl. XIX, fig. 89). M. p. 98; F. p. 119. Arms so short that the body is subpentagonal, a feature which is even more pronounced in the bird's foot star (*Palmipes*), from deeper water. Flattened on the oral side but slightly swollen aborally. The rays are not crested and tend to curl up at the tips when the animal is disturbed. Colour greenish, tinged with yellow and red. $R=2r$. Diameter 6 cm. The smallest British starfish. Amongst algae and on stones between tide marks to deep water. South and west coasts.

Solasteridae.

Disk large, arms rather long, rounded. Often more than five arms. Aboral skeleton reticulate, with paxillae. A single or double series of large marginal paxillae representing the marginal plates. Large, brightly coloured sun stars.

Solaster papposus. The common sunstar. M. p. 112; F. p. 112. Rays eight to thirteen. $R=2r$. Arms rather short, about as long as the diameter of the disk. Scarlet and yellow, with banded arms. To 30 cm. or more in diameter. Occasionally on the shore, but usually in shallow water. Feeds on Molluscs and some Echinoderms.

S. endeca. The purple sunstar. M. p. 115; F. p. 109. Rays usually nine to ten. $R=2{\cdot}5r$. Skeleton much more compact than in the preceding species, so that the animal is hard to the touch. Paxillae small, close set, with short spines. Violet or yellowish red.

40 cm. in diameter. West, north and east but not south coasts. Littoral to shallow water.

ECHINASTERIDAE.

Disk very small. Arms five, long, cylindrical, slender. Spines single or in small groups. No pedicellariae. Podia in two series in narrow grooves.

Henricia sanguinolenta. M. p. 118; F. p. 100, as *Cribella oculata*. In the stiffness of its arms it resembles *Solaster endeca* but can be readily distinguished from the latter by the number of arms. Rays rather slender, hard and tapering. Usually red but may be orange to purple. $R=4\cdot5r$. To 20 cm. diameter. Littoral to shallow water.

PORANIIDAE.

Cushion stars. Aboral surface convex, covered by a thick skin, without spines in many members of the family. No pedicellariae. Papulae numerous on aboral side. Oral surface flat. Marginal plates fairly distinct, partly overlapping, with spines forming a distinct fringe along the edge of the disk.

Porania pulvillus (Pl. XIX, fig. 90). Cushion star. M. p. 90; F. p. 122, as *Goniaster templetoni*. Inhabits shallow water, but occasionally specimens are thrown up on the shore. Disk large, rays short but distinct. $R=1\cdot5$ to $2r$. Aboral surface without spines, scarlet, oral side white. To 11 cm. in diameter. South, west and north coasts.

OPHIUROIDEA

(Pl. XX)

Brittle stars. The disk, containing the body organs, is round and flat, and is sharply marked off from the five (occasionally more) thin, jointed arms. There are no ambulacral grooves, but tentacle-like podia issue from the holes between the joints along the sides of the arms. No anus, no pedicellariae. The mouth is central, on the under side of the disk, and is provided with calcareous teeth. At the root of the arms and on each side of it is a slit leading to a pocket or bursa, probably respiratory, often called the *genital bursa*. The skeleton consists of disk plates and arm plates. The former differ on oral and aboral sides of the disk.

Aboral disk plates:

(*a*) A *central* plate and one or two alternating circlets of five plates each.

(*b*) Ten radial plates, or *radial shields*, usually large and near the margin of the disk.

(*c*) Secondary plates in the form of imbricating *scales*.

Oral disk plates:

(*a*) Five mouth plates or *mouth shields*, of which one bears the madreporite. Usually accompanied by *adoral shields*.

(*b*) Ten lateral mouth plates, each pair forming a "jaw".

(*c*) Ten *genital plates* on the edges of the genital bursae.

(*d*) Imbricating *scales*, spines or granules.

The arm joint or "*vertebra*" is composed of one dorsal, one ventral and two lateral plates. The lateral plates usually bear arm spines.

In addition to the plates above mentioned, movable or immovable calcareous, scale-like papillae may be present:

(*a*) Foot papillae or *tentacle scales* over the pores of the podia.

(*b*) *Mouth papillae* on either side of the teeth on the lateral mouth plates or jaws.

(*c*) *Arm combs* or rows of papillae at the bases of the arms.

(*d*) *Genital papillae* beside the genital bursa.

(*e*) *Tooth papillae*. This term is applied to the compound type of outer tooth composed of several rows of rigid papillae. The real tooth is an enamel-covered structure hidden by the tooth papillae.

Brittle stars are active forms, moving by wriggling snake-like movements of the arms. They feed on small animals such as Molluscs and worms. When irritated they break off one or more arms, but possess great powers of regeneration.

1. Arm spines pressed close against the arm and often reduced. Disk scaly. **Ophiolepidae**, p. 236
   Arm spines erect or outstanding. 2
2. Mouth papillae present. 3
   No mouth papillae. Disk and arms spiny. Radial shields distinct and naked. **Ophiotrichidae**, p. 233
3. Disk granular, sometimes with fine spines and scales. 4
   Disk scaly. Arms long and slender. **Amphiuridae**, p. 235

4. Primary plates usually distinct. One tentacle scale.
Ophiactidae, p. 234

Primary plates usually concealed. Two tentacle scales.
Ophiocomidae, p. 234

It is not easy to identify species of brittle stars when seen for the first time. As a guide, the length of the arm relative to the disk may be employed. Thus the short-armed genera *Ophiura*, *Amphipholis* and *Ophiopholis* have arms not more than four times the diameter of the disk. Brittle stars with arms of moderate length, up to eight times the diameter of the disk, include *Ophiothrix*, *Ophiocomina*, *Ophiopsila*, *Ophiactis*, and *Amphiura*. *Acrocnida* is the only long armed littoral genus, with arms up to fifteen times the diameter of the disk. *Ophiura* is the only genus with arm spines closely pressed against the sides. *Ophiothrix* alone has strong spines on the disk, *Ophiopholis* and *Ophiactis* have a few spines on the disk, but the former has three mouth papillae on the side of the jaw, and short, stout arms, the latter one mouth papilla and a small disk, only 5 mm. in diameter. *Ophiocomina* has numerous mouth papillae and a large disk up to 25 mm. *Ophiopsila* has only two mouth papillae and a disk half that size, and both genera have strong compound teeth. *Amphiura* has only one mouth papilla and very slender arms, *Amphipholis* two mouth papillae, rather short, stout arms and a very small disk (5 mm. in diameter).

OPHIOTRICHIDAE.

Disk with well-developed scales, which may be concealed by spines. Radial shields large and distinct, others hidden by spines. No mouth papillae. Tooth compound, with a group of tooth papillae, and inside these a series of strong, broad teeth. Arm spines erect, often spiny.

Ophiothrix fragilis. (A brittle star in the extreme sense.) M. p. 174; F. p. 60, as *Ophiocoma rosula*. Arm spines long, glassy and thorny, seven to each vertebra, the third from the dorsal side the longest and the lowermost spine often hooked. One small tentacle scale. Colour variable, usually dark purple or reddish, yellow, grey or brown, sometimes spotted or banded. Diameter of disk 20 mm., arms about five times as long. Feeds on worms, Crustacea, Molluscs, etc., and is itself eaten by starfish. Very common on all coasts, on hard ground, from tide limits to deep water.

OPHIOCOMIDAE.

Disk usually granulated, often concealing the scales and radial shields, but not spiny. Arm spines erect. Mouth papillae present. Jaw with tooth papillae and a series of strong teeth. Two tentacle scales.

1. Disk granulated. Tentacle scales simple, subequal. Radial shields concealed. **Ophiocomina**
   Disk not granulated. Tentacle scales unequal, the outer small, the inner very long and crossing its fellow of the opposite side mid-ventrally. Radial shields exposed. **Ophiopsila**

Ophiocomina. Arm spines simple, delicate, tapering, not flattened. Outer tentacle scale only slightly longer than the inner.

O. nigra. Black brittle star. M. p. 178; F. p. 50, as *Ophiocoma granulata*. Arm spines five to seven, smooth, hollow, slender, rather long (as long as two arm joints). Black or brown, sometimes pink or grey, often spotted. Disk diameter 2·5 cm. Arms 12·5 cm. Littoral and in shallow water, south and west coasts and east English coast. Carnivorous, and at times a scavenger.

Ophiopsila. Disk covered with very fine scales but not granular. Arm spines short, outstanding, flattened. Luminescent.

O. aranea. M. p. 180. Radial shields distinct, small, linear, separated. Two pairs of mouth papillae. Arm spines six to eight, rather short. Reddish brown, often spotted with white. Disk 12 mm. in diameter, arms 80 mm. South coast.

OPHIACTIDAE.

Disk sometimes with spines or granules, but the scales are not concealed by them. A single series of square teeth, no tooth papillae. Mouth papillae present. Spines short, outstanding.

1. Dorsal plates of the arms surrounded by small plates. Ventral arm spine hooked. **Ophiopholis**
   No small plates round the dorsal arm plates. No hooks on arm spines. **Ophiactis**

Ophiactis. Disk scaly, generally with some spines. One or two mouth papillae. Arm spines rather short, none of them hooked. Mostly small forms.

O. balli. M. p. 200; F. p. 57, as *Ophiocoma goodsiri*. Radial shields plainly visible, separated by a wedge of scales. One broad tentacle scale. Four or five short conical arm spines. One mouth papilla on each side of jaw. Red, or banded with red, brown and white, variable. Disk 5 mm., arms 30 mm. Under stones, usually concealed. South and west coasts, east coast of England. Littoral to deep water.

Ophiopholis. Disk with granules or small spines which cover the scales but do not conceal the primary plates. Arm spines short and blunt, the lowermost spine hook-shaped and often forked.

O. aculeata. Daisy brittle star. M. p. 204; F. p. 53, as *Ophiocoma bellis*. A robust, common shore form, with rather short, coarse arms. Disk with ten radiating rows of naked primary plates, but radial shield is concealed by small spines and granules. Six or seven short, thick arm spines. One rather large tentacle scale. Three mouth papillae on each side of jaw. Red, often variegated and banded with red and white. Disk diameter 15–20 mm., arms 80 mm. In crevices. Feeds as a scavenger. On all coasts, but locally common. Littoral to shallow water.

AMPHIURIDAE.

Disk scaly, rarely spiny, small. A single series of square teeth. Mouth papillae but no tooth papillae. Arms usually very long, slender and flexible. Arm spines short, outstanding. Most live on muddy bottoms and burrow in it, a few in sand or under stones.

1. One mouth papilla on each side of jaw. Radial plates not contiguous, diverging. 2
   Two mouth papillae. Radial shields contiguous. **Amphipholis**
2. Radial shields with a transverse furrow. Arms fifteen times diameter of disk. **Acrocnida**
   Radial shields without a transverse furrow. Arms eight times diameter of disk. **Amphiura**

Amphiura. Arms long, slender, flexible.

A. chiajei. M. p. 212. Two tentacle scales. Primary plates of disk distinct, radial shields separated. Spines four to six, conical, not hooked. Reddish or greyish brown. Disk diameter 10–11 mm. Arms 80 mm. Low water to shallow. Common. Deposit feeder.

Acrocnida. Marginal and ventral scales tuberculate. Ventral plates of arms furrowed or keeled.

A. brachiata. M. p. 218; F. p. 45, as *Ophiocoma*. This is a sand form, and lies buried with the point of the arms protruding. Disk small, the arms very long and flexible. Primary plates inconspicuous. Eight to ten short arm spines. Two tentacle scales at root of arm but usually one or none near the tip. Bluish or brownish grey. Disk 12 mm., arms 180 mm. Commensal with *Mysella bidentata* (bivalve Mollusc) and *Harmothoe lunulata* (Polychaete worm). South and west coasts.

Amphipholis. Two mouth papillae. Primary plates concealed. Radial shields contiguous throughout their length. Three to four arm spines, short. Arms usually short, tapering.

A. squamata. M. p. 221. A small and delicate species. Two tentacle scales almost equal. Greyish white or bluish grey. Disk 5 mm. diameter, arms 20 mm. Viviparous. Female phosphorescent. On sandy or stony bottom, under stones, amongst algae. Common from tide marks to shallow water.

OPHIOLEPIDAE.

Disk scaly on both sides. Mouth papillae present. No tooth papillae, but a single series of teeth. Arm spines pressed close against the arm, small. Arm short, often with combs of papillae at the base covering the arm slit.

Ophiura. Sand stars. Disk flat. Arm combs well developed. Radial shields visible but other primary plates often concealed.

1. Three to four tentacle scales. Arm comb with about thirty papillae. **O. texturata**
   One tentacle scale except in the first few pore pairs. Arm comb with fewer than twenty papillae. **O. albida**

O. texturata (Pl. XX, fig. 93). Sand star. M. p. 236; F. p. 22. Disk scales coarse. Ventral arm plates with pore-shaped grooves between them. Mouth papillae four to six, pointed. Red or reddish brown, often spotted. Disk diameter 35 mm., arm 10–12 cm. On all coasts. Low tide to shallow water, in sand.

O. albida. M. p. 239; F. p. 27. Disk scales coarse. Arm comb with ten to twelve short papillae. Three to five mouth papillae. Distal pores with one tentacle scale, proximal with two to three. Disk reddish brown, radial shields white. Disk 15 mm., arm 60 mm. On all coasts, to shallow water. Often washed up on shore.

ECHINOIDEA

Sea urchins. Often of large size, with compact, hard tests formed of regularly arranged plates, and bearing articulated spines. Podia in ten series, two to each radius, but not in grooves. Each has a double pore, and at the tip, in regular forms, a calcareous disk for attachment, and spicules. The body is globular, heart-shaped, or flat and cake-like. The area around the circumference of the test at its widest part is called the *ambitus*. The mouth lies on the under side, the anus at the opposite pole or posteriorly. Pedicellariae are well developed, three-rayed and of four different kinds. At the upper pole is the *apical system*, consisting of ten plates arranged in a circle. Each of the five *radial plates* is perforated by a pore for the unpaired tube foot, which does not, as was formerly supposed, bear an eye-spot. (Hence the term ocular applied to the radial plate is a misnomer.) Each interradial plate is perforated by the genital duct, and is often called, therefore, the *genital plate*. One genital plate is larger than the others, and functions as the *madreporite*. Within the circle of ten plates in the regular urchins there is a membrane bearing small, irregularly arranged movable plates, and perforated excentrically by the anus. This area is called the *periproct*. In irregular forms the anus has wandered out of the apical system. The mouth is often surrounded by a soft skinned area, the *peristome*, which may bear small plates, large buccal tube feet, and ten dermal gills. In regular forms and cake-shaped forms the mouth has five calcified teeth, set in an elaborate apparatus called *Aristotle's lantern*. In heart urchins, on the other hand, the mouth is shovel-like and there are no teeth. In habit urchins are sluggish. They walk and climb slowly, but cannot swim. The heart urchins burrow in the sand. Some feed on algae, others on small animals and detritus.

1. Anus within the apical system. Test regular. **Diadematoidea**
   Anus outside the apical system. Test irregular. **2**
2. Mouth central. Tooth apparatus present. **Clypeastroidea** (cake urchins)
   Mouth anterior. No tooth apparatus. **Spatangoidea** (heart urchins)

DIADEMATOIDEA

Regular urchins. The body is regularly round, oval or slightly flattened. The anal aperture is in the centre of the apical system. The mouth is at the lower pole, also centrally, and is provided with five strong teeth set in an elaborate jaw apparatus known as Aristotle's lantern. The peristomial membrane bears plates, often irregular in arrangement, special buccal tube feet and arborescent gills. Both ambulacral (radial) and interambulacral (interradial) plates bear primary tubercles and spines, as well as smaller or secondary ones. The pedicellariae are well developed, with three rays, and are of four kinds, of which one bears a poison gland. The tube feet have disks for attachment and spicules in the stem; they function as the chief organs of locomotion.

1. Glandular (globiferous) pedicellariae with teeth on both sides of the valves. **Echinidae**
   Glandular (globiferous) pedicellariae without side teeth on the valves. **Strongylocentrotidae**

Echinidae.

Tubercles of spines smooth. Ambulacral plates with three or more pairs of pores. Spicules of podia C-shaped.

1. Ambulacral plates with three pairs of pores. 2
   Ambulacral plates with five or six pairs of pores. Deep purple. **Paracentrotus**
2. Peristomial area with closely packed plates. Green urchins. **Psammechinus**
   Peristomial area with few thin plates. Reddish with lilac and white spines. **Echinus**

Psammechinus. Test closely set with spines, the primaries not being very conspicuous. Small forms.

P. miliaris. M. p. 294; F. p. 161, as *Echinus*. A small littoral species, living amongst algae, and feeding on all kinds of small organisms. It is greenish in colour, but its green spines may have violet points. To 50 mm. in diameter. Common on all coasts from shore to shallow water. It often conceals itself by pieces of seaweed stuck on its spines.

Echinus. Primary spines usually conspicuous. Large urchins.

E. esculentus (Pl. XXI, figs. 97*a* and 98). Common sea urchin. M. p. 297; F. p. 149, as *E. sphaera*. Test almost globular, but somewhat flattened at the poles, deep red, tubercles white, spines red and white, with lilac tips. One of the largest Echinoids and often attains 120 mm. in diameter, the largest recorded being 176 mm. in diameter. Feeds on algae and small incrusting animals. Common on nearly all rocky coasts, comes inshore in spring for breeding and extends into shallow water.

Paracentrotus. Test closely set with long spines. Peristomial membrane with few plates. Glandular pedicellariae with only one lateral tooth on each side.

P. lividus. M. p. 306; F. p. 167, as *Echinus*. Test rather low, rounded, with very numerous spines, deep purple, diameter rarely exceeding 70 mm. Usually on rocky shores, in which it may bore. Gregarious, often protecting itself with shells and algae. Littoral to shallow water. On south and west Irish coasts, and less commonly on coasts of Devon and Cornwall.

STRONGYLOCENTROTIDAE.

Tubercles smooth. Ambulacral plates with more than three pore pairs. Valves of glandular pedicellariae forming a closed tube.

Strongylocentrotus drobachiensis. M. p. 313. Test low but not flattened. The primary tubercles form distinct vertical series. Usually five to six pore pairs on each ambulacral plate. Test greenish brown, with greenish or reddish spines, sometimes violet with white points. To 80 mm. diameter. Littoral to shallow water. Often boring holes in rocks. East coast only.

CLYPEASTROIDEA

The cake urchins have an elongated or rounded test, considerably flattened, with the mouth in the middle of the under side, the anus outside the apical system and usually posterior or on the under side. Aristotle's lantern is well developed, but there are no gills in the peristomial area. The spines are small, forming a close velvety coat, and the tube feet and pedicellariae are also small. The ambulacral areas are widened to form "petals". Internally, calcareous struts or buttresses extend between upper and lower sides of the test.

Echinocyamus pusillus (Pl. XXI, fig. 97*c*). M. p. 316; F. p. 175. A small greenish form about 15 mm. long and of a depressed oval shape, found on coarse sand or gravel from extreme low-water mark to considerable depths. Mouth and anus on the under side, petals well developed. Feeds on small organisms and detritus.

SPATANGOIDEA

Heart urchins. The test is elongated, grooved but not flattened. The mouth is on the under surface, towards the anterior side, and the anus outside the apical system posteriorly. Gills, internal supports to the test, and teeth are absent. The ambulacra are simple or petaloid on the upper side and spread out on the under side. The mouth has a shovel-like lip or *plastron* provided by the pòsterior interambulacrum. The spines are not strong, and are modified for various purposes. Those on the *plastron* (posterior interambulacrum) are flat and blade-like for digging. In certain areas, e.g. near the anus, around the petals, etc. a band of small close-set spines called a *fasciole* occurs. The tube feet are also modified, some being tentacle-like (probably respiratory), others fringed at the end and highly extensile for food collecting, some anemone-like, with spicules in the arms.

Spatangidae.

The ambulacra on the upper side of the test are sunk, especially the anterior one, while the remaining four are petaloid. A subanal fasciole is present. The lip formed by the posterior interambulacrum is prominent over the transverse mouth.

1. Only one fasciole, the subanal present. Spines purple. **Spatangus**
   Subanal fasciole, and also one over the anterior ambulacrum and apical system (inner fasciole). Spines yellow. **Echinocardium**

Spatangus. The anterior ambulacrum is not very deep and the four petals are conspicuous. Test strong and both it and the spines purple in colour. A closed subanal fasciole present.

S. purpureus. M. p. 328; F. p. 182. Test broad and low. Violet, with some of the spines whitish. To 120 mm. in length. In coarse sand, gravel or mud. Low tide limit to shallow water. Off all British coasts, but local.

Echinocardium. Two fascioles, a small subanal one, and a larger one including the apical system and anterior ambulacrum (known as the inner fasciole, as distinct from a peripetalous fasciole). Smaller and more fragile than *Spatangus*. Yellow in colour.

E. cordatum (Pl. XXI, fig. 97*b*). Sea potato. M. p. 331; F. p. 190, as *Amphidotus*. The anterior ambulacrum is rather deeply sunk. Spines straw yellow in colour. To 90 mm. long, buried in sand, in a definite burrow communicating with the surface by means of a mucus-lined tube in the sand. Feeds by collecting small animals and plants from the surface of the sand by means of the long tube feet in the anterior ambulacrum. Gregarious. Very common all round the coast from intertidal to shallow water zones.

HOLOTHURIOIDEA

Sea Cucumbers. Elongated, worm-like Echinoderms, without arms, but exhibiting a pentaradiate arrangement of podia despite their bilateral symmetry. The mouth is anterior and is surrounded by a circlet of *oral tentacles*. The anus is posterior. The skeleton is reduced and consists of scattered plates embedded in the skin. These plates are of characteristic shape and are known as wheels, anchors and anchor plates. The skin is thick and leathery. The ventral side is the walking sole and comprises three radii (*trivium*), in which the podia are not fixed in ambulacral grooves but are usually scattered. In some families podia are absent and movement is facilitated by the anchors in the skin. The food consists of minute organisms and detritus.

1. Podia well developed. Tentacles arborescent. 2
   Podia reduced. Tentacles feather-like or digitate. **Apoda.** 3
2. Tentacles shield-shaped, with short cauliflower-like branches, without retractor muscles. **Aspidochirota.** 4
   Tentacles bushy, with retractor muscles. **Dendrochirota.** 5
3. Skin with anchor plates and anchors, the latter clinging to the hand. No podia. **Synaptidae**
4. Skin very thick and leathery, darkly pigmented. **Holothuriidae**
5. Body cylindrical, without definite creeping sole, usually white. **Cucumariidae**

HOLOTHURIIDAE.

There is a definite trivium, with numerous podia, the upper surface or bivium having only reduced, papillose podia. The skin is very thick and bears flat "tables" of calcareous matter. Near the anus internally are tubular cotton glands or Cuvierian organs, and when irritated the thread-like contents of the glands may be shot out through the cloaca, entangling any animal within range.

Holothuria forskali. The nigger or cotton spinner. M. p. 393. Twenty yellow tentacles, each short and shaped like a cauliflower. Skin thick, soft, of a black or dark chocolate brown colour on the back, yellow ventrally. To 20 cm. long. From extreme low-tide marks to shallow water, on south-west coasts of England and Ireland and west coast of Scotland.

CUCUMARIIDAE.

Tentacles arborescent. Podia present, often in five fairly regular series. Skin plates irregular and variable, but usually fenestrated tables. Ten tentacles.

Cucumaria. Body usually curved, five distinct rows of tube feet, mouth and anus terminal, and of the ten tentacles, the two ventral ones are usually smaller than the others. The whitish skin is often stiff with calcareous plates.

1. Skin plates of flat fenestrated tables. **C. saxicola**
   Skin plates fenestrated, nodular. 2
2. Skin wrinkled, tough, with cup-shaped bodies as well. More than 4 cm. long. **C. normani**
   Skin smooth, thick. Small forms with scarce tube feet. To 4 cm. **C. lactea**

Cucumaria saxicola (Pl. XXI, fig. 99). Sea Gherkin. M. p. 401. Body almost cylindrical. Skin rather delicate, with smooth surface, and thin. Plates fenestrated, button-like. White. Tentacles dark. 15 cm. Littoral to shallow water, in rock crevices and under stones. Common on south-west coasts of England and Ireland.

C. normani. M. p. 402. Skin tough, leathery, wrinkled. Skin plates nodular with four holes; also cup-shaped plates. Dirty brownish white to black. 15 cm. Littoral to very shallow water. Not common. South coast of England and west coast of Ireland.

C. lactea. M. p. 402; F. p. 229, as *Ocnus brunneus*. Skin very thick, leathery, smooth. Tube feet scarce. Skin plates nodular with four holes, but variable. White or brown. 4 cm. Littoral, on stony bottoms to shallow water. South, west and east coasts.

SYNAPTIDAE.

Body elongated, worm-like. Skin adhesive owing to presence of anchors which are used in locomotion. No podia.

XIX. ECHINODERMATA

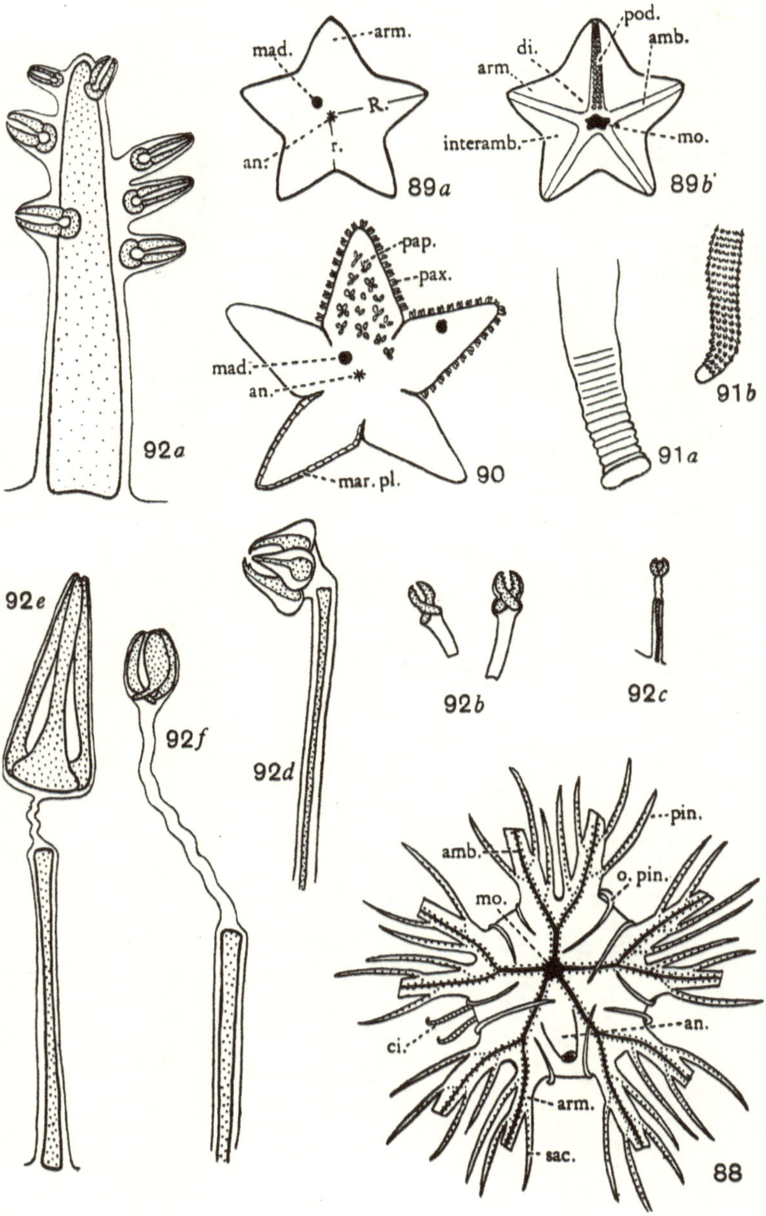

(*facing p.* 243)

1. Anchor plate narrowed at one end. Tentacles with only four branches. **Labidoplax**
Anchor plate rounded. Tentacles much branched. **Leptosynapta**

Leptosynapta. Tentacles usually with numerous branches. The anchor plates of the skin rounded and fenestrated, but not prolonged into a handle. Transparent, usually small. Holothurians burrowing in sand or mud, and feeding on detritus.

L. inhaerens. M. p. 427. Tentacles usually twelve, much branched, the terminal branches the longest. Pink or greyish in colour, to 30 cm. long but usually about half this size. Burrowing in sand or mud. Commensal with *Harmothoe lunulata*, etc. East, south-west and west coasts. Intertidal to shallow water.

Labidoplax. Basal part of the fenestrated anchor plates has a narrow handle so that the whole is like a tennis racket. Tentacles twelve, with few branches.

L. digitata. M. p. 433. Tentacles without terminal prolongations, but with two pairs of finger-shaped lateral branches. Red or brownish. To 30 cm. In mud or sand between tide marks to shallow water. West coasts only.

EXPLANATION OF PLATE XIX

Echinodermata (i), Figs. 88–92

Fig. 88. *Antedon bifida*. Oral view of the calyx and bases of the arms. Only two of the cirri are shown. × 1½.

Fig. 89. *a* and *b*, *Asterina gibbosa*. Oral and aboral views, to illustrate starfish structure. × 1.

Fig. 90. *Porania pulvillus*. Aboral view. One arm shows the bordering paxillae and dermal gills (papulae), another the paxillae only, and a third the marginal plates (hidden within the skin). Note the fusion of the bases of the arms for the enlargement of the disk. × ½.

Fig. 91. *a*, Podium of *Asterias rubens*; *b*, of *Ophiothrix fragilis*. × 12. *a*, the Asteroid and Echinoid type with disk for attachment. *b*, the Ophiuroid and Crinoid type for use as a tentacle. The surface is covered with sensory papillae except at the tip.

Fig. 92. Types of Pedicellariae, illustrated by those of *Asterias rubens* (*a* and *b*), and *Echinus esculentus* (*c*–*f*); *a* and *b* have two jaws, *c*–*f* three. *a*, Straight type on a primary spine; *b*, crossed type; *c*, triphyllous; *d*, globiferous (glandular); *e*, tridentate; *f*, ophiocephalous. Calcareous parts dotted, soft tissues plain. × 15.

amb.	ambulacrum	pap.	papula (dermal gill)
an.	anus		
arm	arm	pax.	paxilla
ci.	cirrus	pin.	pinnule
di.	disk	pod.	podium
interamb.	interambulacrum	R.	distance from centre of disk to tip of arm
mad.	madreporite		
mar. pl.	marginal plate		
mo.	mouth	r.	radius of disk
o. pin	oral (first) pinnule	sac.	sacculus

EXPLANATION OF PLATE XX

Echinodermata (ii), Figs. 93–96

Fig. 93. *Ophiura texturata*. To illustrate the structure of the hard parts in a brittle star. *a*, oral view; *b*, aboral view. The teeth, mouth papillae and pores for the tube feet have not been lettered. × 4.

Fig. 94. Five types of disk in Ophiuroids. *a*, *Ophiothrix fragilis*. Radial shields plain, disk spiny. *b*, *Ophiopsila aranea*. Radial shields long and narrow. Disk with minute scales. *c*, *Ophiactis balli*. Radial shields small and separated by a wedge of scales. *d*, *Ophiopholis aculeata*. Primary plates plain, radial shields hidden by granules and low spines. *e*, *Amphipholis squamata*. Radial shields small, contiguous, disk scaly.

Fig. 95. Two types of mouth apparatus. × 3. *a*, *Ophiothrix fragilis*. Tooth papillae but no mouth papillae. *b*, *Ophiura texturata*. Mouth papillae but no tooth papillae.

Fig. 96. Oblique transverse sections of the arm of *Ophiothrix fragilis*. On the left the spines are shown; on the right, in the gap between lateral and ventral plates, lies a podium, with a single tentacle scale covering the opening through which the podium protrudes. × 3.

ador.sh.	adoral shield	pod.	podium
arm.c.	arm comb	pod.2	aperture for second podium
c.	central plate of disk	r.sh.	radial shield
d.	dorsal plate of arm	rad.v.	radial water vessel
g.	genital plate	s.	scale
gen.b.	genital bursa	sp.	spine
j.	jaw	to.	tooth
l.	lateral arm plate	to.p.	tooth papilla
mo.p.	mouth papilla	t.s.	tentacle scale
mo.sh.	mouth shield	v.	ventral arm plate
musc.	muscle		

XX. ECHINODERMATA

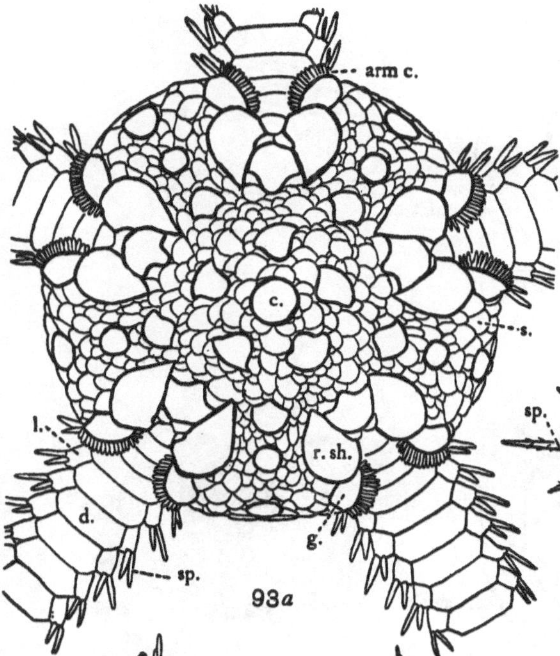

93*a*

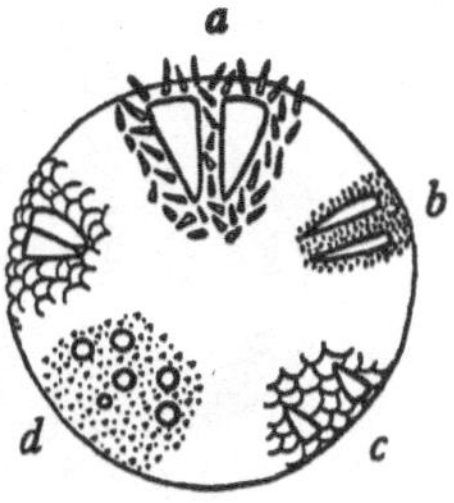

94

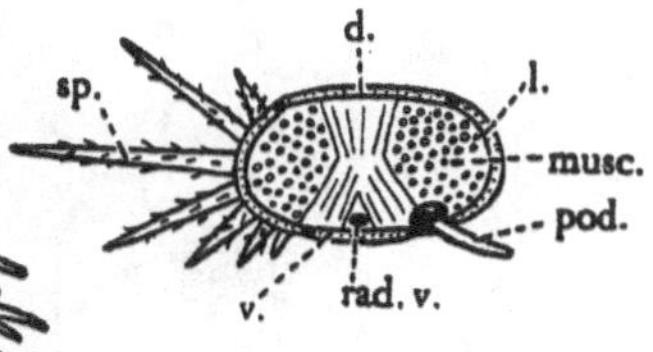

96

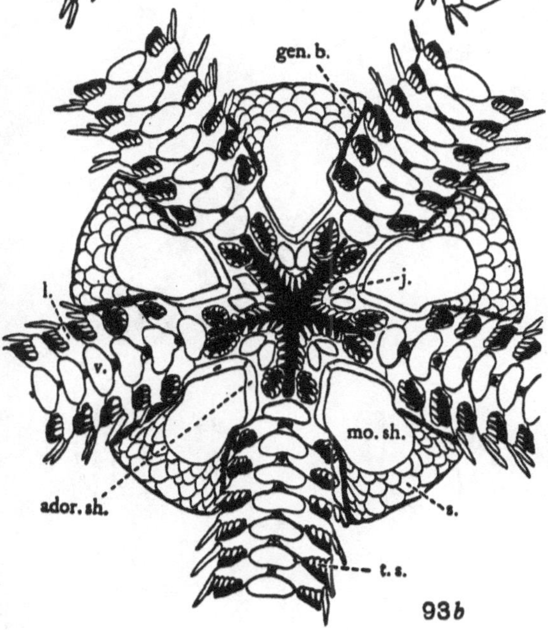

93*b*

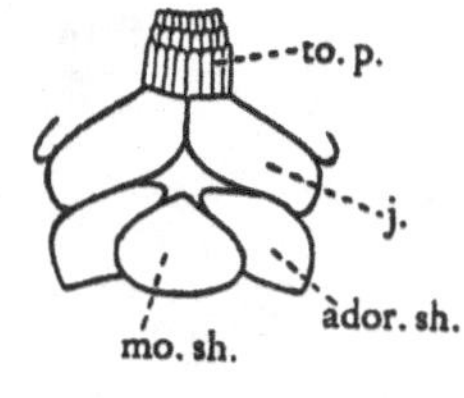

95*a*

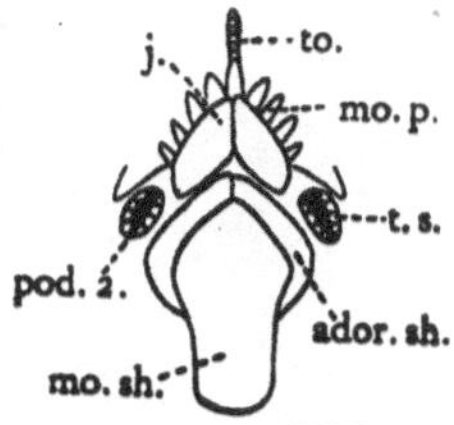

95*b*

(*facing p.* 244)

XXI. ECHINODERMATA

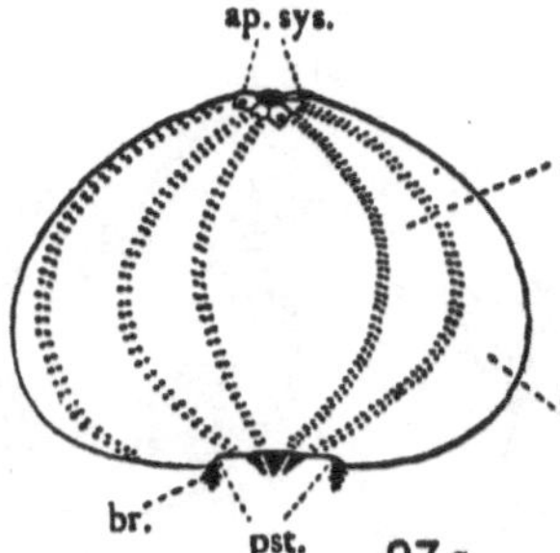

97*a*

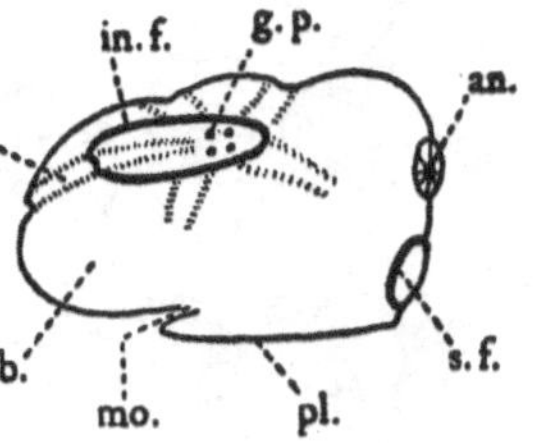

97*b*

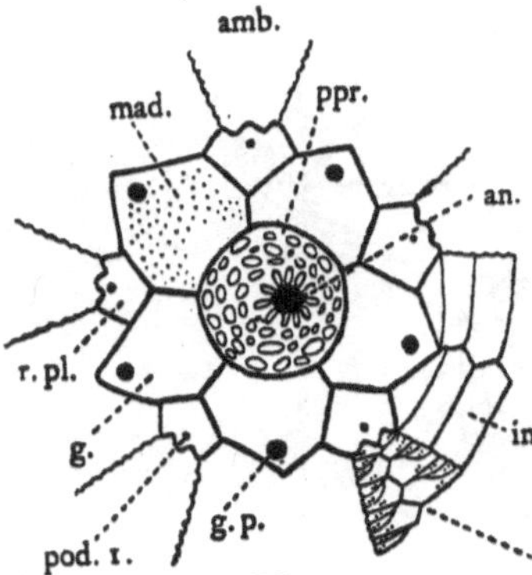

98*a*

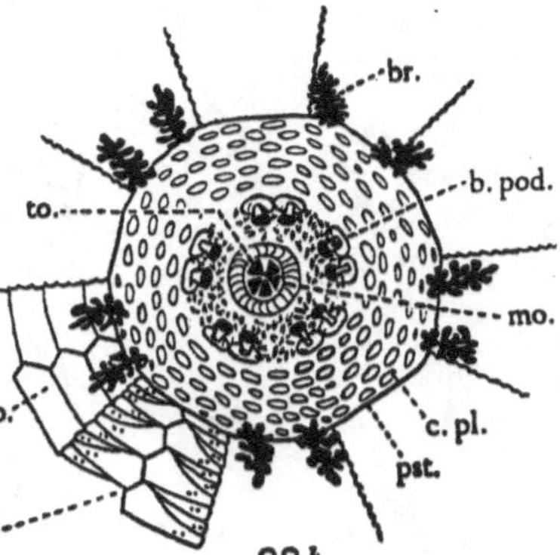

98*b*

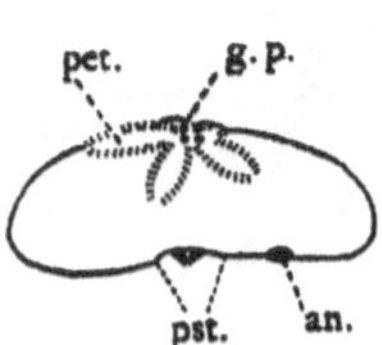

97*c*

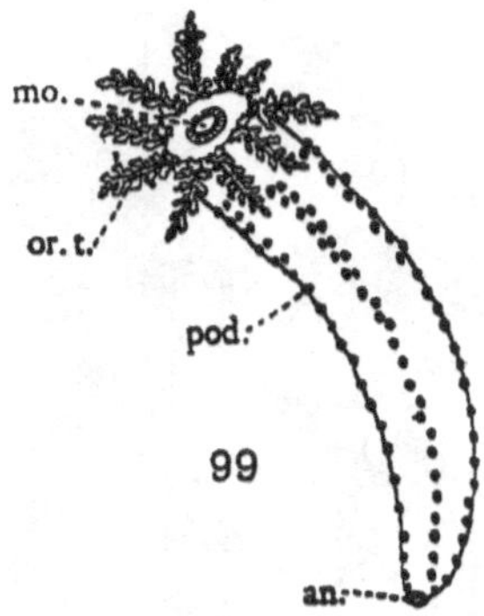

99

100

EXPLANATION OF PLATE XXI

Echinodermata (iii), Figs. 97–100

Fig. 97. The three orders of Echinoids. Lateral view of test after removal of the spines. The dotted lines represent podia of the radii, but not the details of their arrangement. *a*, *Echinus esculentus*, a regular urchin. Anus within the apical system, dermal gills on the peristome. Tooth apparatus present. × ½. *b*, *Echinocardium cordatum*, a heart urchin. Anus outside the apical system. Test irregular. No teeth. × ½. *c*, *Echinocyamus pusillus*, a cake urchin. Anus outside the apical system. Tooth apparatus present. × 4.

Fig. 98. *a*, The apical system; *b*, the peristome of a regular Echinoid (*Echinus*). × 1½. One ambulacrum and one interambulacrum have been sketched. The buccal tube feet pass through calcareous rings and are surrounded by small spines and pedicellariae.

Fig. 99. *Cucumaria saxicola*. Lateral view. The oral tentacles are of the arborescent type. The two short tentacles are ventral in position. The podia are arranged irregularly along the radii. × ½.

Fig. 100. *a*, *b*, *c*. Three types of tentacle in Holothurians. *a*, *Leptosynapta inhaerens*. Feather-like (bipinnate). *b*, *Labidoplax digitata*. Digitate. *c*, *Holothuria nigra*. Shield-shaped.

amb.	ambulacrum	mo.	mouth
an.	anus	or.t.	oral tentacle
ap.sys.	apical system	pet.	petal
b.pod.	buccal tube foot	pl.	plastron
br.	dermal gill	pod.	podium
c.pl.	calcareous plate of peristome	pod.1.	aperture for unpaired tube foot
g.	genital plate	ppr.	periproct
g.p.	genital pore	pst.	peristome
in.f.	inner fasciole	r.pl.	radial plate
interamb.	interambulacrum	s.f.	subanal fasciole
mad.	madreporite	to.	tooth

TUNICATA

(Pls. XXII and XXIII)

AUTHORITIES

ALDER, J. & HANCOCK, A. 1905–12. *The British Tunicata.* Ray Soc. Publ. 3 vols.

BERRILL, N. J. 1950. *The Tunicata.* Ray Soc. Publ. no. 133. London.

GARSTANG, W. 1891. "Report on the Tunicata of Plymouth. Part I." *J. mar. Biol. Ass. U.K.* II, 47–67.

GIARD, A. 1872. "Recherches sur les Ascidies composées, ou Synascidies." *Arch. Zool. exp. gen.* ser. i, I, 501–704.

—— 1873. "Contributions à l'histoire naturelle des Synascidies." *Ibid.* II, 481–514.

HARTMEYER, R. & SEELIGER, O. In Bronn's *Thier-reich.* 1893–1913.

HERDMAN, E. C. 1924. "Botryllus." *L.M.B.C. Memoir XXVI.*

HERDMAN, W. 1881. "Notes on British Tunicata, with descriptions of new species. I. Ascidiidae." *J. linn. Soc.* (*Zool.*), XV, 274–90.

—— 1891. "A revised classification of the Tunicata. Part I." *J. linn. Soc.* (*Zool.*), XXIII, 558–652.

—— 1899. "Ascidia." *L.M.B.C. Memoir I.*

KOTT, PATRICIA. 1952. Compound Ascidians of the Plymouth area. *J. mar. Biol. Ass. U.K.* XXXI, 1, 65–83.

LAHILLE, F. 1890. *Contributions à l'étude anatomique et taxonomique des Tuniciers,* pp. 1–328. Toulouse.

THOMPSON, H. 1930, 1931, 1933, 1934. "The Tunicata of the Scottish Area. Parts 1–4." *Fish. Scot. Sci. Invest.*

The Tunicates are so called because the body is protected by a coat called the *test* or tunic. This is a thick, gelatinous, and often transparent covering, external to the epidermis and the muscular body wall or *mantle.* The general form of the body is that of a leather bottle, viz. an ovate or globular sac with two necks ending in apertures, one median in position and the other usually on one side of the body. The two apertures are the *branchial* (oral) and the *atrial* (cloacal), and it is usual for each to be raised on a short tube or *siphon.* The opening is often lobed, with eyespots (*ocelli*) between the lobes. The siphon is provided with sphincter muscles, lying in the mantle. The *branchial siphon* takes in food and water, separation of the two occurring in the *pharynx*. Between the siphon and the pharynx is a *prebranchial zone*, bounded orally by a circlet of *oral tentacles* (often called the tentacles), and on the pharyngeal side by two parallel ridges, one or both of which may be ciliated, called the *peripharyngeal bands*. One band is complete, the other is broken mid-dorsally and mid-ventrally, at these points turning back at right angles towards the pharynx. In the ventral wall of the pharynx it communicates with a longitudinally placed ciliated and glandular groove called the *endostyle*, in the dorsal wall with a projecting longitudinal shelf, the *dorsal lamina*. The latter may be provided with tentacles or *dorsal languets*. Laterally the walls of the pharynx are divided into rectangular areas or *meshes*, which are usually recessed and are supported by transverse and sometimes by longitudinal bars, within which lie blood vessels. Each mesh contains several gill slits or *stigmata* arranged spirally or longitudinally. Surrounding the pharynx is a peripharyngeal chamber, the *atrium*, which communicates with the pharynx through the stigmata and with the exterior through the *atrial siphon*. The lower end of the pharynx leads, on the dorsal side (i.e. at the lower end of the dorsal lamina), to a short oesophagus, which expands into a globular stomach. Thence an intestine, usually bent in a crossed loop, passes into a rectum, whose anal aperture opens into the atrium. The hermaphrodite gland (*ovo-testis*), renal organ and *heart* usually lie within the intestinal loop. Separate oviduct and vas deferens discharge also into the atrium, whose aperture acts, therefore, as a cloaca. When the water current containing food enters the branchial aperture, the food particles are separated out by the action of the endostyle and the dorsal lamina, and pass into the stomach, while the respiratory current goes through the gill slits into the atrium and out through the anal siphon. To effect this the endostylar glands secrete a mucus string which is passed upwards by ciliary action towards the branchial aperture. On meeting the entering current the food particles are caught in this string, which travels laterally along the two halves of the peripharyngeal band or are thrown across to the dorsal lamina, and down this to the oesophageal opening at its lower extremity.

In the angle between branchial and atrial apertures lies the *nerve ganglion*, the sole nerve centre. Ventral, but occasionally dorsal to the ganglion, is the *neural gland*, which has a communication with the prebranchial zone by means of a ciliated funnel called the *dorsal tubercle*. Neural gland and dorsal tubercle together make up the *hypophysis*. A notochord occurs in the larva but is rarely present in the adult.

Tunicates may be fixed or free, solitary or colonial. All are marine, and most are small. The free-swimming forms are pelagic and belong to the orders Thaliacea and Appendicularia. The former are barrel-shaped, with strong encircling muscle bands, the latter tadpole-like. Littoral forms belong to the Ascidiacea, often known as Sea Squirts, because when touched they squirt water through their siphons.

The ASCIDIACEA may be fixed or free, simple or compound, but are never free-swimming in the adult stage. The test is well developed, the pharynx usually large, with numerous gill slits or stigmata. The musculature of the mantle is irregular and the body sac-shaped. In many colonial forms a stolon arises from the pharynx wall near the lower end of the endostyle. The individual of the colony is called an *ascidiozooid*.

Identification of Ascidians often involves their dissection, since externally only the test and siphons may be visible.

1. Gut on one side of the pharynx, body not divided into regions. **2**
   Gut posterior to pharynx, body divided into two or more regions. **7**
2. Solitary or gregarious Ascidians. **3**
   Colonial. **6**
3. Stigmata curved, usually spiral. **4**
   Stigmata straight, usually longitudinal. **5**
4. Apertures six- and four-lobed. Oral tentacles branched. **Molgulidae**, p. 249
   Apertures eight- and six-lobed. Oral tentacles filiform. **Corellidae**, p. 251
5. Apertures six- and four-lobed. Gonads numerous or lobulated. **Styelidae**, p. 249
   Apertures eight- and six-lobed. **Ascidiidae**, p. 251
6. Individuals with separate branchial and atrial apertures. **Perophoridae**, p. 253
   Individuals grouped in systems, with a common atrial aperture to each system. **Botryllidae**, p. 250
7. Two regions to body—thorax and abdomen. **8**
   Three regions to body—thorax, abdomen and post-abdomen. **Polyclinidae**, p. 253

8. Solitary, or if colonial joined by stolons, with separate apertures to each zooid. 9
   Colonies with a common cloaca to each system. Test often containing spicules. Flat encrusting forms. **Didemnidae**, p. 255
9. No contraction between thorax and abdomen. Pharynx with longitudinal bars. **Cionidae**, p. 252
   Body wall usually contracted between thorax and abdomen. No longitudinal bars in pharynx. **Clavelinidae**, p. 253

MOLGULIDAE.

Solitary or in groups, sessile, embedded in sand or fixed. Test often incrusted with sand which is attached to prolongations of the test. Stigmata curved, often spiral, the spirals forming a cone or infundibulum. Tentacles branched.

Molgula. Spiral of the stigmata incomplete. Globular Ascidians, in many species incrusted with sand, mud, pebbles or shells.

1. Body smooth, free from incrustation. **M. citrina**
   Body incrusted except near siphons. **M. oculata**

M. citrina. A. and H. 1907, II, 62; T. 1930, p. 27; Berrill, 1950, p. 258. Test usually clear, greenish yellow. Oral aperture subterminal, atrial smaller and about one-third of the way down the side. Oral tentacles bi- or tri-pinnate. Viviparous. Small, to 12 mm. long. Littoral, under stones, on rocks, other ascidians, wharf piles, etc.

M. oculata. A. and H. 1907, II, 48; T. 1930, p. 25; B. 1950, p. 252. Globose, adhering slightly at base, sometimes free. Closely incrusted with shells and pebbles except near the apertures. Apertures short and wide, looking like eyes wearing spectacles. About eighteen pinnate or bipinnate oral tentacles. 5–7 cm. diameter. Lower shore, often on shells. South-west coasts.

STYELIDAE.

Body attached, sessile, rarely incrusted, solitary. Branchial and atrial apertures six- or four-lobed. Tentacles simple, unbranched. Gonads often scattered.

1. Apertures four-lobed. 2
   Apertures six-lobed. Ovaries and testes separate. **Distomus**
2. Gonad on right side only, a single branched mass. **Dendrodoa**
   Gonads numerous (polycarps), with central ovary and peripheral testis. **Polycarpa**

Polycarpa. Sessile. Test rough and opaque. Both apertures quadrate, four-lobed. Tentacles simple, linear. Gonads in form of small separate masses (polycarps) scattered over the mantle.

P. rustica. T. 1931, p. 7, as *Styela*. B. 1950, p. 193. Flesh-coloured, cylindrical, attached. Test rough, leathery, slightly mammillated, naked or covered with fine sand. Siphons not prominent, fairly close together, red. Gonads small, ovoid, red or orange. Gregarious. 25–30 mm. long, taller than broad. Low tide to shallow water. Estuarine.

Dendrodoa. Body subcylindrical, attached, not sandy. Apertures anterior, four-lobed. Test thick, rough, wrinkled. Oral tentacles simple. Gonad a single lobulated mass on right side.

D. grossularia (*Styelopsis*). Sea gooseberry. A. and H. 1907, II, 129; T. 1931, p. 26; B. 1950, p. 198. Rusty or bright red to yellow. Adherent singly or in masses, hemispherical. Apertures short, tubular, bright red, set far apart on upper surface. Viviparous. 6–12 mm. in diameter, broader than high. Between tide marks to shallow water. Variable.

Distomus. Gregarious, sessile or shortly pedunculate. Apertures six-lobed, occasionally four-lobed. Test gelatinous, reddish. Gonads separate. Viviparous.

D. variolosus. T. 1931, p. 30; B. 1950, p. 203. In large masses, each individual joined to its fellows up to the middle of the body. Testes orange, ovoid. Ovaries red, spherical. Intense red colour. 8 mm. high. Low tide to shallow water. Local, often on *Laminaria* holdfasts.

BOTRYLLIDAE.

Colonial. Zooids united into systems and forming thin incrustations, or in more favourable positions pendent in thick masses. System of zooids elliptical, circular or branched, with separate rounded branchial apertures, and a common cloacal aperture. Test usually soft, traversed by vessels with terminal knobs or ampullae. Gonads paired. Budding (gemmation) lateral and also from stolons.

1. Systems regular, neither branched nor anastomosing. **Botryllus**
   Systems irregular, branched and anastomosing. **Botrylloides**

Botryllus. Colony thin, incrusting, jelly-like. Systems regular, circular or elliptical, nine or ten zooids in each system, star-like.

B. schlosseri (Pl. XXIII, fig. 106). Star sea squirt. A. and H. 1912, III, 54; T. 1931, p. 36; E. C. Herdman, 1924; B. 1950, p. 216. Numerous colour varieties of this beautiful Ascidian occur. The violet form has a cream star and red spot on each zooid, another form is green and yellow, and a third transparent yellowish green with a red spot on each zooid. Individuals about 2 mm. long. Littoral, very common on stones and weed. The vase-shaped egg cases of the Cowrie (*Trivia monacha*) are found attached to the jelly of this species.

Botrylloides. Colony thin, incrusting, gelatinous. Systems elliptical or elongated, branched and anastomosing, irregular.

B. leachi. A. and H. 1912, III, 77; T. 1931, p. 39; B. 1950, p. 224. Yellow and red, very irregular. Littoral.

Corellidae.

Attached, sessile or pedunculate. Branchial aperture with eight, six or no lobes. Test cartilaginous, gelatinous or horny. Stigmata curved, without papillae or longitudinal vessels.

Corella parallelogramma (Pl. XXIII, fig. 105*b*). A. and H. 1907, II, 25; T. 1933, p. 6; B. 1950, p. 149. Body ovate and compressed, sessile and attached, test soft and transparent. Mantle crimson and yellow, showing through the test. Branchial apertures with eight lobes, atrial with six, both with red ocelli. 25–50 mm. high and about 30 mm. broad. On shells, etc., from low tide to shallow water.

Ascidiidae.

Attached, usually sessile. Test gelatinous. Branchial aperture eight-lobed, atrial usually six-lobed. Ocelli red. Mantle musculature reticulating. Pharynx without folds, with longitudinal bars and usually papillae. Stigmata straight. Stomach and intestine beside the pharynx, gonad close to intestine.

1. Pharynx recurved posteriorly. Test thick, cartilaginous, mammillated. **Phallusia**
   Pharynx not recurved posteriorly. **2**
2. Nerve ganglion close to dorsal tubercle. **Ascidiella**
   Nerve ganglion distant from dorsal tubercle. **Ascidia**

Ascidia. Body naked, or incrusted with sand. Nerve ganglion and neural gland distant from the dorsal tubercle.

1. **Large Ascidians, about 10 cm. or more high. Colour milky, horn-coloured or reddish. Ascidia mentula**
   Small Ascidians, depressed, about 1 cm. high. Colour pale green, transparent. A. conchilega

A. mentula. A. and H. 1905, I, 75; B. 1950, p. 160. Erect, often attached by one side, compressed, pear-shaped. Test free from sand. Apertures distant. Milky, horn-coloured or reddish. 10 cm. or more high. From tide marks to shallow water. South and west coasts.

A. conchilega. A. and H. 1905, I, 111, as *A. depressa.* B. 1950, p. 162. Body depressed, attached usually by left side. Test transparent with small tubercles. Siphons rather long. Pale green. Only about 1 cm. thick, 3–5 cm. long and 15 mm. broad. Tide levels to shallow water.

Ascidiella. Attached, not incrusted. Apertures not far distant. Nerve ganglion close to dorsal tubercle.

1. **Attached by posterior end. Stigmata about five in a mesh. A. aspersa**
   Attached by left side. Stigmata about twelve in a mesh. A. scabra

A. aspersa (Pl. XXIII, fig. 104). Herdman, 1881, p. 281; B. 1950, p. 154. Test with branched projections, rough. Colourless or reddish. About 5 cm. long. Estuarine, intertidal to shallow water, especially south coast. On stones, algae, etc. Often in clusters.

A. scabra. A. and H. 1905, I, 128, as *Ascidia*; T. 1933, p. 28; B. 1950, p. 158. Siphons long, diverging. Test tough, transparent, depressed. 18–30 mm. Intertidal to shallow water, on stones, shells, etc. Common in northern areas.

Phallusia. Erect, attached. Test cartilaginous, opaline.

P. mammillata (*Ascidia*). A. and H. 1905, I, 72; T. 1933, p. 10; B. 1950, p. 171. Test thick, white or yellowish, opaline, divided into areas with a rounded eminence in the centre of each (mammillated). Oral tentacles few, small, sixteen to twenty-four. 7–12 cm. long. Adhering to stones, etc., in shallow water beyond the tidal zone, but sometimes washed up on the shore.

CIONIDAE.

Body elongate, cylindrical, attached posteriorly. Test thin and gelatinous. Branchial aperture eight-lobed, atrial six-lobed. Mantle musculature mainly longitudinal. Stigmata straight, meshes rectangular, dorsal lamina with languets, which are long, numerous and slender. Stomach, intestine and gonad below the branchial sac.

Ciona intestinalis. A. and H. 1907, II, 9; T. 1933, p. 34; B. 1950, p. 131. A large, tubular sea squirt, with the two apertures anterior, unequal and not far distant from one another. Green or grey. 7–10 cm. tall, 2–4 cm. wide. On stones, etc., from extreme low-tide mark to shallow water.

PEROPHORIDAE.

Colonial Ascidians, the zooids being distinct and attached by short peduncles to creeping stolons. Test thin and transparent but not investing the stolon. Branchial aperture six-lobed, atrial five-lobed, rather distant. Four rows of straight stigmata.

Perophora listeri (Pl. XXII, fig. 101). Garstang, 1891, p. 58. B. 1950, p. 145. Body short, laterally compressed. About 1 cm. high. On stones and algae in shallow water and in tidal pools. Colony resembles a small bunch of grapes. Viviparous.

CLAVELINIDAE.

Colonial Ascidians, with a gelatinous or cartilaginous test, which is in some individual, in others common. Body divided into thorax and abdomen, connected by a slender oesophageal region. Apertures terminal, close together, smooth edged. Stigmata straight.

Clavelina lepadiformis (Pl. XXII, fig. 102). A. and H. 1907, II, 152; Garstang, 1891, p. 51; T. 1934, p. 4; B. 1950, p. 70. This beautiful Ascidian is very common locally. Colony compact, test very clear and gelatinous, through it appearing bright yellow and brown pigment lines situated on the thorax. Ovary like a bunch of grapes. Viviparous. 1–2 cm. high. On rocks and stones, from low-tide marks to shallow water.

POLYCLINIDAE.

Colony usually massive and fleshy, in some incrusting, in some lobed or pedunculated. Zooids wholly immersed in a common envelope, elongated antero-posteriorly, usually with three regions—thorax, abdomen and post-abdomen. When systems of zooids are present, there is a common cloaca, usually inconspicuous. Branchial aperture six- or eight-lobed, atrial aperture often with a languet. Test gelatinous or cartilaginous, often with embedded sand grains but without spicules. Dorsal lamina with languets. Oral tentacles small, few. Colonies reproduce by gemmation.

1.	Stomach smooth-walled.	**Polyclinum**
	Stomach areolated or channelled.	**2**
2.	Stomach channelled with longitudinal folds.	**3**
	Stomach areolated, with mulberry-like markings.	**Morchellium**
3.	Branchial aperture with six teeth, atrial with a languet.	**Aplidium**
	Branchial aperture with eight teeth.	**Sidnyum**

Polyclinum. Systems simple or compound. Branchial aperture with six teeth. Atrial languet well developed. Stomach simple, without folds or channels. Male and female follicles separate in gonad.

P. aurantium. A. and H. 1912, III, 2; T. 1934, p. 20; B. 1950, p. 87. Systems compound. Colonies subglobular, flattened on top, contracting to a narrow base. Ascidiozooids in elliptical curves round the large smooth cloacae. Colonies 6–25 mm. across. Yellow. On undersides of stones, between tide marks to shallow water.

Morchellium. Colonies mushroom-shaped, with enlarged heads on slender stalks, occasionally sessile. Branchial aperture eight-lobed. Four red pigment spots, two beside endostyle, two by dorsal tubercle. Post-abdomen large, but not distinctly separated from the abdomen, and about double the length of thorax and abdomen together. Simple atrial languet.

M. argus (Pl. XXII, fig. 103). Herdman, 1891, p. 625; T. 1934, p. 24; B. 1950, p. 95. This very common Ascidian is vermilion or yellow in colour and reaches 35 mm. in height and 12 mm. in breadth. It may be incrusted with sand. An atrial languet present. On stones, etc., between tide marks.

Aplidium. Often confused with *Morchellium*, but distinguished by the longitudinal folds in the stomach wall. Colony massive, lobed or pedunculated, fleshy. Branchial aperture six-lobed, atrial with a long languet close to the branchial aperture. Viviparous.

1.	Atrial languet simple. Colony reddish to yellow.	**A. proliferum**
	Atrial languet three-lobed. Colony transparent, with an orange spot on the anterior end of the endostyle of each zooid.	**A. punctum**

A. proliferum. Kott, 1952, p. 73. This includes the forms known as *nordmanni* and *densum*. B. 1950, pp. 105 and 108. Variable. Colonies fleshy, sessile or shortly stalked, and may be sand encrusted. 20–40 mm. high. Zooid with short, broad post-abdomen. Systems with few, regularly arranged zooids, or numerous, closely packed zooids. South and west coasts. Lower shore to shallow water.

A. punctum. Giard, 1873, p. 495; T. 1934, p. 33; B. 1950, p. 100. Resembles *Morchellium argus* in appearance, but each individual has a single red spot which lies at the upper extremity of the endostyle. Atrial languet trilobed. Stomach simply channelled with six folds. Yellow to bluish white. About 12 mm. in diameter. Under rocks to shallow water, often on algae. South-west coasts.

Sidnyum. Colony massive or lobed, with compound systems. Post-abdomen separated by a constriction from the abdomen. Branchial aperture eight-lobed, atrial with a median lobe, and usually a languet. Seven to nine rows of stigmata. Stomach wall thickened irregularly.

S. turbinatum. Giard, 1872, p. 639, as *Circinalium concrescens*; T. 1934, p. 22; B. 1950, p. 112. Colony lobed, arising from stolons, each system a truncated cone. Branchial aperture with eight teeth. Atrial languet present. Post-abdomen short, ovary bilobed, testis with numerous follicles. Yellow or orange with white pigment round the branchial aperture, often with red spots on the surface. On *Zostera* and under rocks. Tidal to shallow water.

S. elegans. Giard, 1872, p. 638, as *Fragarium*. B. 1950, p. 110. Flower-like. Colony consisting of a few zooids, with several large cloacal apertures, toothed and surrounded by a toothed collar. Branchial apertures with about eight teeth. Rich rose colour, with white tipped branchial teeth. Somewhat like *Aplidium proliferum*. On *Zostera* and under rocks.

Didemnidae.

A family of thin, flat, incrusting lichen-like compound Ascidians. The systems are often inconspicuous and irregular, but the common cloacae are usually plain and without lobes. The ascidiozooids are small, divided into thorax and abdomen, without post-abdomen. Branchial aperture six-lobed, atrial plain or with a languet. The test usually contains stellate calcareous spicules. Stigmata usually in three or four, rarely five rows. Stomach united to the thorax by a narrow oesophagus, and usually smooth-walled. Certain Gastropods belonging to the families Cypraeidae and Lamellariidae browse on Didemnid Ascidians, and lay their eggs on the test or in the Ascidian's cloaca. The veliger larva of *Lamellaria* was called *Echinospira* before its identity was discovered.

1. Branchial aperture six-lobed. Vas deferens convoluted or spiral. **2**
   Branchial aperture circular or minutely toothed. No anal languet. Vas deferens straight. Test gelatinous. **Diplosoma**
2. Brightly coloured, with spicules. Two anal languets. Vas deferens convoluted. **Trididemnum**
   Gelatinous or opaque, with or without spicules. Vas deferens spiral. Cloaca slit-like. **Didemnum**

Diplosoma. Test soft and gelatinous, usually transparent, sometimes pigmented but without spicules. Branchial sac with four rows of stigmata.

D. listerianum. Lahille, 1890, p. 107, as *D. gelatinosum* var. *listeri*; T. 1934, p. 17; B. 1950, p. 125. Colony thin and translucent, without pigment, ascidiozooids plainly visible through the test. Colonies of ten to fifty zooids. On weed, to shallow water. *Trivia monacha* (Cowrie) lays its eggs on this Ascidian. The eggs are small vase-shaped structures and are attached by stalks to the test.

Trididemnum. Spicules present in the test. Well-developed atrial siphon, one pair anal languets. Sixteen oral tentacles in two rows.

T. tenerum. Giard, 1872, p. 648, as *Didemnum cereum*; T. 1934, p. 10; B. 1950, p. 116. Colony 25–75 mm. long, straight, with a few cloacae. Brilliantly coloured yellow, orange, bluish or greenish grey. Spicules evenly distributed and not massed round the apertures. Surface smooth. Stomach globular, vas deferens with twelve convolutions. Variable, either forming thin crusts or irregular lobed fleshy masses. On *Zostera*, from coast to shallow water. The Gastropod *Lamellaria perspicua* lays its eggs in the cloaca of this species, and its veliger larvae hatch there.

Didemnum. Colony fungiform, thick and fleshy, or thin and incrusting, with a hard firm test, resembling a lichen. With or without calcareous spicules. Cloaca without lobes, slit-like, with or without anal languet. Eight oral tentacles. Ovary single, lateral, only one egg maturing at a time. Testis with spiral vas deferens.

XXII. TUNICATA

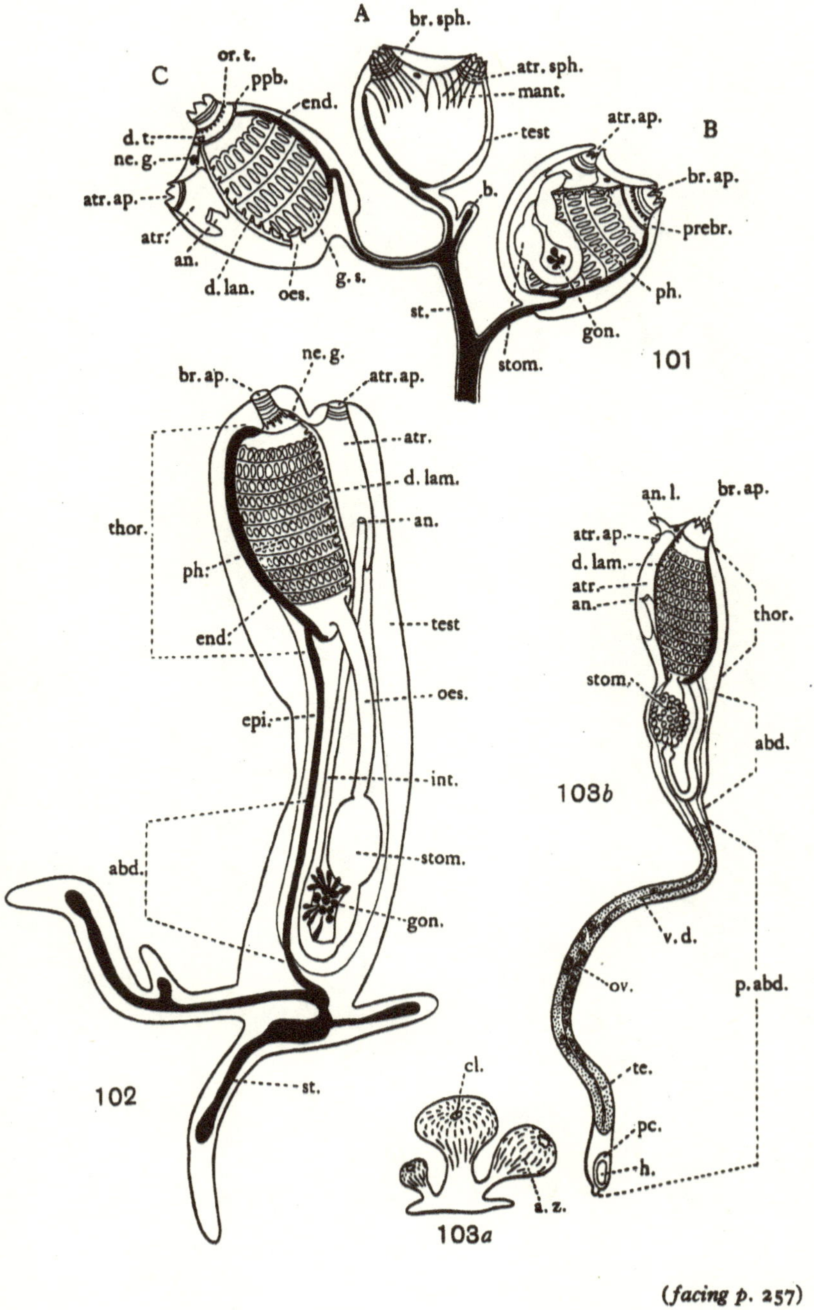

(*facing p.* 257)

1. Test gelatinous, with few or no spicules. **D. gelatinosum**
   Test an opaque crust filled with spicules. **2**
2. Colour grey or blue, with purplish lines. **D. maculosum**
   Colour yellow or white, not lined but covered with conical papillae. **D. asperum**

D. gelatinosum. A. and H. 1912, III, 32; B. 1950, p. 122. Incrusting, thin, viscous and without colour. May have a few spicules round the gut but none near the surface. 25–50 mm. in diameter. On stones, shells, etc. Littoral. South coast.

D. maculosum. A. and H. 1912, III, 44, as *Leptoclinum*. B. 1950, p. 125. With calcareous spicules in the thin, tough, leathery crust. Whitish grey or bluish colonies marked with branched purplish streaks, 25 mm. or more across. Common cloacae rather inconspicuous. On holdfasts of *Laminaria* in littoral zone on English coasts.

D. asperum. A. and H. 1912, III, 45, as *Leptoclinum*. Incrusting, leathery, thin, covered with conical papillae, one or more near each branchial aperture. Dull yellowish white. On stones, holdfasts of *Laminaria*. May be a variety of *maculosum*.

EXPLANATION OF PLATE XXII

Tunicata (i), Figs. 101–103

Fig. 101. *Perophora listeri.* To illustrate the general structure of a Tunicate. The colony consists of numerous ascidiozooids budded off from a stolon, which is covered with ectoderm but is not enveloped in test. Three zooids are shown. A shows test and mantle with simple musculature, B the general structure, with the stomach and intestine beside the pharynx, C the more detailed structure of the pharynx. ×10.

Fig. 102. *Clavelina lepadiformis.* Colonial form with separate apertures to each ascidiozooid. Budding by stolons. Body divided into thorax and abdomen, constricted between the two. The mantle is not shown. ×7.

Fig. 103. *Morchellium argus.* *a*, Three systems on an incrusting base. ×1. *b*, A single ascidiozooid. Note the three regions of the body—thorax, abdomen and post-abdomen; also the mulberry-like stomach swellings, typical of the genus. ×15.

a.z.	ascidiozooid	b.	bud
abd.	abdomen	br.ap.	branchial aperture
an.	anus	br.sph.	branchial sphincter
an.l.	anal languet	cl.	cloaca
atr.	atrium	d.lam.	dorsal lamina
atr.ap.	atrial aperture	d.lan.	dorsal languet
atr.sph.	atrial sphincter	d.t.	dorsal tubercle

end.	endostyle	p.abd.	post-abdomen
epi.	epicardium (leading to stolon)	pc.	pericardium
		ph.	pharynx
g.s.	gill slit	ppb.	peripharyngeal band
gon.	hermaphrodite gland	prebr.	prebranchial zone
h.	heart	st.	stolon
int.	intestine	stom.	stomach
mant.	mantle	te.	male part of hermaphrodite gland
ne.g.	nerve ganglion		
oes.	oesophagus	test	test
or.t.	oral tentacle	thor.	thorax
ov.	female part of hermaphrodite gland	v.d.	vas deferens

EXPLANATION OF PLATE XXIII

Tunicata (ii), Figs. 104–106

Fig. 104. *Ascidiella aspersa*. Test, epidermis and mantle removed from the apertures and pharyngeal regions to show more detailed structure of these regions. ×15.

Fig. 105. The chief types of pharynx wall in Tunicates. *a*, *Ascidia mentula*, transverse bars, longitudinal bars with papillae, straight longitudinal stigmata. ×35. *b*, *Corella parallelogramma*, free longitudinal bars, without papillae. Stigmata spiral, on elevated cones. ×35.

Fig. 106. *Botryllus schlosseri*. A single system resting on the incrusting corm. The system contains six ascidiozooids discharging into a common central cloaca by six atrial apertures. Four of these apertures are figured; these are normally covered by atrial languets similar to those shown in the remaining two zooids. Gonads are figured on two zooids; in the lower zooid the gonad is predominantly male, in the upper predominantly female. The branchial sphincter is shown in varying degrees of contraction. On the right a zooid has pushed its way out of the circle and has already given rise to two buds. From these a new system will be formed. ×9.

amp.	ampulla	gon.f.	gonad, predominantly female
an.	anus		
atr.ap.	atrial aperture	gon.m.	gonad, predominantly male
atr.sph.	atrial sphincter		
b.	bud	l.b.	longitudinal bar
br.ap.	branchial aperture	mesh	mesh of pharynx
br.l.	branchial lobe	ne.g.	nerve ganglion
br.sph.	branchial sphincter	ne.gl.	neural gland
corm	corm	oc.	ocellus
d.lam.	dorsal lamina	or.t.	oral tentacle
d.t.	dorsal tubercle	pap.	papilla
e.k.	end knob of vessel	ppb.	peripharyngeal band
end.	endostyle	prebr.	prebranchial zone
g.s.	gill slit	t.b.	transverse bar

XXIII. TUNICATA

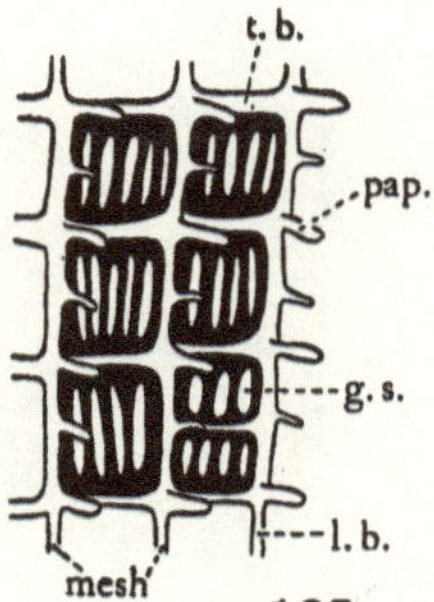

105*a*

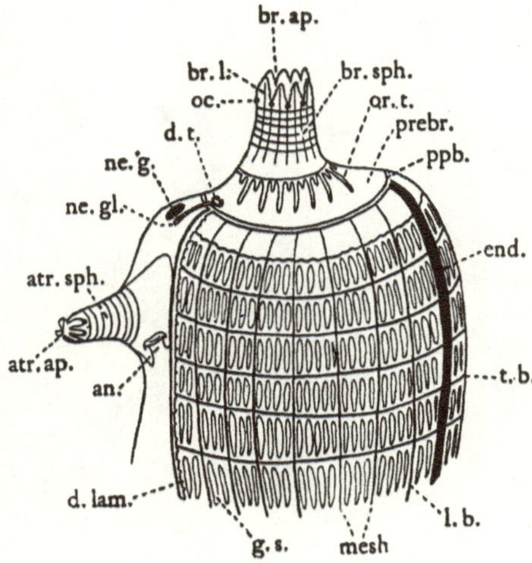

104

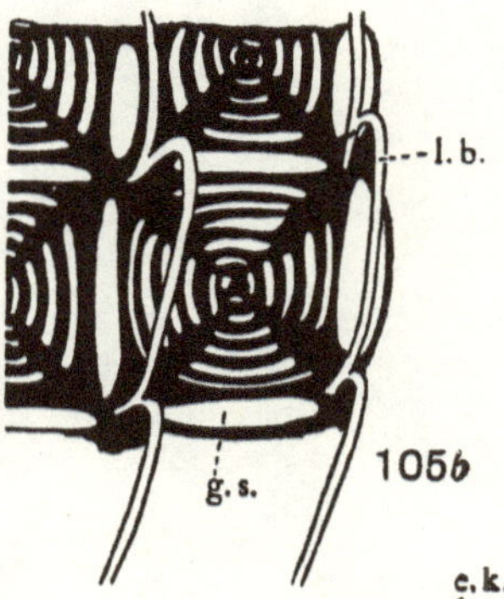

105*b*

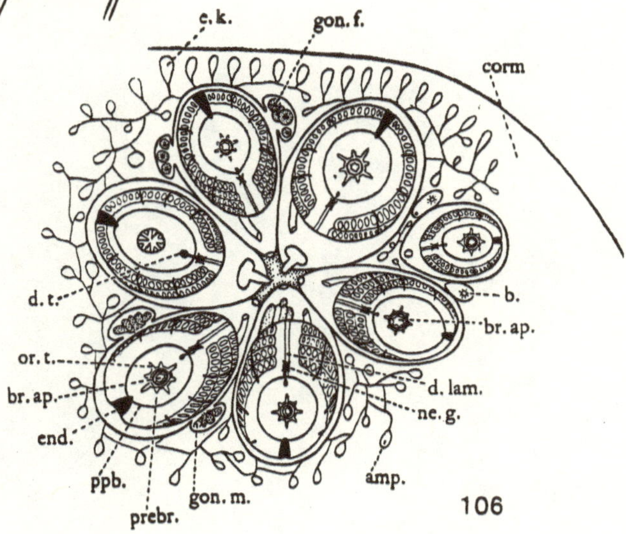

106

A small marine group with isolated distribution. In British seas they are represented by the order Enteropneusta, with four genera. The larval form (Tornaria) is common all round the British coasts, adults have been recorded in the Channel Islands, the Menai Straits and a few other places. *Balanoglossus* (usually a species of *Glossobalanus* or *Saccoglossus*) lives buried in the sand and swallows large quantities for the sake of its organic content, ejecting a casting of a denatured sand in much the same way as *Arenicola* does. The body is unsegmented, elongated, soft and worm-like. It is divided into three regions: proboscis, collar and trunk. The proboscis is shaped like an elongated acorn, and is a hollow organ used for burrowing, with a small opening near the base called the proboscis pore. A marked constriction separates it from the collar, which stands up round the body immediately posterior to the proboscis, its edges projecting at both ends. There are two collar pores. The small mouth, which cannot be closed, lies beneath the anterior collar rim on the ventral side, in the incision between proboscis and collar. The trunk is the longest part of the body and is depressed, with lateral wing-like processes in some genera. Anteriorly it is perforated by numerous pairs of curved gill-slits, opening from the pharynx. The sexes are separate, the gonads numerous and paired, each opening by a dorso-lateral pore. Posteriorly the trunk contains irregular hepatic diverticula which fade out towards the tail. The anus is termnial.

Saccoglossus ruber, Tattersall 1905, *Ann. Rep. Fish. Ireland*, 1902–3, p. 213. Proboscis pinkish orange, elongated, with a shallow dorsal longitudinal groove and a ciliated band surrounding its stalk. Collar brick red, with dark circular bands anteriorly and posteriorly. Trunk dull orange to yellow, long and spirally coiled, very fragile. Number of gill pores variable, but may reach 80 or more pairs. Branchial and genital regions overlapping; no genital wings. Hepatic region dull yellow with red and buff spots. The gut may show through as a green streak. Burrows U-shaped, irregularly spiral, lined with yellowish green mucus, 5–20 cm. down, in fine clean sand at low water spring tide level. Worm about 75 mm. long, proboscis 15 mm. but much more when extended. Breeding in April. Coast of North Wales.

See also: C. Burdon Jones and A. M. Patil, The Genus *Saccoglossus* in British Waters, *Proc. Zool. Soc. Lond.* **134**, 635, 1960.

PISCES

(Pl. XXIV)

AUTHORITIES

DAY, F. 1880–4. *The Fishes of Great Britain and Ireland.* 2 vols. Vol. I, 1880–2; Vol. II, 1882–4. London.

JENKINS, J. TRAVIS. 1925. *The Fishes of the British Isles.* London: Warne.

Plymouth Marine Fauna List. 1931.

British Museum (Natural History) List of British Vertebrates. 1935.

The classification is based mainly on the Plymouth list.

The true fishes are aquatic Chordates extracting their oxygen from the water which bathes them. For respiratory purposes a current of water is passed into the pharynx and makes its exit laterally through paired perforations, called gill slits, in the pharynx wall. The sides of the slits are highly vascular, with a folded mucous membrane, and it is through this membrane that oxygen is absorbed into the blood stream. The folded membranes are called gills and are supported on gill arches of cartilage or bone.

The body of the fish is adapted for rapid movement in a watery medium. It is usually blunt in front, with either a terminal or a ventral mouth, and tapers posteriorly to a tail which is normally the organ of propulsion. The contour of the body gives a cigar-shaped outline in most fishes; apart from the fins there are neither excrescences (such as ears) nor incisions (such as a neck). The skin is usually covered with scales, whose structure and shape differ in the orders of fishes. A few have smooth skins, and nearly all exude mucus from the skin.

The body is divided into *head*, *trunk* and *tail*. The head bears the mouth, the nostrils, the eyes and the hidden ears; the trunk has the gills and the anus, urinary and genital apertures; the tail is very muscular. Each of these areas passes insensibly into the next, but the head lies anterior to the first gill opening and the tail commences posterior to the anus. Trunk and tail bear fins, as follows:

(1) One or more median *dorsal* fins.
(2) One pair of *pectoral* or forelimb fins.

(3) One pair of *pelvic* or hindlimb fins (sometimes called *ventral*). They may be abdominal, thoracic or jugular in position. If thoracic, they lie posterior to the pectorals but high up on the trunk. If jugular, they are anterior to the pectoral fins.

(4) A median *anal* fin.

(5) A median *caudal* fin set vertically.

A fin is a flattened expansion of the body, sometimes permanently expanded, sometimes folding like a fan. It is supported by skeletal fin-rays. Each set of fins has a definite function to perform during swimming. The caudal fin and the tail act together as the propelling organ. Alternation of the segmental muscles of the right and left sides of the tail causes strong lateral movements in contraction. The median and paired fins are used for steering and balancing. It is difficult, however, to separate these fins as regards function, because they act together. The number of bony rays in the fins is usually fairly constant for the species, so that it is possible to write a *fin formula* in terms of rays. Thus:

Trachinus vipera. D. 5/21–24, P. 14, V. 1/5, A. 25–26, C. 12, means: anterior dorsal fin of five rays; posterior dorsal fin of twenty-one to twenty-four rays; pectoral fin of fourteen rays; ventral (pelvic) fin of six rays, of which the first is separable; anal fin of twenty-five to twenty-six rays; caudal fin of twelve rays.

On the head and along each side of the body are pores belonging to a sense organ of little known function. This is the *lateral line* system.

The mouth is provided with jaws of bone or cartilage, usually beset with teeth. Certain bones of the roof of the mouth also bear teeth. No mastication of the food takes place, however; the food is bolted whole and the teeth serve merely for gripping.

Fishes are divided into two sub-classes: the Elasmobranchii (cartilaginous fishes, with at least five pairs of separate gill slits) and the Teleostomi (bony fishes, with the gills covered by a gill cover or *operculum*).

The littoral fishes are rather specialised members of the subclass Teleostomi, order Teleostei. Some burrow, others adhere by means of suckers to rocks, or squirm into clefts.

KEY TO SHORE FISHES

1. Pelvic fins present. 4
   No pelvic fins. 2
2. One dorsal fin and a caudal peduncle. Body very slender and stiff. Pipe fishes. **Syngnathidae**, p. 264
   One dorsal fin. Body flexuous. 3
3. Dorsal, caudal and anal fins continuous. Eels. **Anguillidae**, p. 263
   Distinct caudal fin. Sand eels. **Ammodytidae**, p. 266
4. Pelvic fins small, abdominal in position. Anterior dorsal fin spiny. Sticklebacks. **Gasterosteidae**, p. 264
   Pelvic fins thoracic, jugular, or forming part of a sucker. 5
5. Pelvic fins modified to form a sucker. 6
   Pelvic fins not sucker-like. 9
6. Body scaly. Eyes protruding. Two dorsal fins. **Gobiidae**, p. 271
   Body scaleless. 7
7. Skin tuberculated in adult, first dorsal fin reduced to tubercles in adult. Lumpsuckers. **Cyclopteridae**, p. 276
   Skin smooth, only one dorsal fin. 8
8. Skin loose. Dorsal fin long, extending to tail. **Liparidae**, p. 277
   Skin not loose. Dorsal fin usually short. **Gobiesocidae**, p. 270
9. Pelvic fins thoracic (i.e. posterior to pectorals). 10
   Pelvic fins jugular (i.e. anterior to pectorals). 13
10. First three rays of pectoral fins free and feeler-like. **Triglidae**, p. 274
    No free rays on pectorals. 11
11. One dorsal fin. Mouth with very thick lips. Scales cycloid. **Labridae**, p. 278
    Two dorsal fins. 12
12. Scales ctenoid. Anal fin with three spines. **Serranidae**, p. 279
    Scaleless. **Cottidae**, p. 275
13. Fish lying on one side, both the eyes on the upper side. **Pleuronectidae**, p. 273
    Not a "flat fish". 14
14. Anterior dorsal fin, operculum and pelvic fins spiny. **Trachinidae**, p. 269
    No spines on operculum. 15
15. Body covered with bony plates. Feeler-like outgrowths on lower jaw. **Agonidae**, p. 275
    No bony scutes on body. 16
16. Pelvic fins reduced, with not more than five rays. Dorsal fin very long, not separated into several fins. **Blenniidae**, p. 267
    Pelvic fins with more than five rays. Barbels and nasal tentacles frequent. **Gadidae**, p. 266

ANGUILLIDAE.

Eels. Elongated fish, usually almost cylindrical, with reduced scales or none at all. No pelvic fins. Dorsal fin confluent with the reduced caudal fin.

1. Scales embedded in the skin. Dorsal fin not reaching forwards to level of pectoral fin. **Anguilla (anguilla)**
   No scales. Dorsal fin reaching level of pectoral fin. **Conger (conger)**

Anguilla anguilla. Day, 1880–4, II, 241; Travis Jenkins, 1925, p. 267 (*A. vulgaris*). The common eel inhabits both salt and fresh water. There is marked sexual dimorphism, the olive and yellow females attaining a length of a metre or more, the bronze metallic males rarely exceeding 50 cm. Moreover, as the female approaches maturity it becomes laden with fat, and its jaw muscles broaden. In this stage it is called a yellow eel, broad-nosed eel or frog-mouthed eel. In autumn the yellow eel leaves the rivers and shores, becomes sharp-nosed and silvery, and migrates to the deep sea. It is now called the silver eel, and starts on its long journey across the Atlantic to its breeding ground in the Sargasso Sea. The young return to Europe as pelagic, transparent, laterally compressed, leaf-like larvae known as *Leptocephalus* larvae. Arrived off the coast at the age of three years, they ascend the rivers, but not before a considerable metamorphosis has occurred. They become rounded and opaque, acquire pigment, and are called elvers. In this stage they are about 7 cm. in length. Thus, in the littoral zone may be found an occasional eel, and in early spring the elvers, especially in estuaries. Fin formula D.+A.+C. together 480–500, P. 17–18.

Conger conger. Day, 1880–4, II, 250; T. J. 1925, p. 273, *C. vulgaris*. The conger is larger and more heavily built than the fresh-water eel. Eyes large, scales absent. Purely marine and inhabits shallow water, but may occur in pools along the coast. Spawning occurs in deep water, and the *Leptocephalus* stage migrates towards the coasts, but does not ascend rivers. The young congers are pink in colour, later they become deeply pigmented, especially along the back. The adult reaches several metres in length. D.+A.+C. 525–546, P. 16–19. Under weeds on sandy bottoms with rocks, to shallow water. South and west coasts chiefly.

GASTEROSTEIDAE.

Sticklebacks. Small fish with elongated, depressed head, small terminal mouth and spiny rays in the anterior dorsal fin. The body contracts to a narrow peduncle at the root of the tail. Pelvic fins reduced, abdominal. No scales but bony scutes present. There are three species of stickleback, two inhabiting fresh water, viz. the three-spined and ten-spined, while one, the fifteen-spined, occurs in the sea.

Spinachia vulgaris. Fifteen-spined stickleback. Day, 1880–4, 1, 246, as *Gasterosteus*; T. J. 1925, p. 125. Body elongated, with about forty large keeled scutes arranged longitudinally on each side, forming a single raised row about the middle. The head resembles that of the pike. Caudal peduncle long and slender. D. 15/6–7, P. 9–10, V. 1/1, A. 6–7, C. 12–14. Olive dorsally, silvery ventrally, with dark and silvery bands on the head and sometimes on the body. Colour variable. Reaches 15 cm. in length. Under stones or seaweed and in rock pools between tide marks. They may be caught in shrimp nets on all coasts. The nesting habits of this species are well known. The male becomes blue in the breeding season, the female remaining green. The male constructs a nest for the eggs by fastening together fronds of seaweed in some rocky pool until an ovate structure about 10 cm. long is formed. The male guards the eggs deposited by the female in this nest for an incubation period of about three weeks.

SYNGNATHIDAE.

Pipe fish. Body extremely elongated, with hard scutes instead of scales. Fins very small. Pelvic fins absent, and in some genera the pectorals also. One dorsal fin. A long caudal peduncle precedes the minute caudal fin (even this fin is absent in *Nerophis*). The male incubates the eggs and has a marsupial pouch for the purpose.

1.	Body ridged. Pectoral and caudal fins present.	**Syngnathus**
	Body smooth, rounded, very slender. No pectoral fins.	**2**
2.	Caudal fin present.	**Entelurus**
	No caudal fin.	**Nerophis**

Syngnathus. Body ridged longitudinally, with strong scutes in several transverse rows. Tail not prehensile.

1. Body spotted with light spots ventrally. **S. typhle**
Body banded and spotted with dark marks. **S. acus**

S. typhle. Broad-nosed pipe fish. Day, 1880–4, II, 257; T. J. 1925, p. 208; both as *Siphonostoma*. D. 38–46, P. 15, V. 4, C. 9–10. Dark brown, spotted with lighter colour ventrally. 25–30 cm. Not common. On all coasts but rather local.

S. acus. Greater pipe fish. Day, 1880–4, II, 259; T. J. 1925, p. 208. D. 40–44, P. 13, V. 3, C. 3–10. Brown with a black stripe over the eye. Body with dark bands, about twenty in number. Dorsal fins with bands of spots and a black spot at the base of the first fin ray. 30–40 cm. long. Offshore, but occurs also in shallows, on rocky ground and amongst weed. On all coasts.

Nerophis. Body rounded with indistinct ridges. Tail tapering to a point. Male incubates the eggs in a marsupium.

1. Dorsal fin with more than thirty rays. **N. ophidion**
Dorsal fin with fewer than thirty rays. **N. lumbriciformis**

N. ophidion. Straight-nosed pipe fish. Day, 1880–4, II, 262; T. J. 1925, p. 209. Eyes at about the middle of the length of the head. Body smooth, stiff. Green, with white dots, spots and blotches on the under side, and a black band across the operculum. 25–30 cm. long. On all coasts, under stones between tide marks to shallow water. Not very common.

N. lumbriciformis. Worm pipe fish. Day, 1880–4, II, 263; T. J. 1925, p. 210. Eyes in anterior half of head. Olive brown, with white lines and dots ventrally. To 25 cm. long. In rocky pools, under stones and weed, between tide marks. Does not extend into deeper water. On all coasts, especially on the west.

Entelurus. Similar to *Nerophis*, but with a small caudal fin of six rays.

E. aequoreus. Snake pipe fish. Day, 1880–4, II, 261 (*Nerophis*); T. J. 1925, p. 209. Body rather rounded, not showing the ridges until the fish is dried. Dorsal fin with thirty-seven to forty-four rays. Caudal fin with six rays but said to be absent in large specimens. Olive with narrow bluish transverse bands. Dorsal rays yellow. Clinging to *Zostera* by the tail, but mostly in deeper water. Usually about 40 cm. long but may be much longer. Off all coasts.

AMMODYTIDAE.

Sand eels. Body elongated, compressed and silvery, covered with small cycloid scales. There is one long regular dorsal fin distinct from the caudal fin. No pelvic (ventral) fins. The mouth is protractile and shovel-like, without mandibular teeth; the anus placed far back. The gill opening is wide. The sand eels are believed to live in deep water during the winter and to migrate inshore to sandy bays in spring. Here they bury themselves in the sand in shoals and may be dug up at low tides.

1. Dorsal fin starts posterior to apex of pectoral fin. Large vomerine teeth present. More than 15 cm. long. **Ammodytes lanceolatus**
   Dorsal fin starts anterior to apex of pectoral fin. Minute vomerine teeth. Less than 15 cm. long. **A. tobianus**

A. lanceolatus. Greater sand eel. Day, 1880–4, I, 329; T. J. 1925, p. 166. A pair of vomerine teeth. Very voracious fish. Olive green, with a bright broad longitudinal band laterally, abdomen silvery. To 30 cm. Not common on the shore. Used as food on west coast and as bait.

A. tobianus. Lesser sand eel. Day, 1880–4, I, 331; T. J. 1925, p. 166. Minute deciduous vomerine teeth. More silvery than the above species. To about 15 cm. Very common on shore and used for bait or food. On sandy coasts to shallow water.

GADIDAE.

This very large family includes the cod, hake, ling, haddock, whiting and other food fishes.

The body is symmetrical and elongated, with wide gill openings and large eyes. The fin rays are soft and without spines, the scales small, cycloid and smooth. One, two or three dorsal fins are present, extending along the greater part of the length of the fish. One or two anal fins, and usually a distinct rounded caudal fin. Pelvic (ventral) fins jugular in position (anterior to the pectorals). The rocklings belong to a subdivision of the family characterised by the presence of two dorsal fins, the anterior sunk in a groove on the back. Barbel-like processes occur on the nasal openings and the lower jaw.

Onos mustelus (*Motella*). Five-bearded rockling. Day, 1880–4, I, 314; T. J. 1925, p. 163. Five barbel-like structures on the head.

Of these, four are dorsal, viz. a pair of long ones on the anterior nasal openings, and a shorter pair on the edge of the upper lip. A single ventral one on the lower lip. Anterior dorsal fin is set in a groove which bears taste buds and acts as a food locating organ. Caudal fin rounded. The body bears no spots, but is shaded from a dark olive or chestnut brown on the back to a light brown on the ventral side. The paired fins are often orange at the base. D. 47–55, P. 15, V. 7–8, A. 40–46, C. 20–21. 30–40 cm. In rock pools on all coasts.

O. tricirratus (*Motella tricirrata*) (Pl. XXIV, fig. 108). Three-bearded rockling. Day, 1880–4, I, 317; T. J. 1925, p. 162. Differs from the former species in the following points: (i) only one pair of barbels on upper lip instead of two pairs, (ii) marked by black spots on head, back and sides of body and usually on dorsal and pectoral fins, but these may be absent when young. D. 55–57, P. 16–18, V. 6–7, A. 48–50, C. 21. 30–45 cm. long.

BLENNIIDAE.

The blennies are small littoral fishes, some of which extend into brackish water. Head short, with terminal mouth and large eyes. The body is not very shapely, being thick in the tail region. It is usually spotted or banded, with or without scales. The dorsal fin may be continuous or may be divided into two or three parts; it is always very long, usually extending along the entire length of the body. The anal fin is single. The pectoral fins are rather large and rounded, but the jugular pelvics consist of fewer than five rays and are occasionally completely aborted.

1. Scales present. Dorsal fin long and undivided. **Chirolophis**
   Scales minute or absent. **2**
2. Dorsal and anal fin continuous. No caudal fin. **Zoarces**
   Caudal fin present. **3**
3. Pelvic fin reduced to one spine and one ray. A row of black spots at intervals along the base of the long dorsal fin. Body of the same depth throughout. **Centronotus**
   Pelvic fins small, but with one spine and two rays. Body tapering. **Blennius**

Blennius. Snout short, body elongated and compressed, with narrow mouth and usually with a tentacle near the orbit. Dorsal fin continuous, some of the rays being soft and others spiny. Pelvic fins small, consisting of one spine and two rays. Caudal fin distinct. No scales.

1.	One pair of orbital tentacles present.	2
	No orbital tentacles.	**B. pholis**
2.	First dorsal fin elevated, with an eye-like spot.	**B. ocellaris**
	First dorsal fin not elevated.	**B. montagui**

B. montagui. Montagu's blenny. Day, 1880–4, I, 200; T. J. 1925, p. 111 (*B. galerita*). Head with a pair of small red fringed tentacles posterior to the rather small eyes, and a soft erectile fold of skin forming a crest. Dorsal fin divided into almost equal halves, both semi-circular, the anterior lower than the posterior. D. 12–13/15–17, P. 12, V. 2, A. 17–18, C. 11. Body covered with bluish white spots, with white blotches. Ground colour brown or yellowish grey. Pectoral fin with a black band at its base. 5–7·5 cm. Littoral, under stones and amongst weed; and in rock pools.

B. ocellaris. Butterfly blenny. Day, 1880–4, I, 201; T. J. 1925, p. 111. A black eye-like spot ringed with white on the elevated first dorsal fin, which is much taller than the posterior dorsal. Head very short and blunt, with a pair of clubbed tentacles over the eyes. D. 11–12/14–16, P. 12, V. 2, A. 16–18, C. 11. Grey, with six or seven vertical lighter bands on the body. To 15 cm. On south and west coasts, amongst algae to shallow water.

B. pholis. Common blenny or shanny. Day, 1880–4, I, 203; T. J. 1925, p. 112. The commonest of the blennies. Dorsal fin continuous, though notched to form anterior and posterior portions. D. 11–13/18–20, P. 13, V. 2, A. 18–20, C. 13. Colour variable, greenish or yellowish, banded, spotted or blotched with black. To 15 cm. In rock pools on the shore, and on weed. Can climb or jump out of the water, and will bite the fingers with its strong teeth. Feeds on Barnacles and shellfish. Eggs yellow ovoid, with flattened underside, attached, in masses, guarded by male. June.

Chirolophis. Dorsal fin long and spiny, small scales cover the body.

C. galerita. Yarrell's blenny. Day, 1880–4, I, 206 (*Carelophus ascanii*); T. J. 1925, p. 112. A well-developed erectile, fringed supraorbital tentacle over each eye, and a shorter one in front of it, as well as a nasal tentacle. Dorsal fin long and neither divided nor notched. D. 50–52, P. 14, V. 1/3, A. 36–40, C. 17. Reddish brown, lighter beneath, banded transversely or mottled. Usually about 10 cm., but specimens up to 15 or 18 cm. are found. Within tide marks in northern districts, but in deeper water on south-west coasts.

Zoarces. Dorsal fin long, continuous, formed of soft rays except for a small depressed area near the end of the tail, where spines occur. No caudal fin, dorsal and anal fins continuous posteriorly. Pelvic fins with three or four rays. Scales vestigial.

Z. viviparus. Viviparous blenny. Day, 1880–4, 1, p. 211; T. J. 1925, p. 115. Caudal fin present in young stage but atrophies later. D. 76–80/10/20–25, P. 19, V. 3, A. 84–89. Viviparous, producing numerous young. Colour olive, with irregular spots and bands. Living within tide marks, under stones and weed, but retiring to deeper water in summer. It is said to reach 60 cm. in length.

Centronotus. Body elongated and compressed, ribbon-like. Dorsal fin long and formed entirely of spines. Pelvic fins reduced or absent. Caudal fin distinct. Scales very small.

C. gunnellus. Butterfish or gunnel. Day, 1880–4, 1, 208; T. J. 1925, p. 114 (*Pholis gunnellus*). A common shore fish shaped like a compressed snake, golden brown in colour, with a row of about ten to thirteen white-ringed dark spots along the base of the dorsal fin, which is low, regular and spiny. Caudal fin rounded. Pectorals small, pelvics very small. Skin very slimy, making the fish very difficult to hold in the hand. D. 75–82, P. 11–12, V. 1/1–2, A. 39–45, C. 15. 15 cm. long. Between tide marks, under stones or weeds. On all coasts. Common. The yellow egg masses as large as a walnut are also found on the shore in spring, the fish sometimes being curled up round them, the male alternating with the female in guarding them.

TRACHINIDAE.

The Weevers are the only harmful fishes on the shore, and should not be handled because of their habit of striking and wounding the skin by means of their poison spines. The opercular spines and the five spines of the anterior dorsal fin inject the poison, which causes rapid and painful swelling of the limb lasting about twelve hours, but with no more serious effect as a rule. Since the habitat of the fish is restricted to the sandy zones where shrimps and prawns are abundant, care should be exercised when handling the animals of this region in case a weever fish is amongst them. The fish can be recognised by the following characters: The mouth is directed obliquely upwards, there are spines on the head and operculum, the dorsal fin has a five- or six-rayed anterior and many-rayed posterior portion, the former spiny, the latter soft, the anal fin is soft and equal in size and shape to the posterior dorsal, the caudal fin large and square cut, the pectorals rounded or triangular, the pelvics jugular in position. The scales are very small. The two British species, known as the greater and lesser weever, belong to the genus *Trachinus*. Both sting by means of the opercular and dorsal spines, but the lesser weever is the more venomous species.

Trachinus draco (Pl. XXIV, fig. 107). The greater weever. Day, 1880–4, I, 79; T. J. 1925, p. 59. Two small spines on the upper angle of the orbit, several spines, including one large one, on the operculum. D. 5–6/29–31, P. 16, V. 1/5, A. 31, C. 12. Grey or yellow, darkest along the back, head spotted with grey, brown and yellow. 30–40 cm. long. On shrimping grounds, usually in deeper water, but at times, chiefly on the west coast, it may be found in the littoral zone, either buried in the sand or on its surface. It comes inshore for spawning in early summer.

T. vipera. Viper weever, lesser weever or stingfish. Day, 1880–4, I, 81; T. J. 1925, p. 60. This vicious little fish often lies buried in the sand near low-tide mark, with only the top of the head and the tips of the dorsal poison spines protruding. It feeds on shrimps. It resembles the previous species but does not exceed half the size, being usually from 10 to 15 cm. in length. There are no orbital spines, but opercular spines are present, and five or six poison spines in the anterior dorsal fin. D. 6/21–24, P. 14, V. 1/5, A. 25–26, C. 12. The colour is grey, paler on the abdomen. Confined to the coastal and littoral areas on sandy shores. Common on Irish and English coasts, less common in Scotland.

Gobiesocidae.

Sucker fish. Small sluggish fish provided with a large adhesive apparatus or sucker whose margin is formed in part from the pelvic fins and in part from the coracoid area. The body is soft, scaleless and spineless, the head depressed. Only one dorsal fin, with very few rays and at the posterior end of the body. The pectoral fins are rounded, the pelvics wide apart in the subjugular position, with one spine and four or five rays supporting part of the sucking disk. Caudal fin small and pointed or rounded. Shape tadpole-like.

Lepadogaster.

1.	Caudal fin distinct.	2
	Caudal fin confluent with dorsal and anal fins.	**L. gouanii**
2.	Dorsal fin with more than ten rays.	**L. candollii**
	Dorsal fin with fewer than ten rays.	**L. bimaculatus**

L. gouanii (Pl. XXIV, fig. 109*a*). Cornish sucker. Day, 1880–4, I, 189; T. J. 1925, p. 107. Snout spatulate, depressed, body thick, tapering to a pointed tail. D.+C.+A. 44–50, P. 20–25, V. 1/4. Red or purplish, lighter on ventral surface, occasionally brown or green. Two large oval spots, black surrounded by lighter rings, on the back of the head. Body banded or striped, but variable. To 10 cm. Common on west coasts, less common on east coast. In rock pools and under stones between tide marks.

L. candollii. Connemara sucker. Day, 1880–4, I, 191; T. J. 1925, p. 107 (*L. decandolii*). Head broad, snout spatulate. No nasal tubercles. Dorsal fin distinct, caudal fin rounded. D. 14–16, P. 25, V. 1/4, A. 8–11, C. 18. Red with crimson fins. Body with light spots, but variable. To 6 cm. On west and south coasts. Under stones between tide marks. Not very common.

L. bimaculatus. Two-spotted sucker. Day, 1880–4, I, 192; T. J. 1925, p. 107. The smallest species, not exceeding 5 cm. in length. Snout depressed but not so strongly spatulate as in the other two species. Dorsal fin small, separated by a considerable distance from the wedge-shaped caudal fin. D. 5–7, P. 17, V. 5, A. 4–6, C. 12. Red and orange, with reddish V-shaped bands on head and body. A black or purple eye-like spot just behind each pectoral fin, giving the species its name. Adults are said to inhabit deeper water but young ones are found on the shore. On all coasts.

Gobiidae.

The gobies are small, gregarious, goggle-eyed fishes with rounded fins and usually scaly bodies. There are two dorsal fins separated by a short interval, the anterior fin with three or four flexible spines, the posterior fin soft. The pelvic fins are small, set close together, and in many form a single fan-like fin which acts as a sucker. Sexual dimorphism is often marked. The males guard the pear-shaped eggs in a nest and are usually pugnacious.

Gobius. Pelvic fins united to form a disk attached only by its base, each fin consisting of one spine and three or four rays.

1. First dorsal fin with seven spines. **G. ruthensparri**
   First dorsal fin with six spines. 2
2. Dorsalmost rays of pectoral fins separated and silk-like. 3
   Dorsalmost rays of pectoral fins not separated. 4
3. Anterior dorsal fin with a dark edge, its middle spines longest. **G. niger**
   Anterior dorsal fin with a light edge, its anterior spines longest. **G. paganellus**
4. A row of black spots along the base of the dorsal fins. **G. pictus**
   A black or blue spot at posterior end of the dorsal fin. **G. minutus**

G. ruthensparri (*flavescens*). Spotted goby. Day, 1880–4, 1, 160; T. J. 1925, p. 94. Seven (instead of six) spines in the anterior dorsal fin, a black spot at the root of the tail. Scales angular and toothed on the posterior end of the body, but anterior to the dorsal fin there are none. D. 7/10–12, P. 19, V. 1/5, A. 10–12, C. 15. Yellow to olive brown, with pale markings dorsally and bands of red or pink on the yellow dorsal fins. The adult male is more brilliantly coloured than the female. To 6 cm. long. In rock pools amongst seaweeds, often swimming in shoals. Common.

G. paganellus. Rock goby. Day, 1880–4, 1, 162; T. J. 1925, p. 92. Dorsal fins almost touching, the pectorals nearly as long as the head, the caudal fin rounded. Head scaleless, a few scales on the operculum, the rest of the body with toothed (ctenoid) scales. D. 6/14–16, P. 20, V. 1/5, A. 12–14, C. 13. Often confused with *G. niger*, but can be distinguished by the light edge to the anterior dorsal fin and by the shape of this fin, which has the two to five spines almost equal in length. Its body is also spotted or banded, and not so darkly coloured as in the black goby. Length 10 cm. Common in rock pools, under stones and in shallow water. On all southern coasts but not in north Scotland. Breeding April to June.

G. niger. Black goby. Day, 1880–4, 1, 163; T. J. 1925, p. 91. Dorsal fins almost continuous, but their bases do not quite join. The pectorals have a few of the dorsalmost rays separate and silk-like. No scales on the head and operculum, the body scales large and ctenoid. D. 6/12–13, P. 18, V. 1/5, A. 10–12, C. 13. Colour variable, dull brown to black, with blotches and marblings. The fins have dark bands on a lighter base. 12 cm. On rocky ground and in rock pools, on all coasts.

G. minutus. Common one-spotted goby (Pl. XXIV, fig. 109*b*). Day, 1880–4, 1, 165; T. J. 1925, p. 93. A small species, variable in size, colour and markings, and adapting itself to varying conditions of salinity. A considerable space between the two dorsal fins, of which the first is lower than the height of the body. Caudal rather slender, rounded. D. 6/10–12, P. 21, V. 6, A. 10–12, C. 11. Colour light brown to grey, with dark blotches on the sides and a deep blue or black spot at the posterior end of the dorsal fin. Female lighter than the male. Length 8 cm. In shallow and sandy bays, often caught in shrimp nets; or in rock pools. Gregarious. On all coasts.

G. pictus. Painted goby. Day, 1880–4, 1, 167; T. J. 1925, p. 95. Resembles *G. minutus*, but brighter in colour, with red bands and black spots on the dorsal fins. D. 6/10, P. 20, V. 1/5, A. 9, C. 13. Very small. 4 cm. On coarse sand, from tide marks to shallow water. South and west coasts.

PLEURONECTIDAE.

The flat-fishes include plaice, dabs, brill, etc., and although these inhabit the shallow waters off the coast, small specimens of several species may often be caught at low tide in shrimping areas. The body is laterally compressed, the fish lying on one side with its head twisted to the opposite side so as to bring the eyes uppermost. The mouth, on the other hand, is obliquely twisted towards the under side. The upper side is pigmented, the under or eyeless side is often white. The dorsal and anal fins are very long, and the dorsal fin continues to grow forwards after the head twist occurs during development so that it never is truly dorsal (i.e. between the eyes) on the head. The pelvic fins are jugular in position, the caudal fin is symmetrical and rounded. In the genus *Pleuronectes* the fish lies on its left side, with the right side uppermost. The head twist is therefore towards the right and is described as dextral. The teeth are better developed on the blind side. Scales are present but are small.

1. Lateral line nearly straight; bony tubercles between the eyes. Red spots on skin. Caudal fin with seventeen or eighteen rays. **P. platessa**

   Lateral line curved anteriorly. Caudal fin with fourteen rays. **2**

2. Scales brown, rough. Bony ridge between eyes smooth. **P. limanda**

Rough tubercles along base of fin. Scales reduced. **P. flesus**

P. platessa. Plaice. Day, 1880–4, II, 25; T. J. 1925, p. 171. Young specimens can be caught in shrimping nets on sandy shores. The scales are small and are embedded in the skin, there are bony knobs on the head between the eyes, and the dark, greyish black or brown skin is relieved at intervals by bright red spots. In immature specimens the red spots are placed directly on the dark ground, in mature fish each spot is surrounded by a white ring. The colour changes with the background. D. 66–77, P. 10–11, V. 6, A. 50–57, C. 17–18. Plaice migrate inshore for spawning during the spring, then return to deeper water. Very large specimens have been known to attain a length of over 75 cm. and a weight of 15 lb. but the usual size is about 30 cm. Most abundant in North Sea and Irish Sea.

P. limanda. The dab. Day, 1880–4, II, 31; T. J. 1925, p. 185. Smaller than the plaice, with rougher scales, without bony tubercles between the eyes, and if spots are present they are dull in colour. The lateral line is strongly curved anteriorly. D. 65–78, P. 10–11, V. 6, A. 50–62, C. 14. A coastal fish, usually attaining about 25 cm. in length. The smallest species of the genus. Common in sandy bays in September.

P. flesus. The flounder. Day, 1880–4, II, 33; T. J. 1925, p. 187. Upper side dark olive brown, under side white. Scales small, cycloid. Lateral line slightly curved anteriorly. Rough, rounded tubercles along the bases of the rays of dorsal and ventral fins. D. 60–62, P. 10, V. 6, A. 39–45, C. 14. The flounder ascends the river estuaries in search of food, but unlike the two previous species, goes into deep water to spawn. Length about 22 cm.

TRIGLIDAE.

Gurnards. Three feeler-like rays on the pectoral fins. These rays are separate from the fin proper and are used for walking. The head is strongly armed with bony scutes, some bearing spines, and there are often spines on the lateral line and at the base of the first dorsal fin. The thoracic pelvic fins are narrow and slender. The anal fin is usually equivalent in size to the second dorsal fin, the caudal fin is large and triangular.

Trigla.

1. Colour brownish, with large pectorals deep blue behind. **T. lucerna**
   Colour slate grey, with white spots. **T. gurnardus**

T. lucerna. Yellow gurnard. Day, 1880–4, 1, 59; T. J. 1925, p. 50 (*T. hirundo*). Tub fish, sapphirine gurnard. Pectoral fins very large and coloured a rich blue on their posterior sides. Caudal fin emarginate. D. 9/16–17, P. 10, +3, V. 1/5, A. 15–16, C. 12. Red to brownish red, second dorsal, pectoral and caudal fins violet, the pectorals blue underneath. Coloured like a butterfly. To 60 cm. long. West and south-west coasts, rare on east. Young on shrimping grounds.

T. gurnardus. Grey gurnard. Day, 1880–4, 1, 62; T. J. 1925, p. 52. The commoner species. Grey with white spots. Utters a grunt when removed from the water. Lateral line scales armed with spines, other scales small. Caudal fin broad and short. D. 8–9/19, P. 10, V. 1/5, A. 18–19, C. 15. To 75 cm. On all coasts.

Agonidae.

Armed bullheads. The body is small, elongated and almost cylindrical, and is covered with bony scales. The head is strongly armed with spiny plates. There are two dorsal fins, a small anal fin, and rounded caudal fin set on a long slender caudal peduncle. The pectoral fins are well developed, the pelvics small and jugular in position.

Agonus cataphractus. Armed bullhead or pogge. Day, 1880–4, 1, 67; T. J. 1925, p. 55. The lower jaw has feeler-like outgrowths. Head wide and depressed, with protruding eyes and short concave snout. D. 5/5–6, P. 16, V. 1/2, A. 6, C. 12. Light grey, tinged with yellow, banded on the sides with black, and often spotted. Fins usually yellow but variable. About 15 cm. long. From coast to deep water, coming inshore to spawn in spring. Egg masses straw-coloured in *Laminaria* holdfasts.

Cottidae.

Bullheads or sea scorpions. Small, prettily marked, coastal fish, with broad, depressed, spiny head, eyes on upper surface, rounded snout, subcylindrical body without scales. Two large dorsal fins, the first spiny with separate rays, anal fin similar to second dorsal, pectoral fins rounded, pelvic fins thoracic, caudal fin rounded with forked rays. A strong spine on the pre-operculum. When disturbed the fish utters a grunting noise and erects the spines on its head.

Cottus.

1. Five spines on the operculum, one very strong. **C. bubalis**
 Four spines on the operculum. **C. scorpius**

C. scorpius. Father lasher, Greenland bullhead. Day, 1880–4, I, 49; T. J. 1925, p. 47. Four spines on the operculum, but one is especially well developed though shorter than the eye length. A spine over the eye. D. 9–10/13–14, P. 16–17, V. 2–3, A. 9–13, C. 11. Grey, mottled and banded with black, ventral surface white and yellow. Fins yellow, with oblique dark bands and spots. 15–30 cm. Under stones and weed and in rock pools. Common. Eggs in large coloured masses, December–March.

C. bubalis. Long-spined sea scorpion. Day, 1880–4, I, 51; T. J. 1925, p. 48. Usually smaller than the preceding species, about 15–22 cm. long, and more highly coloured. The large spine on the operculum is longer than the eyeball and there are four other spines in this area. D. 8/11–12, P. 16, V. 1/3, A. 9, C. 10. Colours bright, yellow, red and green being predominant, with darker markings on the body, but not on the posterior fins as a rule. Under stones, in pools and in deeper water, often estuarine. Common. Egg masses orange.

CYCLOPTERIDAE.

The lumpsuckers have rounded bodies, with short thick head. There are two dorsal fins, the first reduced to a rayless stump, the second similar to the anal in shape, but with forked rays. Pelvic fins with a spine and a few (five) reduced rays, forming the centre of a suctorial disk with a cutaneous margin. The skin is naked in parts, but there are longitudinal rows of strong tubercles in the adult. *Cyclopterus* has a viscous skin and the first dorsal fin is replaced by stout tubercles.

Cyclopterus lumpus. Lumpsucker, Sea hen. Day, 1880–4, I, 179; T. J. 1925, p. 101. While primarily a deeper water species, the lumpsucker comes inshore for spawning in spring, and may be left stranded at the low spring tides. After egg-laying the female returns to deeper water, leaving the male to guard the eggs, which are attached to rock and weed between tide marks in irregular masses, containing thousands of eggs. Four rows of large tubercles along the body between the head and tail, and scattered smaller tubercles. Young specimens smooth, with distinct anterior dorsal fin. D./10–11, P. 20–21, V. 1/5, A. 9–10, C. 10–11. Colour usually dark brown or bluish black, with bright red ventral side and fins, particularly in the male during the breeding season. Length up to 60 cm. A typically northern form, though it occurs on all coasts.

LIPARIDAE.

Smooth suckers. Small, clumsily built fishes with smooth, naked, movable skins, and a ventral sucker formed around the fused pelvic fins as in the Cyclopteridae. The body is subcylindrical anteriorly, but is compressed posteriorly. The gill openings are narrow. There is a single long dorsal fin.

Liparis. Sea snail. Dorsal fin longer than the anal, both linked with the narrow triangular caudal. Head smooth and rounded, with an almost terminal mouth and small eyes.

1. Body with irregular longitudinal bands. Dorsal fin with more than thirty rays. **L. liparis**
   Body spotted. Dorsal fin with not more than thirty rays. **L. montagui**

L. liparis. Common sea snail. Day, 1880–4, I, 184; T. J. 1925, p. 105, as *L. vulgaris*. Head broad and wide. Dorsal fin long, commencing at about the middle of the pectoral and extending to the tail. Anal commencing below the sixth dorsal ray and extending also to the tail. Pectorals very broad, extending forwards and downwards under the throat, the six ventral rays longer than the others so that the fin appears notched. Pelvics with six rays each, surrounding the disk. D. 34–36, P. 28, V. 1/5, A. 27–28, C. 12–14. Pale grey or brown, with purplish longitudinal bands. Pectoral and caudal fins yellowish, spotted. Variable. To 15 cm. long. Near low-water mark, often in estuaries, under rocks, etc. On all coasts. Egg masses attached to weed or stone, January–July.

L. montagui. Montagu's sea snail (Pl. XXIV, fig. 109*c*). Day, 1880–4, I, 186; T. J. 1925, p. 105. Very similar to the above, but smaller, and spotted on the body. D. 26–30, P. 30, V. 6, A. 24, C. 14. Purplish or yellowish brown, with dark spots. Fins yellow, often spotted. 8 cm. long. Under stones at low tide, in rock pools, amongst *Fucus*, etc. Often estuarine.

LABRIDAE.

Wrasses. Rather small fishes, often brilliantly coloured or iridescent, but of little or no economic importance. The body is heavily built and the lips very thick and folded. There is a single long dorsal fin, which may be partly spiny and partly soft-rayed. The caudal fin is broad and rounded. The pectorals are small, the pelvics thoracic in position, i.e. rather far forward but not in front of the pectorals. Strong crushing teeth are present on the jaw but not on the palate. The scales are cycloid.

1. Three anal fin spines. 2
   More than three anal fin spines. Conical teeth in a single row in the jaws. **Centrolabrus**
2. Teeth in a single row. 3
   Teeth in a band, with an outer series of stronger conical teeth. **Ctenolabrus**
3. Lateral line scales more than forty. **Labrus**
   Lateral line scales fewer than forty. **Crenilabrus**

These four genera were formerly included in the genus *Labrus*, and the differences between them are slight.

Labrus. Dorsal fin of many rays of which at least thirteen to twenty-one are spiny but not much elongated. Anal fin with three spines and numerous soft rays. Scales rather large. Lips very thick.

L. bergylta. Ballan wrasse. Day, 1880–4, I, 252; T. J. 1925, p. 128 (*L. maculatus*). A red or greenish fish, spotted with white, but variable, and brown, blue and green may occur. Each scale has a white centre. The unpaired fins bear yellow or orange rings. D. 20–21/9–11, P. 14–15, V. 1/5, A. 3/8–10, C. 14. To 40 cm. On rocky coasts, coming inshore to spawn in spring. The "nest" is made in seaweed in a crevice, but is not guarded.

Crenilabrus. Resembling *Labrus* but with serrated pre-operculum and fewer than forty scales along the lateral line.

C. melops. Corkwing. Day, 1880–4, I, 260; T. J. 1925, p. 129. With numerous varieties. Usually reddish purple along the back, lighter ventrally, with spots beneath the dorsal fin, a spot near the posterior end of the lateral line, and two transverse rows of spots along the caudal fin. D. 14–17/8–9, P. 14, V. 1/5, A. 3/9–10, C. 14.

Rarely exceeds 22 cm. Gregarious, in rock pools and in rocky crevices, on sand and under weed, where it builds nests. On all coasts especially in the south.

Ctenolabrus. Dorsal fin of twenty or more rays, of which sixteen to eighteen are spiny and none is elongated. Scales of moderate size. Teeth in the form of a band of conical teeth, those of the outer row the largest. Three anal fin spines.

C. rupestris. Goldsinny wrasse. Day, 1880–4, I, 264; T. J. 1925, p. 130. Spinous portions of dorsal fin shorter than soft portion. Caudal fin square. D. 16–18/8–10, P. 14, V. 1/5, A. 3/8, C. 15. Pink or golden, darker dorsally. Fins red, the dorsal with a dark mark anteriorly and usually with a black spot at the root of the tail. 15 cm. long. Usually occur in deeper water but sometimes cast ashore after gales. East and south coasts.

Centrolabrus. More than three anal spines. Dorsal fin of more than twenty rays, of which sixteen to twenty are spinous and none elongated. Teeth in a single row on the jaws.

C. exoletus. Rock cook. Small-mouthed wrasse. Day, 1880–4, I, 267; T. J. 1925, p. 131. Anal fin with five spines, the first short. D. 18–20/6, P. 14, V. 1/5, A. 5/7–8, C. 15. Rich brown dorsally, yellow and silvery on the underside. A dark spot over the eye and two blue bands from eye to angle of mouth. Fins silver and yellow. A row of dark marks along spinous dorsal fin. Caudal fin with a white rim and dark base. 10–12 cm. In rock pools and in deeper water. On east and south-west coasts. Not common.

SERRANIDAE.

Two distinct dorsal fins, separate or united only at the base. Anterior dorsal fin spinous, posterior dorsal with some soft rays. Pectoral fins small. Pelvics thoracic, narrow. Spines and serrations on operculum. Teeth on jaws, vomer, palatine and tongue. Scales ctenoid. Estuarine fishes.

Morone. Teeth all alike. Anterior dorsal fin with nine spines, posterior dorsal partly spiny, partly soft-rayed. Anal fin with three spines.

M. labrax. The bass. Day, 1880–4, I, 8, as *Labrax lupus*; T. J. 1925, p. 34. Two opercular spines. Anterior dorsal fin elevated in the centre, the shortest spines being the first and last. D. 8–9/12–13, P. 16, V. 1/5, A. 3/10–11, C. 17. Grey, silvery ventrally. A dark spot on the operculum. Dorsal, anal and caudal fins greyish, paired fins yellow. 30–45 cm. or more. Voracious and gregarious, appearing inshore and in estuaries in spring. West and south coasts. Rare on east.

EXPLANATION OF PLATE XXIV

Pisces, Figs. 107–110

Fig. 107. ***Trachinus draco.*** Lateral view to illustrate the structure of a fish. The scales are not shown. $\times \frac{1}{4}$.

Fig. 108. *Onos tricirratus.* Lateral view of head and trunk. $\times \frac{1}{2}$.

Fig. 109. The three types of sucker in Sucker fish. All in ventral view. *a*, *Lepadogaster gouanii.* Pelvic fins, the coracoid region and the skin posterior to the pelvic fins take part in the formation of a frilled sucker. $\times 1\frac{1}{2}$. *b*, *Gobius minutus.* The right and left pelvic fins fuse to form a cravat-shaped sucker. $\times 1\frac{1}{2}$. *c*, *Liparis montagui.* The pelvic fins are fused and greatly reduced. They form a circular sucker. $\times 1\frac{1}{4}$.

Fig. 110. Two types of scales. *a*, *Onos mustelus.* Cycloid. $\times 30$. *b*, *Gobius minutus.* Ctenoid. $\times 12$.

a.d.f.	anterior dorsal fin	n.	nostril
an.	anus (urinary and genital apertures are close to this)	operc.	operculum
an.f.	anal fin	p.d.f.	posterior dorsal fin
caud.f.	caudal fin	pect.f.	pectoral fin
d.bar.	dorsal barbel	pelv.f.	pelvic fin
l.l.	lateral line	sp.	poison spine
mo.	mouth	su.	sucker
		v.bar.	ventral barbel

XXIV. PISCES

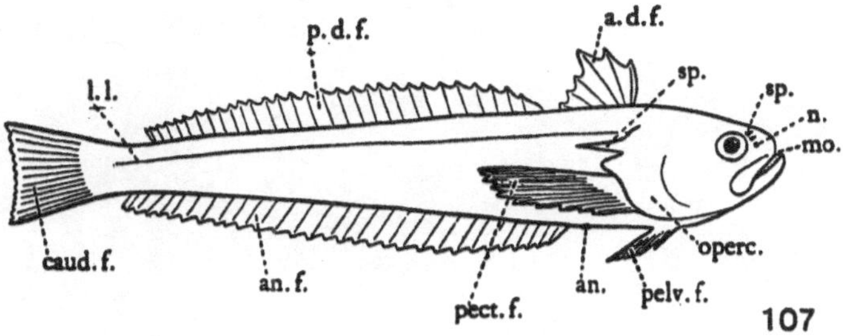

107

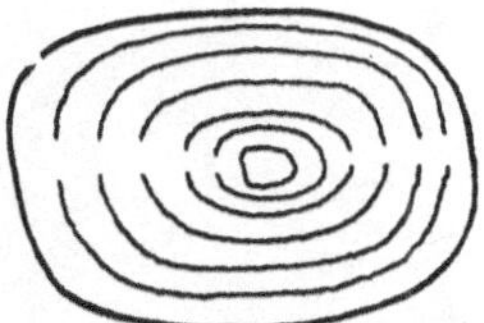

110*a*

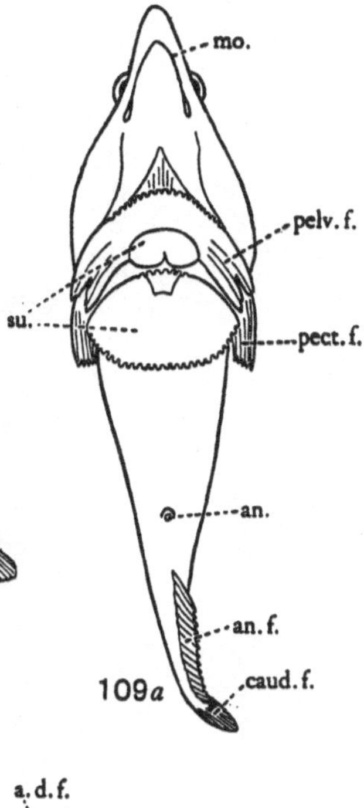

109*a*

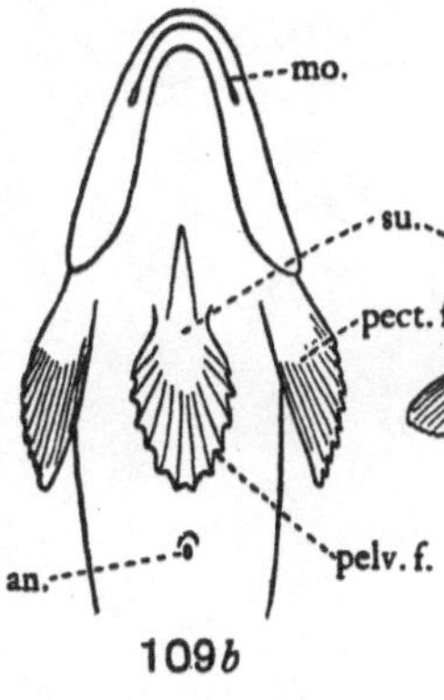

109*b*

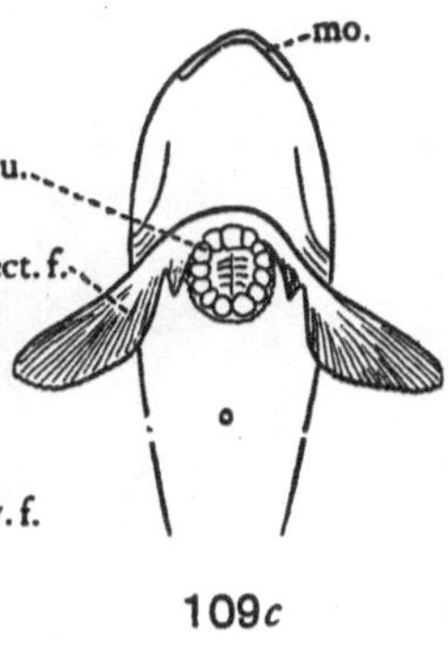

109*c*

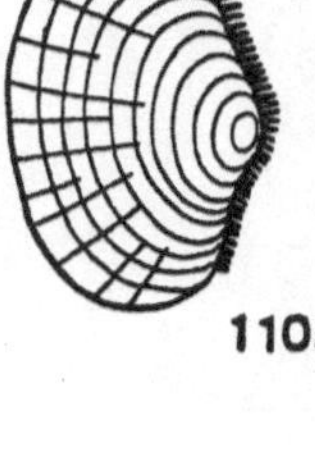

110*b*

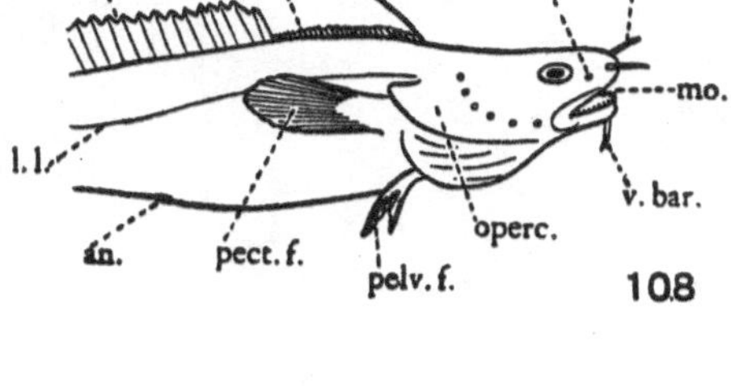

108

(*facing p.* 280)

APPENDIX

1. PRACTICAL HINTS ON SHORE COLLECTING

For shore collecting the best times in the year are the Spring and Autumn Equinoxes, for at these times the greatest area of shore is exposed at low water, but the fauna at the two periods will vary considerably. In the spring some species come inshore to spawn (e.g. Nudibranchs), many colonial animals are beginning to form their colonies (e.g. Coelenterates, Tunicates), whereas in autumn the offshore migration has already occurred and the colonial forms have usually passed their zenith.

1. Choose the ground according to the weather conditions. The ideal is a fair, calm, warm day, but these three factors are rarely co-existent at the Spring and Autumn Equinoxes. The most troublesome factor is wind, and shelter from the prevailing wind is imperative, not merely for the comfort of the collector, but because the wind, blowing against the outgoing tide, splashes the water about, making visibility difficult, and preventing the tide from attaining its recorded lowest limit. If therefore an east wind be blowing, choose a collecting ground facing west, or at least sheltered from the east. For example, in the Isle of Man one cannot collect at Gansey Point, Port St Mary, in a strong easterly wind, but conditions are likely to be good on the *east* side of Port Erin Bay, which is protected from the east by the cliffs.

2. Wear suitable clothes. Old mackintoshes, short leather coats, old shoes for paddling, Wellingtons, etc. have been found most useful. Do not wear clothes that will get heavy when wet.

3. Take the following general equipment:

An unbreakable vessel, such as an enamelled milk can, for use as a collecting jar.

A few specimen tubes, enclosed in a box or tin, for special small finds.

A sharply pointed stainless knife for digging specimens from crevices.

A hand lens, tied on.

A sponge bag for shell fish, sponges, weed, etc.

Rucksacks, pockets and straps are useful.

Equipment for special operations is given below.

4. Be on the collecting ground, near the water's edge, ready to begin work, at least *half-an-hour* before the time of low tide. This allows about an hour for observation and collecting. The walk or climb to the ground may be long and arduous. Great rocks covered with slippery wet seaweeds, pools that look temptingly shallow, narrow crevices, and gullies that involve difficult detours require care and time, and allowance should be made for slow progress over such an area. The slogan for rock climbing is "Prehension by three appendages". Hold on, and balance yourself, by two feet and one hand or two hands and one foot; test the next hold before transferring your weight to it.

5. Before beginning work, study the general conditions of the shore. Note whether it is exposed or sheltered, washed by a swift current or by slack water, muddy or sandy, with healthy seaweed or partly rotting weed, and so on. All these alternative conditions will affect the fauna, and should be recorded along with the animal population.

The methods employed in collecting vary with the nature of the shore. The region with the richest fauna is the rocky shore with a mixture of small movable rocks, large rocks with overhanging surfaces covered with weeds, and rock pools in the hollows. Turn over the small rocks and search in the cavity beneath, then examine the overturned surface, lifting animals off with your knife. *Finally put the stone back in its place, under side downwards as before*, otherwise the sedentary forms growing on it will be killed. Raise the weed hanging from the big rocks, and look on the weed itself and on the rock surface for crawling and sedentary forms. Lie face downwards on a rock and peer into a rock pool, moving the weed very gently. Fishes, Crustacea and other quickly moving animals lurk among the weed. Nudibranchs creep slowly along, or swim in a reversed position on the surface, while zoophytes and anemones expand their tentacles on the bottom of the pool. Study the zoophytes in the pool, remove a colony and examine it in the contracted condition, so that you will recognise it when you see it on a rock or weed that is not covered with water. A small net or sieve is useful for catching the swimming inhabitants of the pools.

Collect *variety* rather than quantity, and put small specimens in separate receptacles. Put a little damp weed in the bottom of the jar and place the animals upon it. Water in a collecting jar rapidly becomes harmful to marine animals, most of whom can endure temporary removal from water better than immersion in foul water.

Always take back bunches of the plant and animal growths, and scrapings from the rocks, in the sponge bag, and when you get back to the Laboratory turn them into shallow dishes containing plenty of fresh sea water. Creeping forms such as Nudibranchs, Worms, Isopods, Amphipods, etc. are often found by this method, whereas searching for them directly may be futile owing to their protective colouring.

The procedure on a sandy or muddy shore is different because few of the animals are exposed, although some may indicate their presence by holes, castings and tracks visible on the surface. For these animals digging is necessary. Turn up the sand or mud smartly with a fork or a spade and examine for worms, fish, molluscs, sand urchins and sand anemones. Remove the catch to a bucket or jar, and take back plenty of sand also, so that you can watch the animals burrowing at your leisure. Another good method is to transfer the contents of the spade to a large sieve and wash out the sand or mud immediately, by shaking the sieve under water. If the bottom of the sieve be the ordinary perforated zinc, most of the smaller worms, Crustacea and Molluscs will be left behind in the sieve when the sand and mud are washed away.

Arrived back at the laboratory with the haul, fill large dishes with freshly drawn sea water, and place them in a shaded place. Tip out the contents of the collecting jar, and sort out the specimens as far as possible into phyla, e.g. place the Coelenterates in one dish, the Molluscs in another, the worms in a third. Put the weed in a large dish by itself, and later examine it, and the water beneath it, for small animals. Return to the sea animals you do not want; do not leave them to rot in the laboratory. *Use the keys for tracing the animals as far as possible, then look up the book to which reference is made and verify with the help of the fuller description.* Repeated use of the keys should enable a student to omit many of the earlier portions, e.g. he will soon learn to recognise a Sertularian hydroid, a Polynoid worm or a Portunid crab from the other members of the same class. *Record every animal found*, giving data regarding environment, as in the following example:

Myriothela cocksi.

Coelenterata. Gymnoblastea—Myriothelidae. 21/3/36, Wembury Bay. LWEST.* Lower shore, under stones and on rocks. Sporosacs ripe. Abundant.

It is a good plan to keep named type specimens. See below on Preservation of Marine Animals.

In practice, a student cannot tackle all the phyla, and it has been the author's method to allot a chosen phylum to one student, or to two students working together. After the sorting of the material into phyla, the identification is left to the students specialising on any particular phylum. Subsequently the animals are exhibited on a demonstration table, with an explanatory label attached to each specimen, based on the example given above. Every student spends part of his time examining other people's catches, and the student specialising on a group explains his specimens to the others.

Many marine animals can be kept alive for considerable periods. Anemones have been kept for more than 30 years in a salt water aquarium. Conditions in the sea are more uniform than in fresh water, and an aquarium should be as cool as possible, in an even but not strong light. Constant changing of the water is not necessary, but uniform salinity and aeration are important. Mark the level of the water, and as evaporation proceeds from day to day, fill up to the level mark with fresh non-chlorinated water. Tap water can be boiled, cooled and aerated for this purpose.

Arrange stones and shelter for the animals. Seaweed cannot be used in a marine aquarium as it decomposes very readily. Predatory and herbivorous animals will not live together, and a preliminary study of feeding habits is therefore important. Minute green algae may cause trouble after a time and are difficult to remove; a limpet may assist for a while, but changing the water is imperative if the green film becomes too dense.

Carnivorous forms may be fed with very small pieces of mussel, fish, etc. Crabs soon become tame and will take food from forceps; anemones pass food to the mouth if small pieces are placed on the disk or in contact with the tentacles. Remove at once any food that is not eaten.

* LWEST means low water of the Equinoctial spring tide.

The *quantitative*, as opposed to the *qualitative* study of the littoral fauna can provide data on the range, size distribution, frequency, feeding habits and life histories of the species chosen, and is best performed by a number of people working as a team. Two examples are given. The first deals with the numbers and distribution of a single species on a sandy shore free from weed, the second with the numbers and range of several species in relation to weed growth on a rocky shore.

Station records and size distribution of Tellina tenuis *on a sandy shore**

Apparatus required: Square wooden frames, each side half a metre long, and about 20 cm. in depth, buckets, spades, sieves with a mesh of 2 mm., collecting jars.

The count should be made at low water of an equinoctial spring tide, if possible, as the range of this species extends into shallow water. Taking extreme high water mark as zero, mark four stations on the sand at 80, 100, 150 and 200 yards below high water mark. The distances may be measured by pacing. At each station a squad presses the frame into the sand to a depth of about 15 cm., digs out the contained sand and transfers it to the buckets, the contents of which are then sieved under water, washing out the sand and leaving the shells. The living specimens of *Tellina tenuis* are put into a collecting jar and marked with the number of the station.

Since the frame is square and has a side of half a metre, the area dug will be a quarter of a square metre.

The specimens are taken to the laboratory and sorted into sizes, rising by 2 mm. from the smallest (about 3 mm. length) to the largest (about 19 mm. length). The number in each group is counted and plotted on a graph, or tabulated as follows:

Station	3 mm.	5 mm.	7 mm.	9 mm.	11 mm.	13 mm.	15 mm.	17 mm.	19 mm.	Total
I. 80 yd.	—	1	4	4	12	19	12	3	—	55
II. 100 yd.	1	2	2	12	20	43	16	7	4	107
III. 150 yd.	1	4	13	68	106	113	37	25	5	372
IV. 200 yd.	52	27	89	297	112	62	102	—	—	741

* The example taken is based on that worked out by Dr A. C. Stephen at Kames Bay, Millport, Firth of Clyde, where *Tellina tenuis* is the dominant Lamellibranch species. *Journ. Mar. Biol. Ass. U.K.* 15, no. 2, 1928. The figures quoted were obtained in a class study under the direction of the author.

From the above figures it can be inferred that the maximum number of littoral individuals lives near extreme low water mark, and here also are the greatest number of young, but a relatively small number of older specimens. The largest specimens are found at stations II and III. The numbers of all sizes diminish rapidly as the distance from low water increases. Thus the size frequency is in inverse relation to the number of individuals occupying a zone. The upward range is just above station I, and this corresponds with high water of the neap tides (78 yards below zero in Kames Bay). Records made in shallow water indicate that the lower range of this species is in about 4 fathoms, and that spat is deposited in the sublittoral zone during the late summer. First year individuals attain a size of 5 mm., and though they migrate inshore they rarely ascend above low water mark of the neap tides (180 yards from HWES). In their second year they measure 9 mm. and keep low on the beach or even descend to shallow water again (for spawning?). Third year specimens up to 13 mm. migrate upshore again and decrease in numbers as they grow older; they occupy the region between HWN and LWN. The range of the species is thus between HWN and 4 fathoms depth.

Belt Transect of a rocky shore

Apparatus required: String or measuring tapes; Metre or yard sticks; Collecting jars; Note-books for records on the beach.

Choose a portion of the shore with as gentle a slope as possible, avoiding, if possible, sudden drops or deep pools, which may be studied separately. Run a string *AB* straight down the beach from extreme high water mark to extreme low water mark of the spring tide. Place a second string *CD* parallel with *AB*, but at a distance of 1 metre from it. The area between the strings can be marked off in square metres in ladder formation. Divide the class into squads, instructed beforehand to recognise the species which they are to record.

Squad 1. Working outside the ladder, plot the ranges in metres of the chief seaweeds, e.g. *Enteromorpha intestinalis*, *Pelvetia canaliculata*, *Fucus spiralis*, *Ascophyllum nodosum*, *Fucus vesiculosus*, *F. serratus*, *Laminaria digitata*. Enter up the data in the note-book.

Squads 2–5, or more if required. Between the strings *AB* and *CD* work out the range and count the number of individuals per

XXV. ZONATION OF CERTAIN LITTORAL SPECIES

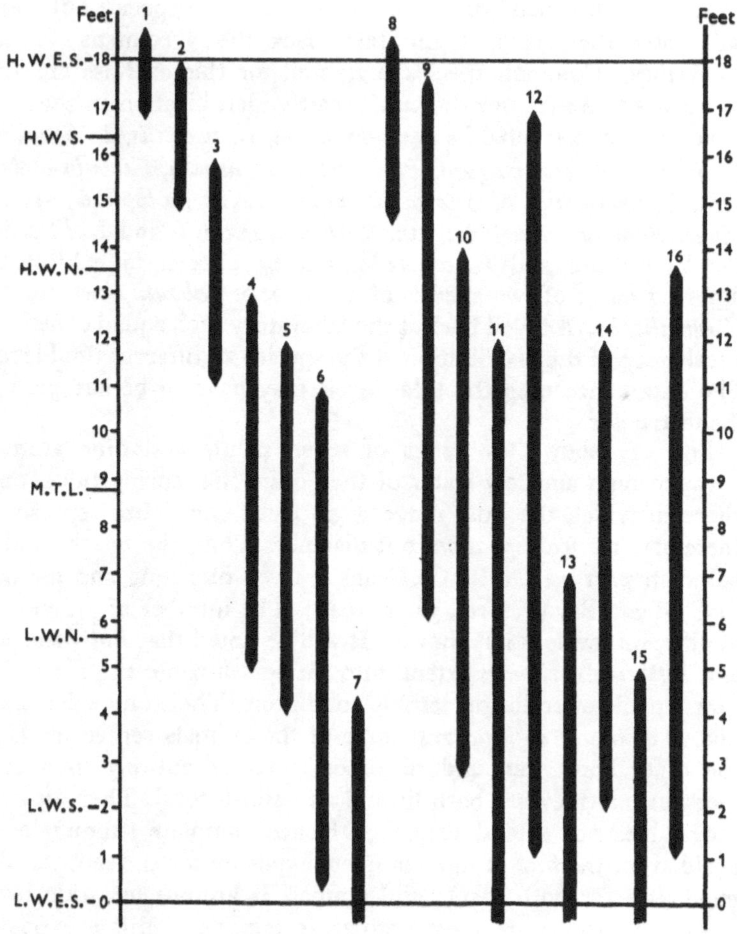

Fig. 111. Vertical zonation of plants and animals between tide marks. The extreme tidal range is taken as 18 feet and the tidal contours are given for an area such as Cardigan Bay or Port Erin Bay. 1. *Enteromorpha intestinalis*. 2. *Pelvetia canaliculata*. 3. *Fucus spiralis*. 4. *Ascophyllum nodosum*. 5. *Fucus vesiculosus*. 6. *F. serratus*. 7. *Laminaria digitata*. 8. *Littorina neritoides*. 9. *L. saxatilis*. 10. *L. littoralis*. 11. *L. littorea*. 12. *Patella vulgata*. 13. *P. aspera* (*athletica*). 14. *Gibbula umbilicalis*. 15. *G. cineraria*. 16. *Nucella lapillus*. After R. G. Evans.

square metre of species selected for their abundance on the site chosen. Each squad will be responsible for one species only, and will enter their records and take back the specimens to the laboratory. Common species suggested for this exercise are *Littorina neritoides* (rather difficult because identification in the field is not easy and it inhabits deep crevices), *L. saxatilis*, *L. littoralis*, *L. littorea*, *Patella vulgata*, *P. depressa*, *P. aspera*, *Gibbula umbilicalis*, *G. cineraria*, *Calliostoma zizyphinum*, *Nucella lapillus*, *Mytilus edulis*, *Balanus balanoides*, etc. *Littorina saxatilis* and *L. littoralis*, *Patella vulgata* and *P. aspera* might be chosen to exhibit the differing range of two species of a genus, or *Gibbula cineraria* and *G. umbilicalis*. Arrived back at the laboratory each squad constructs a scale map of the distribution of the species at different tidal levels. The data concerning the tidal levels may have to be supplied by the instructor.

Fig. 111 shows the range of seven plants and nine animals between high and low water of the equinoctial spring tides, on a shore in which the tidal range is 18 feet. The figure represents, therefore, *vertical zonation*, not distances along the beach, and is based in part on Dr R. G. Evans' survey of plants and animals in Cardigan Bay (*J. Ecol.* 34, 2, 1947). The number of individuals at different levels is not shown. It will be noted that *Enteromorpha* and *Littorina neritoides* extend into the splash zone, and are therefore capable of enduring aerobic conditions. Their range is a short one. *Laminaria digitata* and three of the animals represented, on the other hand, can endure almost total or entirely total submersion, i.e. they are both littoral and sub-littoral. Their upward range does not extend very far. Hence complete submersion is avoided for the first group, complete exposure for the second. The maximum for both plants and animals is around mid-tide level, where conditions are most regular as regards alternate exposure and submersion. The limits of the ranges, however, will vary from one shore to another, and even from one part of a given beach to another part, according to conditions.

Other suggested studies are:

1. The population rate per square metre of a single gregarious species, e.g. *Balanus balanoides*, *Patella vulgata*.
2. A comparison of the population rate of a gregarious species in sheltered, partly exposed and fully exposed situations.

2. PRESERVATION OF MARINE ANIMALS

The methods employed are so numerous that only a few can be mentioned here.

(1) *Dry Methods*

Specimens can be dried by exposure to air at ordinary temperatures, or the desiccation accelerated by placing the specimen in a warm air chamber such as a low temperature incubator.

The method is possible for small specimens possessing hard parts. The soft tissues dry out, leaving the skeleton, which in time becomes very brittle. It is advisable to pin larger specimens to a board in the position required, in order to prevent uneven shrinkage during drying. Can be used for Foraminifera, some sponges, Hydroids, Crustacea, Mollusc shells, Bryozoa and Echinoderms. In the Urchins, the test can be prepared by scrubbing off the spines and pedicellariae after drying.

(2) *Wet Methods*

Specimens are immersed in a preserving (or fixing) agent for a definite period, then transferred, usually after washing free from the fixative, to Alcohol or Formalin.

(*a*) *Formalin* is a cheap fixative and works well for marine animals in 5 % neutral solution. It hardens tissues after prolonged immersion, has poor penetrating powers, and becomes gradually acid by oxidation of the formaldehyde into formic acid. The hardening effect can be counteracted by the addition of a little glycerine. Suitable for soft delicate organisms, e.g. most worms, Coelenterates, Tunicates, etc. Some Nudibranchs and Alcyonarians preserve well in *hot* formalin. Make up in sea water.

(*b*) *Alcohol* is a more rapid fixative and penetrates better than formalin, but decolorizes the tissues and causes shrinkage. It also tends to precipitate the salts from sea water on the surface of the animals, thus impeding the entrance of the fixative. Specimens should be preserved for 24 hours in 30 % alcohol, then transferred to higher strengths by gradual stages up to 70 %. Alcohol does not attack calcareous structures as formalin and the acid fixatives do.

(*c*) *Alcohol acetic.* A mixture of alcohol and acetic acid, made up with 99 parts of 70% alcohol and 1 part of glacial acetic, is a useful general fixative, where preservation of calcareous parts is not required. Thus it can be used for a siliceous, but not for a calcareous sponge, for a Holothurian but not for other Echinoderms. The acidulation of the alcohol prevents the deposition of salts from the sea water on the surface of the animal. After fixation keep in 70% alcohol.

(*d*) *Formalin and alcohol.* Equal parts of a 5% solution of formalin and 50% alcohol. The specimens can be stored in the mixture.

(*e*) *Bouin's picro-formol fixative* is excellent for marine animals, especially when made up in sea water. After fixation for from 4 to 16 hours, the specimens must be washed in 50% alcohol or treated with lithium carbonate until all traces of the yellow picric acid disappear.

Picric acid, saturated solution in sea water ...	75 parts
Formalin	20 parts
Acetic acid (glacial)	5 parts

It is especially suitable for soft-bodied animals (e.g. worms, jellyfish), and the specimens can afterwards be used for histological work. Ctenophores must not be kept in it.

(*f*) Corrosive sublimate mixtures (Gilson's, Susa and Zenker's fluids), Bi-chromate, Osmic and Nitric acid mixtures. For directions in the use of these, reference should be made to books on preserving methods, such as Lee's *Microtomist's Vade Mecum*, and Guyer's *Animal Micrology*.

Special Methods of Fixation and Preservation

1. Immerse the animal in *fresh* water, and as soon as it is dead replace the water by preservative (e.g. 30% alcohol). Sometimes used for Ophiuroids, Holothurians and some worms.

2. Heat the sea water containing the animals and when dead transfer them to preservative. Some Nemerteans fix well by this method.

3. Pour *warm* fresh water on the animals suddenly. Small Nemerteans may be compressed between two glass slides, flooded with hot water and then preserved. A large Nemertean may be

wrapped round a stick, the ends of the animal held with the fingers and the whole plunged into a tube containing hot water.

4. Pour the preservative, either hot or cold, suddenly on the animal. This method may be used for slowly contracting animals such as some Tunicates and Medusae.

5. Narcotise the animals in sea water. This method is invaluable for strongly retractile forms such as Hydroids, Bryozoa, Holothurians and Molluscs. The specimens should be placed in a large clean bowl containing an abundant supply of clean fresh sea water. Put the bowl in a cool place away from direct sunshine, and do not allow it to be moved or shaken during the process of narcotisation. After a time the animals will expand. The narcotic is added to the water either by sprinkling slowly soluble crystals on the surface or by pouring it in drop by drop in liquid form. After a few minutes to a few hours, according to the narcotic used, the specimens will become immobile, and will not contract when touched with a needle. As soon as this stage is reached, they should be removed to preservative. Care must be taken that they are not left too long under the action of the narcotic or the tissues will decompose rapidly. In other words the animal should be insensible but not dead when placed in preservative.

The chief narcotics employed for marine animals are formalin, alcohol, menthol, cocaine, chloral hydrate and eucain, chloretone, magnesium sulphate and magnesium chloride.

The cocaine group are difficult to obtain, but the others can be purchased from most chemists. Chloretone is made up in crystal form; Parke Davis and Co. supply it in gelatin capsules.

(*a*) *Menthol.* Sprinkle a few crystals on the surface of the water. A slow narcotic, sparingly soluble, taking from 12 to 24 hours, hence the tissues are liable to macerate if specimens are left too long in the narcotising mixture.

(*b*) *Chloretone.* Method as for menthol. A slow narcotic, good for some species of worms.

(*c*) *Magnesium sulphate.* Add a few crystals every ten minutes until a concentration of about 1% is obtained after 30 minutes, though this strength may be varied according to the species to be narcotised. The animals should become immobile in from 1 to 5 hours.

(*d*) *Magnesium chloride, crys.* An 8% solution in *tap water* is isotonic with sea water, and narcotises an animal placed in it, in from 20 minutes to 1 hour. Add strong formalin and transfer to alcohol if desired.

(*e*) *Formalin.* Add a few drops of 1% formalin to the water containing the expanded animals and repeat at intervals of 15 minutes. The formalin acts slowly but may cause violent reaction at first. Hence it is not good for Coelenterates but has been found useful for certain Nudibranchs.

(*f*) *Alcohol.* Can be employed in the same way as formalin.

6. Inject the preservative into the body of large animals, such as starfish, Simple Ascidians, worms, Opisthobranch Molluscs, etc., then immerse the specimen in the fixative. If a hypodermic syringe is not available, use a glass cannula with rubber bulb attachment. Make a small slit in the body of the animal, or pierce it with the point of the syringe, and squeeze gently, allowing the preservative to fill internal cavities and bring about the extension of the body. This gives good internal preservation.

3. ARTIFICIAL FERTILISATION

The developmental stages of many Invertebrate and Protochordate animals such as *Echinus*, *Pomatoceros* and *Ascidiella* can be studied as the result of artificial fertilisation of the eggs in a pure culture. Directions are given for *Echinus*, but the method can be adapted for other animals.

Thoroughly clean several large jars (breffits) and fill with filtered sea water. Cut an *Echinus* into two along the ambitus and examine the gonads in the aboral half. Ripe female gonad exhibits a pinkish, granular, translucent appearance; ripe male is opaque and milk-white. Examine microscopically ripe and unripe ova and active sperm. Suspend the half of a female over the mouth of one of the jars, gonads upwards, and the half of a male over a second jar. If mature, genital products will be discharged through the genital pores into the water beneath within a few minutes; unless this occurs reject the specimen. Ripe ova come down in small groups of isolated granules which float, sperms in milk-white streams which soon render the water turbid. Having obtained one jar containing ova and another sperm, remove the half *Echinus* from the top of each, pipette up *a few drops* of the sperm suspension, add it to the water containing the ova, and stir well. After a time eggs that have been fertilised will sink to the bottom. Decant off the surface water with unfertilised eggs and surplus sperm, fill up with filtered sea water or distribute the fertile eggs by removing to glass finger bowls, putting a few in each bowl. Keep the bowls in a cool, shaded place. Take samples at frequent intervals and examine microscopically the fertilised egg, the blastula and gastrula stages, and the early pluteus stages after hatching, noting the order of appearance of the arms, etc. Keep a record of times and temperatures of the water.

Diatom cultures for feeding the larvae may be purchased from the Marine Biological Laboratory at Plymouth.

4. GLOSSARY OF DESCRIPTIVE TERMS

Aboral. On the region of the body opposite to the part where the mouth lies.

Adductor. A muscle which, by its contraction, draws a part towards the main axis of the body, or bends upon itself a relaxed part.

Annulated. Ringed.

Arborescent. Branching repeatedly, as a tree does.

Areolated. Divided into small distinct areas; often implies the formation of pockets in the wall of an organ.

Articulated. Jointed.

Auriculate With ear-shaped appendages.

Avicular. Shaped like a bird's head.

Bifid. Divided into two by a cleft.

Bifurcate. Divided into two forks or branches.

Biramose. Having two rami or branches.

Biserial. Arranged in two rows or series.

Caducous. Falling off easily.

Capitate. Having a knob-like head.

Carinate. Keeled or ridged.

Cephalic. Belonging to the head.

Chelate. With pincer-like claws.

Cirriform. Applied to processes that are slender and usually curved.

Colonial. Forming an aggregate of individual animals or plants connected by structural and physiological links.

Commensal. Applied to animals living together and sharing the same food. See *parasitic* and *symbiotic*.

Compressed. Flattened laterally. See *depressed*.

Crenate. Having the edge scalloped or notched, with rounded projections. Cf. *denticulate*.

Ctenoid Having a comb-like edge (scales of fishes).

Cycloid. Circular in form, with concentric growth lines (scales of fishes).

Decumbent. Lying down.

Dendritic. Branching, arborescent.

Denticulate. Having tooth-like projections.

Depressed. Flattened dorso-ventrally. See *compressed*.

Dextral. On the right side of the body, or curving towards the right. Opposite of *sinistral*.

Diaulic. Having separate male and female apertures in a hermaphrodite animal.

Dichotomous. Branching into two.

Digitate. Shaped like a finger, or having finger-shaped processes.

Dioecious. Having the two sexes in separate individuals.

Distal Distant from the axis or centre of the body. Opposite of *proximal*.

Divaricator. An opening muscle.

Emarginate. Notched or indented at the margin.

Epizoic. Living upon the surface of.animals.

Excentric. Not centrally placed.

Falciform. Sickle-shaped, curved and hooked.

Fenestrated. Having small perforations or openings like windows.

Filamentous or

Filiform. Thread-like.

Fimbriated. Fringed.

Fusiform. Spindle-shaped.

Geniculate. Bent like a knee.

Gibbous. Humped.

Gregarious. Living in groups of the same species, but without structural continuity of the members. Cf. *colonial*.

Hispid. Finely papillose or hairy.

Imbricating. Overlapping like roof tiles.
Involute. Spirally coiled, usually so closely as to hide the axis of the spiral.

Lamellar or
Lamellate. Consisting of thin plates.
Lamelliform. Having the shape of a thin plate.
Lanceolate. Resembling a spear-head, narrow and pointed.
Linear. Like a thread, very narrow and about the same breadth throughout.
Lobulated. Composed of small rounded processes or lobes.

Mammillated. Conical, with a rounded apex.
Manubrium. A handle, haft or stalk.
Monaulic. Having a single aperture for male and female organs in a hermaphrodite animal.
Monodactylous. Having only one finger or claw, as in subchelate Crustacea.
Monoecious. Having the sexes combined in one animal, hermaphrodite.
Mucronate. Terminating abruptly in a short sharp point.
Muricated. Furnished with sharp points or hard excrescences as in a *Murex* shell.

Nacreous. Pearly.
Natant. Swimming.
Natatory. Adapted for swimming.
Nautiloid. Coiled like a *Nautilus* shell, with the convolutions of the spiral in a single plane.

Occlusor. A closing muscle.
Opaline. Of a clear milky white transparency.
Operculate. Possessing a lid or cover called the operculum.
Orbicular. Round or shield-shaped.
Ovate. Egg-shaped. If attached, fixed by the broader end. See *pyriform*.
Ovigerous. Capable of carrying ova, e.g. ovigerous legs.
Oviparous. Reproducing by means of ova.

Pallial. Pertaining to the mantle, e.g. in Molluscs.

Palmate. Divided into lobes like the palm and fingers of a hand.

Papillate or *papillose.* Covered with small projections from the surface.

Paragastric. Non-flagellate cavities or passages in a sponge through which the water current passes on its way to the osculum.

Parasitic. Depending on another organism for food, etc.

Pectinate. Comb-like, having branches on one side of the axis.

Pedicellate or

Pedunculate. Supported on a stalk.

Pellucid. Transparent.

Penultimate. In a series, the one before the last.

Peripheral. Distant from the centre or axis.

Pinnate. With branches arising on each side of the axis. If the branches are also pinnate, the structure is said to be *bipinnate*; and if these are again pinnate, it is *tripinnate.*

Plumose. Feather-like.

Porcellanous. Pearly.

Prehensile. Adapted for holding.

Protandrous. In hermaphrodite individuals, producing and discharging the sperms before the ova.

Protogynous. In hermaphrodite individuals, producing and discharging the ova before the sperms.

Protractor. A muscle which pushes out or extends a part of the body.

Proximal. Near the axis or centre of the body, opposite of *distal.*

Punctate. Having the surface covered with dots or points.

Pustulated. Covered with blister-like projections.

Pyriform. Pear-shaped. If attached, fixed by the narrow end. See *ovate.*

Ramose. Branched.

Reflected. Turned back upon itself, folded back.

Reptant. Creeping or crawling.

Reticulate. Shaped like a network.

Retractor. A muscle which draws back a part when it contracts.

Rhomboidal. Lozenge-shaped.

Rudimentary. Imperfectly developed or arrested at an early stage of development, but not degenerate. See *vestigial.*

Scalariform. Ladder-like.
Sedentary. Resting on the substratum.
Serrated. With the edge notched like a saw.
Sessile. Fixed by a base, but without a stalk or pedicel.
Setose. Supplied with setae, bristly.
Sinistral. On the left side of the body, or curving towards the left. Opposite of *dextral.*
Spatulate. Spoon-shaped.
Squamous. Covered with scales.
Strigose. Bristly or ridged.
Styliform. Bristle-shaped.
Subchelate. Imperfectly chelate. In this type of claw the dactylopodite or terminal joint bends back to form a hook-like structure, but the propodite does not send up a fixed process to meet it, as in the true chela.
Symbiotic. Applied to two or more organisms living together for their mutual benefit. Cf. *commensal.*

Tactile. Specially adapted for touch.
Triaulic. Having three genital apertures in a hermaphrodite animal—the male, the female and the copulatory apertures.
Tricuspid. Having three cusps or points.
Triramose. Having three rami or branches.
Truncate. Terminating abruptly as if an end were cut off.
Tuberculate, Tuberculous. Having knob-like projections or tubercles.
Tubicolous. Inhabiting a tube.
Tumid. Inflated, e.g. whorls of a Gastropod shell.
Turreted. Tower-like.

Unciform, Uncinate. Hook-like.
Uniramose. Having only one portion, e.g. parapodium of some Annelids. See *biramose.*
Uniserial. With branches on one side of the axis only.

Vermiform. Shaped like a worm.
Verrucose. Covered with warts.
Vestigial. Degenerate, usually applied to an organ that is reduced in size and structure.
Viviparous. Bringing forth young alive, the ova developing within the body.

PLATE A

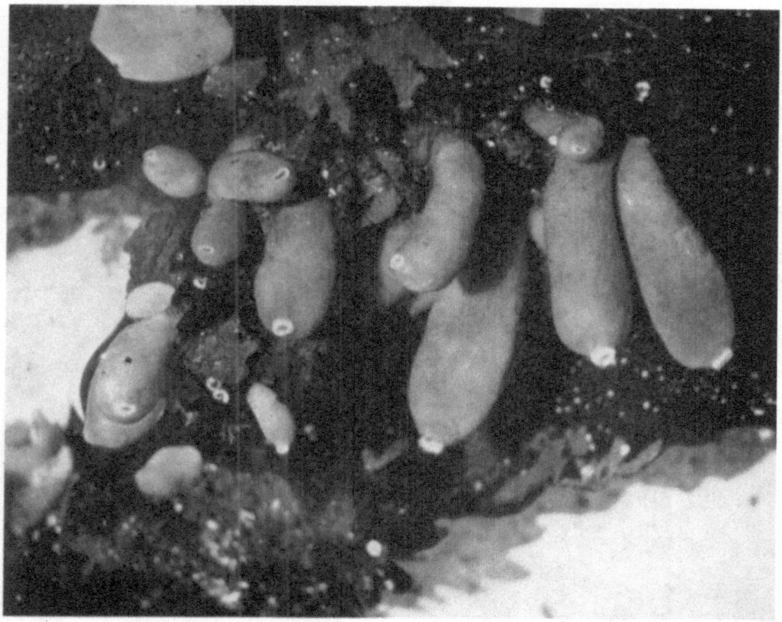

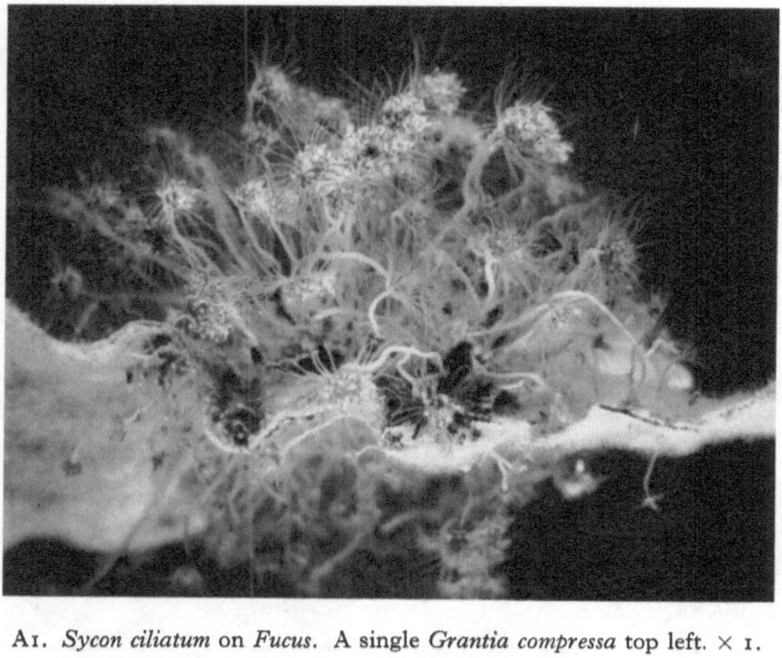

A1. *Sycon ciliatum* on *Fucus*. A single *Grantia compressa* top left. × 1.
A2. *Tubularia larynx*. Colony growing on Bryozoan. × $\frac{1}{2}$.

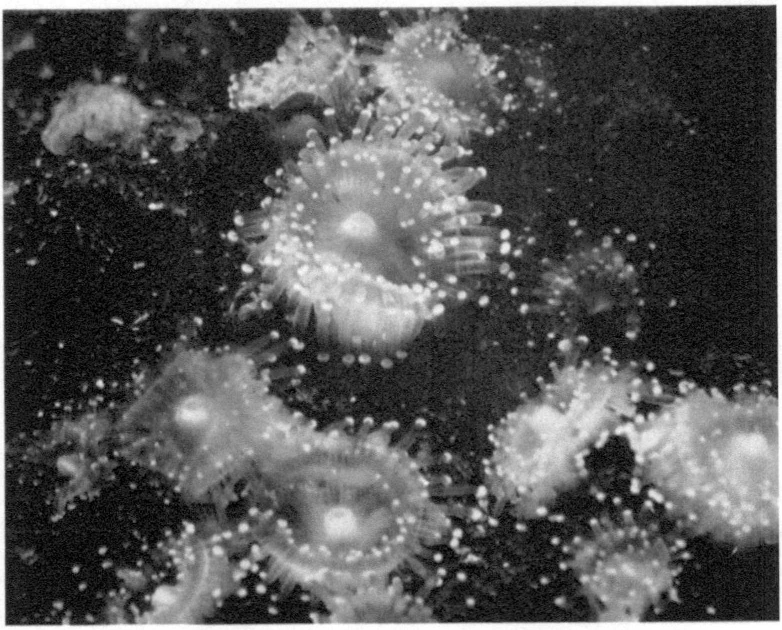

B1. *Cereus pedunculatus*. Daisy Anemone. Natural size.
B2. *Corynactis viridis*. A gregarious Anemone. × 3.

PLATE C

C1. *Caryophyllia smithi*. Devonshire Cup Coral. On the right a corallum, left a partially expanded zooid, centre a fully expanded zooid. × 3.

C2. *Cancer pagurus*. Edible Crab. × ½.

PLATE D

D1. Rock with barnacles (*Balanus balanoides*), Limpets (*Patella vulgata*) and Dog Winkles (*Nucella lapillus*). Note the homing areas of the limpets. × $\frac{1}{6}$.

D2. *Eupagurus bernhardus*. Hermit crab in a whelk shell covered with the hydroid *Hydractinia echinata*. Natural size.

PLATE E

E1. *Pecten septemdentatus*. The scallop is lying on its convex valve; the mantle is expanded to show the tentacular fringe, the ocelli, and the approximated mantle edges forming the velum. × ½.

E2. *Buccinum undatum*. The whelk. The large foot, with its horny operculum, the siphon, and one tentacle with the eye are shown. × ¾.

PLATE F

F1. *Aplysia punctata*. The Sea Hare. Oral tentacles, rhinophores, eye, parapodia and position of the anal spout. × $\frac{3}{4}$.

F2. *Dendronotus frondosus*. Seven groups of branched cerata are well developed. × 4.

G1. *Asterias rubens.* Common Starfish. The line of spines along the middle of each ray, the central anus, calcareous madreporite and terminal eye spots are shown. × $\frac{1}{3}$.

G2. *Solaster papposus.* Sun Star. Juvenile with nine arms. Paxillae (grouped spines) on the aboral surface and edges of the arms. × $\frac{1}{2}$.

PLATE H

H1. *Antedon bifida.* Feather Star. The cirri cling to a rock. × 1.

H2. *Echinus esculentus.* Sea Urchin. Specimen with expanded podia, a few of which are adherent to the bottom of the tank. × ½.

INDEX TO FAMILIES AND GENERA

The bold numerals are figure numbers of illustrations; other numerals are page numbers

Eight plates of photographs of living animals may be found between pages 298 and 299

For EU product safety concerns, contact us at Calle de José Abascal, 56–1°, 28003 Madrid, Spain or eugpsr@cambridge.org.

www.ingramcontent.com/pod-product-compliance
Lightning Source LLC
LaVergne TN
LVHW090305210926
848480LV00001B/134
* 9 7 8 0 5 2 1 2 9 4 7 0 6 *